M000294353

Membrane Processes
for Water Reuse

About the Author

Anthony M. Wachinski, Ph.D., has more than 35 years of experience in the water and wastewater industry and is currently senior vice president and technical director for Pall Corporation, Water Processing Group. He has worked in government, academia, and private industry as a professional engineer, consultant, associate professor of civil engineering, principle research investigator, and expert witness. Dr. Wachinski has experience in research and development, program and project management, new product development, technology acquisition and transfer, and water product certification and verification. He holds numerous patents and has authored more than 40 professional papers, along with books for AWWA and CRC Press, and is a certified forensic examiner for the Department of Homeland Security.

Membrane Processes for Water Reuse

Anthony M. Wachinski

New York   Chicago   San Francisco
Lisbon   London   Madrid   Mexico City
Milan   New Delhi   San Juan
Seoul   Singapore   Sydney   Toronto

The McGraw·Hill Companies

Cataloging-in-Publication Data is on file with the Library of Congress.

McGraw-Hill books are available at special quantity discounts to use as premiums and sales promotions, or for use in corporate training programs. To contact a representative please e-mail us at bulksales@mcgraw-hill.com.

Membrane Processes for Water Reuse

1 2 3 4 5 6 7 8 9 0   DOC/DOC   1 9 8 7 6 5 4 3 2

ISBN 978-0-07-174895-7
MHID 0-07-174895-4

The pages within this book were printed on acid-free paper.

Sponsoring Editor	**Proofreader**
Larry S. Hager	Cenveo Publisher Services
Acquisitions Coordinator	**Production Supervisor**
Bridget Thoreson	Richard C. Ruzycka
Editorial Supervisor	**Composition**
David E. Fogarty	Cenveo Publisher Services
Project Manager	**Art Director, Cover**
Harleen Chopra,	Jeff Weeks
Cenveo Publisher Services	
Copy Editor	
Jeff Anderson	

Contents

Preface

This book is about how membrane processes can be used to recycle and reuse the effluent from wastewater treatment plants—secondary effluent. With water scarcity and changes in global climate cycles, recycle and reuse of secondary effluent is a viable alternative, together with the brackish and seawater desalination, for providing water for low and high water needs.

Global water supply is limited and its availability is variable. Freshwater sources and population density are not evenly distributed. Many countries including the United States and China have sufficient water, but in the wrong place, at the wrong time, and in the wrong amounts. At the writing of this book, almost the entire United States is experiencing drought. Unlike oil, coal, and natural gas though, water is renewable by natural processes. It can also be recycled and reused, but today only 4% of wastewater is recycled globally. The United States recycles 12%, Kuwait on the other hand recycles 91%, and Israel 87%.[1]

This book is written at a time, in my opinion, when the ability of technical books to provide unique knowledge is diminishing. In writing *Membrane Processes for Water Reuse*, I asked myself "Does this book bring value to its reader? Is the cost of this book justified? Can the reader get the information from the Internet?"

The book brings value to the reader because much of the information is not available on the Internet or in the literature. I share my close to 40 years of experience in this industry. The book is structured so that the reader can zero-in on a topic and learn about that topic. Each chapter is stand-alone. I wrote the book to bring value to the reader. Time, of course, will tell.

This book is written for process engineers, regulators, graduate students, and water industry managers. It assumes the reader has, or is willing to gain, a more-than-basic understanding of wastewater treatment. Process engineers can use this book to select and size appropriate membrane processes, pretreatment, and posttreatment to meet reuse water quality. Graduate students will discover what

---

[1]Global Water Summit, 2010.

they know and don't know about wastewater treatment, and understand membrane technology as the kingpin technology in wastewater reuse. Regulators will be able to make informed decisions on how to achieve desired levels of wastewater quality. Municipal wastewater treatment plant management will be able to explore reuse and recycle alternatives to recoup treatment costs to meet or exceed regulatory requirements, and make cost-effective decisions. Industry plant owners will be able to make cost-effective decisions on how to best handle their discharges, i.e., discharge to stream, treat and sell effluent, or recycle effluent.

Flow schemes showing integrated membrane processes—pretreatment and posttreatment—are presented in the book. An extensive glossary is provided to familiarize the reader with definitions of membrane and water reuse terminology.

This book does not cover wastewater treatment plant design or membrane bioreactor design. For these topics, many books and short courses are available. In my opinion, this book complements those texts in defining membrane's role in wastewater reuse.

Anthony M. Wachinski

Acknowledgments

My sincere thanks to the many people who helped me write and publish this book. Special thanks to Dr. Charles Liu, one of the top experts in the world, on integrity testing, and the theoretical aspects of low pressure membrane properties and performance. He was kind enough to let me publish our paper (he was primary author), in this book, with some modification, defining the properties of high performance membranes. He also gave valuable insight to the sections on integrity testing.

Special thanks to Joseph Swiezbin, and his daughter Katie. To Joe, for his insight, in chapter content, and to Katie for providing drawings for two figures.

This project was the brainchild of Larry Hager. Thanks Larry for your guidance and patience.

I offer my sincere thanks to Ms. Harleen Chopra, project manager, for her timely and always professional efforts in getting the book published.

Acronyms

AOP	Advanced Oxidation Process
ASTM	American Society for Testing and Materials
AWWA	American Water Works Association
AWWARF	American Water Works Association Research Foundation
BNR	Biological Nutrient Removal
BOD	Biochemical Oxygen Demand
CA	Cellulose Acetate
CAS	Conventional Activated Sludge
CEB	Chemically Enhanced Backwash
CFT	Conventional Filtration Technology
CIP	Clean-In-Place
COD	Chemical Oxygen Demand
CSTR	Continuous Stirred Tank Reactor
CT	(Disinfectant Residual) Concentration (mg/L) × (Contact) Time (minutes)
CWA	Clean Water Act
CWS	Community Water System
DBP	Disinfection By-Product
DOC	Dissolved Organic Carbon
ED	Electrodialysis
EDR	Electrodialysis Reversal
EFM	Enhanced Flux Maintenance
EGSB	Expanded Granular Sludge Bed
GFD	Gallons per Square Foot per Day
HAAs	Haloacetic Acids
HFF	Hollow Fine Fiber
HPC	Heterotrophic Plate Count
IMS	Integrated Membrane System
LPM	Low Pressure Membranes

MBR	Membrane Bioreactor
MCB	Membrane-Coupled Bioprocesses
MCL	Maximum Contaminant Level
MCLG	Maximum Contaminant Level Goal
MF	Microfiltration
MFT	Mobilized Film Technology
MGD	Million Gallons per Day
MLD	Million Liters per Day
MLSS	Mixed Liquor Suspended Solids
MO	Methyl Orange alkalinity
MMF	Multimedia Filter (filtration)
MWCO	Molecular Weight Cut Off
NA	Not Applicable
NEPA	National Environmental Policy Act
NF	Nanofiltration
NIPS	Non-Solvent Induced Phase Separation
NOM	Natural Organic Matter
NPDES	National Pollutant Discharge Elimination System
NSF	National Sanitation Foundation
NTU	Nephelometric Turbidity Unit
OH alk	Hydroxide alkalinity
Palk	Phenylthaline alkalinity
PA	Polyamide
PAC	Powdered Activated Carbon
PAN	Polyacrylonitrile
PES	Polyethersulfone
PP	Polypropylene
PS	Polysulfone
PSI	Pounds per Square Inch
PVDF	Polyvinylidene Fluoride
RO	Reverse Osmosis
SBR	Sequencey Batch Reactor
SCADA	Supervisory Control and Data Acquisition
SDI	Silt Density Index
SUVA	Specific Ultraviolet Absorbance
SWTR	Surface Water Treatment Rule
TCF	Temperature Correction Factor
TDS	Total Dissolved Solids
TFC	Thin film Composite
ThOD	Theoretical Oxygen Demand

TIPS	Thermally Induced Phase Separation
TKN	Total Kjeldahl Nitrogen
THM	Trihalomathane
TMP	Transmembrane Pressure
TN	Total Nitrogen
TOC	Total Organic Carbon
TOX	Total Organic Halides
TS	Total Solids
TSS	Total Suspended Solids
TTHM	Total Trihalomethanes
TTAL	Treatment Technique Action Level
TVS	Total Volatile Solids
TDS	Total Dissolved Solids
UASB	Upflow Anaerobic Sludge Blanket
UF	Ultrafiltration
USEPA	United States Environmental Protection Agency
UV	Ultraviolet (light)
WEF	Water Environment Federation

Membrane Processes
for Water Reuse

CHAPTER 1

Water Reuse Overview

Introduction

The focus of this book is how membrane technology can be used alone, in integrated membrane systems, or coupled with aerobic or anaerobic processes—membrane coupled bioprocesses—to process treated municipal effluent or industrial wastewater for discharge, recycling, or reuse.

This book does not describe in any technical detail the biological processes used to treat municipal and industrial wastewaters for the removal of dissolved organics and the nutrients nitrogen and phosphorus. A suggested reading list is included at the end of this chapter. The book incorporates an extensive glossary as a reference and as a jumping-off point for the reader to research particular topics.

This book does not recommend or endorse any membrane manufacturer or membrane brand, nor should any conclusions be drawn from the source of tables or figures. One objective of the book is to provide engineers and designers with information to select or recommend a membrane solution for solving a reuse application.

This chapter previews the book, discusses the drivers for reuse—water scarcity and economics—and overviews water reuse technologies and the role membranes have played in the past and will play in the future.

Chapter 2 reviews the basic chemistry needed to understand membrane technologies and provides a detailed description of reuse water characteristics and parameters related to source water quality and product water quality. Drinking water and wastewater standards are presented. A brief description of analytical techniques is discussed. Measuring "what's in water" drives the technology to remove it.

Chapter 3 provides the reader with the basic principles needed to understand membrane applications. This chapter covers in depth the fundamentals of membrane filtration for low pressure and diffusive membranes.

1

Chapter 4 concentrates solely on low pressure membrane filtration, microfiltration (MF), and ultrafiltration (UF). MF is the kingpin technology in making water reuse work. The removal of dissolved constituents, dissolved inorganic solids, dissolved organics, and dissolved organic carbon from membrane treated settled secondary effluent is accomplished using the pressure driven processes of nanofiltration (NF) and reverse osmosis (RO). In Chapter 5, NF and RO configurations are discussed. Topics include feed water composition, indicators of RO feed water quality, membrane fouling, pretreatment considerations, and posttreatment blending.

Membrane coupled bioprocesses (MCBs)—see list of acronyms—marry membranes with biological processes; new and disruptive MCBs are the topic of Chapter 6. Membrane bioreactors (MBRs) are covered in great detail in many texts on the market today. As such they are treated as a subset of MCBs and given only cursory treatment.

Chapter 7 ties in Chapters 1 through 6 and discusses the design, sizing, and selection of membrane technologies for water recycling and reuse applications. A myriad of topics are discussed, from source water considerations to system certification and system reliability. Examples and case studies are included.

The final chapter looks into future trends and challenges.

Water Scarcity

The major driver to further treat domestic and municipal secondary effluent for reuse or recycling is water scarcity. The global water supply is limited and its availability variable. At the same time, the demand for reliable water supply is growing globally. Freshwater sources and population density are not evenly distributed. Many countries, including the United States and China, have sufficient water, but this water is available in the wrong place at the wrong time, and in the wrong amounts. This uneven distribution and consequent mismatch between the demand for water and the supply of water is termed *water scarcity* by Global Water Intelligence (2010).

Water scarcity is caused by floods and droughts. It is caused by agricultural demand, industrial demand, and municipal demand. It is affected by infrastructure to transport the water to where it is needed when it is needed and to treat the water to the quality needed. Unlike oil, which has only a one-time use as fuel, water is renewable, recoverable, and reusable.

The following statistics are adapted from Global Water Intelligence (2010). Renewable freshwater is divided between groundwater and surface water. The annually recharged groundwater volume is said to be 3.0 B gal (11,358 km^3). The annually renewable surface water volume is 10.7 B gal (40,594 km^3). The Global Water Intelligence report notes that according to Aquastat in the Global Water Intelligence report, 600 to 700 km³ (0.15 to 0.2 B gal) of groundwater is believed to be withdrawn annually. The remainder of the 3802 km³ (1.0 B gal) of water withdrawn annually is surface water.

Water shortages, increasing population, and more stringent environmental regulations have created a compelling need for recycling and reuse of wastewater in the United States. The United States Environmental Protection Agency's (USEPA) guidelines on discharge limits of wastewater are forcing industrial users to consider recovery and reuse options. Mercury and selenium, for example, are now regulated to parts per trillion in industrial discharge in many states, forcing industry to consider recycle instead of discharge.

The limited capacity of existing publicly owned treatment plants continues to encourage wastewater recycling while investigating MCB processes to expand wastewater treatment capacity without changing footprint.

Water Supply

Water supply and availability are only partially understood. Estimates vary widely on the amount and availability of fresh water. Global Water Intelligence (2010) estimates that two-thirds of the earth is covered by water—366 trillion gal (1.385 B km³). About 2.5% of this water is freshwater available for use, but 69% is ice glaciers and polar caps and is not available. Thirty percent is fresh groundwater, and about 0.3% is available as surface freshwater, lakes, and rivers. The remaining 0.07% is unusable water, i.e., swamp water, soil voids, general moisture, and frost. The total renewable freshwater available per year is fixed 11.4 B gal (43,250 km³).

Of the freshwater available, on average about 10% is withdrawn each year. Seventy percent of that withdrawn water is estimated to be used for agriculture, 16 to 20% for industry, and 9 to 14% for municipal use. The water withdrawn varies among countries, with many withdrawing more than can be replenished by natural processes, 100%. Global water demand is estimated to increase 40% by 2020 (Global Water Intelligence, 2010).

Water Demand

To better understand the challenges of water scarcity and its complement water availability—of which there are many definitions, we must consider the global water supply and the overall water demand, i.e., agricultural demand, industrial demand, and municipal demand. Water supply is not the issue: water availability in the right place at the right time is the issue. Over the past century, global water withdrawals have increased by a factor of 6 (Global Water Intelligence, 2010).

Agricultural Demand

Water for agriculture is the water used to irrigate crops. Agricultural use of water outpaces industrial and municipal use by almost 5 times. In the United States today, 3000 lpcd (792.6 gpcd) of water is required to produce the country's food. One kilogram (0.45 lb) of grain requires 1 m^3 (264 lb) of water (Global Water Intelligence, 2010).

Industrial Demand

Water scarcity in the United States, together with stringent environmental regulations, has created a compelling need for industry to recycle and reuse wastewater. Industrial demand for water grew about 280% between 1905 and 2011, with power generation, refineries, and pulp and paper as the top three water users (Global Water Intelligence, 2010).

The volume of water needed to manufacture products varies. The volume of water needed to produce 1 L (0.26 gal) of gasoline is 10 L (2.64 gal); generating 1 kW of electricity requires 32 L (8.35 gal) of water. Producing 1 kg (0.45 lb) of steel requires an average of 95 L (25 gal) of water. Manufacturing 1 kg (0.45 lb) of paper requires 324 L (85.6 gal) of water. The figures are not fixed. The more water reused in the plant, the less total volume required (Wachinski, 2009).

Over the years there have been considerable improvements in the water efficiency of industry. Industrial water use per capita fell from 927 L (245 gal) in 1950 to 450 L (118.9 gal) in 2000. Much of this improvement has been achieved as a result of the switch away from single-pass cooling systems for power stations in Europe and North America. China also claims to have controlled industrial water demand by closing down industrial facilities which are wasteful of water. Despite the trend towards greater water efficiency in some parts of the world, industrial water demand is still expected to grow (Global Water Intelligence, 2010).

Regulations governing water treatment tighten yearly. At the same time, new generations of more sensitive water monitoring and measurement devices reveal that previously undetected contaminants exist in drinking water supply systems. Conventional gravity-dependent technologies such as clarifiers and loose media filters do not reliably remove many of the target pathogens that can affect public health, and the performance of the older technologies is also heavily dependent on appropriate applications of water-treatment chemicals.

Industries dependent upon water spend millions of dollars per year for chemicals to treat water. It is easily one of the highest costs for many treatment systems. In addition, the public's awareness of the potential health effects of some water-treatment chemicals is growing. Because of the escalating costs for freshwater and stringent discharge standards, manufacturers are looking for cost-effective methods to treat wastewater and reuse it when possible. Water reuse represents a practical and reliable means of extending water supplies in areas with water shortages.

Domestic Water Demand

The United Nations put the minimum daily demand for drinking, washing, and household use at around 50 L per capita per day (13 gpcd; Global Water Intelligence, 2010). Average per capita use in the United States is around 378.61 pcd (100 gpcd). The domestic water demand is a function of population, income, and water quality; for example, per capita consumption in a small village in Thailand quadrupled after installation of an MF system. Per capita consumption also increases as people install flush toilets, bathtubs, showers, washing machines, and—for those with high incomes—swimming pools and lawn irrigation.

Water Reuse Demand

Water reuse is treating wastewater in such a way that it can be reused. Wastewater is used drinking water with added organic matter, solids, and nutrients, such as nitrogen and phosphorus. Wastewater reuse can also be termed *wastewater mining*. We are not mining coal or gold, we are extracting nutrients, salts, energy, and inorganic compounds. In the future we will be able not only to extract nutrients, salts, energy, and inorganics, but to then produce "designer water," specialized water for special uses.

Total reuse in 15 years will be 33% of treated wastewater, about 80 million m^{3}d. Kuwait and Israel lead the world in reuse at 91% and 85% respectively. Global reuse is now at 4% (Global Water Summit, 2010).

Water Scarcity Solutions

Water, unlike oil, natural gas, and coal, is a renewable resource. It can be treated and then reclaimed or reused. A number of options are available to combat water scarcity. These include but are not limited to

1. Conservation
2. Location and development of new water sources
3. Aquifer storage and recovery
4. Wastewater reuse and recycling
5. Desalination of seawater and brackish water

Conservation

Water conservation should always be practiced, especially in industry, where it is directly tied to the cost of water needed to produce a product.

New Water Sources

Finding and developing new water sources is a viable option until all sources have been discovered and no new water sources can be brought online.

Recovery

The concept of aquifer storage and recovery (ASR), as explained by Rolf Herrinan, is to reinject (treated) water into certain geologic formations and create a large "bubble" displacing the native water below the surface (Global Water Intelligence, 2010). It is a controlled recharge because technicians inject through wells, and can know exactly what the volume going in is and where it goes, and the water can be recovered when it is needed. Aquifer storage provides a strategic reserve where it is possible to store very large volumes of water for up to 10–20 years, and can be used to balance consumption. Reclaimed water ASR systems are more prone to clogging than potable ones are, so further and more costly levels of treatment are needed, especially if injection wells are used. However, water quality improves during storage—it is a natural aquifer treatment, an interesting concept. The main challenge associated with reclaimed water ASR is striking a balance between level of treatment and associated water quality.

Wastewater Reuse and Recycling

The topic of this book! In areas of water scarcity, the upgrading of municipal wastewaters for indirect potable and direct industrial reuse, as well as internal industrial recycling, has become an attractive means of extending existing water supplies. In many of these applications, opportunities exist for the incorporation of membrane technologies.

Figure 1-1 summarizes water reuse today and projected wastewater reuse capacity through 2015. Kuwait leads the world in wastewater reuse today (91%), with Israel a close second at 85%. The United States is seventh in the world, behind Singapore (35%), Egypt (32%), China (14%), and Syria (12%). Global water reuse is in the low single digits (4%), but is expected to grow to 25% by 2025. The opportunity to expand water reuse globally is significant, and membrane-coupled bioprocesses will play a major role in this expanding water reuse market. Membranes can be used to further treat wastewater discharge to surface waters to treat wastewaters to a quality enabling beneficial reuse, to a quality for use in boilers and industrial process waters.

Table 1-1 shows the cost of water in key cities around the world. In one full-scale application, RO membrane technology is used at Water Factory 21 in Orange County, California, as part of a system that indirectly reclaims treated municipal effluent via groundwater infiltration. Reclamation for direct potable reuse is practiced

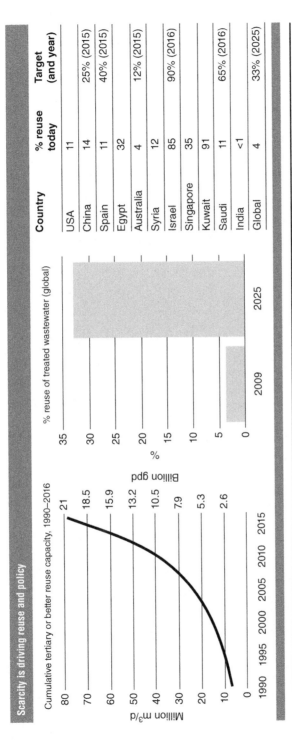

Scarcity is driving reuse and policy

Cumulative tertiary or better reuse capacity, 1990–2016

% reuse of treated wastewater (global)

Country	% reuse today	Target (and year)
USA	11	
China	14	25% (2015)
Spain	11	40% (2015)
Egypt	32	
Australia	4	12% (2015)
Syria	12	
Israel	85	90% (2016)
Singapore	35	
Kuwait	91	
Saudi	11	65% (2016)
India	<1	
Global	4	33% (2025)

FIGURE 1-1 Global water reuse trends. (*From Markoff, 2010.*)

7

Location	$/m³	$/1000 gal
Aarhus (Denmark)	8.59	32.51
Acapulco (Mexico)	0.58	2.20
Adelaide (Australia)	3.02	11.43
Baltimore (USA)	1.88	7.12
Bangalore (India)	0.17	0.64
Bejing (China)	0.54	2.04
Birmingham (UK)	3.86	14.61
Boston (USA)	3.05	11.54
Cleveland (USA)	1.88	7.12
Copenhagen (Den.)	9.07	34.33
Dallas (USA)	2.00	7.57
Denver (USA)	1.39	5.26
Detroit (USA)	2.98	11.28
El Paso (USA)	1.58	5.98
Fort Worth (USA)	2.34	8.86
Ho-Chi Minh City (Vietnam)	0.36	1.36
Houston (USA)	2.12	8.02
Indianapolis (USA)	2.35	8.89
Jeddah (Saudi Arabia)	0.05	0.19
London (UK)	3.46	13.10
Louisville (USA)	2.31	8.74
Manila West (Philippines)	0.42	1.59
Manila East (Philippines)	0.28	1.06
New York City (USA)	2.11	7.99
Philadelphia (USA)	3.21	12.15
San Diego (USA)	4.36	16.50
San Francisco (USA)	3.14	11.88
San Jose (USA)	3.41	12.91
Seoul (S. Korea)	2.35	8.89
Shanghai (China)	0.31	1.17
Singapore (Singapore)	3.56	13.47
Tokyo (China)	1.96	7.42
Washington DC (USA)	2.11	7.99

*The cost of water (selected water tariffs).
Source: Adapted from *Global Water Market 2011* (Global Water Intelligence, 2011).

TABLE 1-1   Cost of Water (2009)*

in Windhoek, the capital of Namibia. Technology for producing high-quality potable water from secondary municipal wastewaters has also been demonstrated in various pilot and demonstration scale facilities. The most notable of these was conducted at the demonstration water reclamation facility in Denver, Colorado, which included RO for organic pollutants. In both the Denver and Orange County facilities, RO is used only as a polishing step for highly treated wastewater. However, RO has been demonstrated to be an effective treatment process for reclamation when pretreatment is designed only to provide suitable water for the RO unit rather than to accommodate reclamation needs.

Desalination

Over 99% of the earth's water is seawater. The oceans and seas are literally drought proof. About 60% of the earth's population lives along ocean coastlines. Desalination of seawater and saline groundwater has become a major source of water in regions of the Middle East, which boast about two-thirds of the worlds desalting capacity. The American Water Works Association Research Foundation, Lyonnaise des Eaux, and the Water Research Commission of South Africa, in *Water Treatment Membrane Processes* (1996) reported the following on desalination capacity:

> Distillation dominated the desalination scene until about 1970. Since then, improvements in RO and *electrodialysis* (ED) technologies have resulted in substantial increases in their application. In 1988 there were 1742 RO plants—that is 49.4% of the total 3527 desalination plants in the world. ED now accounts for 564 plants or 16% of the total. On the basis of installed capacity, RO and ED account for approximately 23 percent and 5 percent, respectively, of the world's desalination capacity. By far the largest RO plants have been installed in Bahrain: a 45,420 m/d plant in Ras Abu-Jarjur, which desalinates highly brackish water, and a 56,000 m/d plant at Al Dur. *Ultrafiltration* (UF) and *microfiltration* (MF) have also been demonstrated as effective pretreatment technology.

Water Reuse Technology Overview

This book is about using membranes to reuse or recycle municipal or industrial wastewater treated to some degree. This book does not describe in detail conventional filtration technology or the biological treatment processes used to treat these waters.

The reader is directed to the following excellent texts:

- Water Environment Federation—*Biological Nutrient Removal (BNR) Operation in Wastewater Treatment Plants,*

- Metcalf & Eddy—*Water Reuse: Issues, Technologies, and Applications* (chapters 6 and 7).

- Metcalf & Eddy—*Wastewater Engineering: Treatment and Reuse* (all five editions),
- Water Environment Federation—*Design of Municipal Wastewater Treatment Plants,*
- Water Environment Federation—*Wastewater Treatment Plant Design,* American water works Association Research Foundation—*Water Treatment,* and
- *Water Environment Federation Manuals of Practice.*

Conventional Water Treatment Technology

Conventional water treatment technology is based on Stokes's law. A typical conventional plant incorporates the processes shown in Fig. 1-2. Source water is oxidized, if necessary, to convert dissolved iron or manganese to particulate (suspended) form. Inorganic coagulants such as alum, lime, iron salts, or organic polyelectrolytes are added in the rapid mix basin. After rapid mixing, the water is slowly mixed to flocculate and coagulate it and form large particles that settle. The coagulated water is sent to either final sedimentation or dissolved air flotation to reduce suspended solids to about 20 mg/L. Dissolved air flotation can be used in many applications, including removal of algae and floatables, thickening the backwash from depth filters or more efficient removal than sedimentation, of low-density materials. This effluent is fed to conventional media filtration before disinfection, and then to the distribution system.

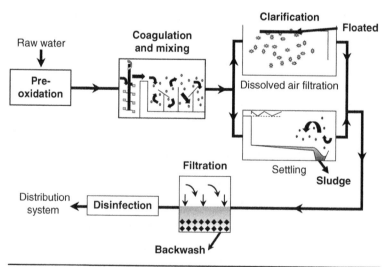

Figure 1-2 Conventional water treatment.

Although chemical precipitation—often a pretreatment step for UF membranes—is normally selected as a treatment method because of its efficiency in softening, it is also useful for the removal of other contaminants. It is particularly effective at removing iron and manganese, heavy metals, radionuclides, dissolved organics, and viruses.

Conventional water and wastewater filtration in this text can mean any depth or surface filtration technology used to remove suspended solids and any associated organic material from a water source or from secondary effluent. It can, in some cases, be used as pretreatment for a low pressure membrane process, i.e., MF or UF. The effluent from conventional filtration is termed *tertiary treatment*. In Fig. 1-2, the filtration effluent prior to disinfection is tertiary treatment.

A brief summary of conventional filtration is offered to the reader, as the term is used in the process flow schemes described in the remainder of the text. Metcalf and Eddy/AECOM (2007) provide an excellent treatment of conventional depth and surface filtration and serve as the source for the depth and surface filtration description offered here.

The mechanism of solids removal in a depth filter is sieving. Water or wastewater is passed through a bed of solids. Depth filtration is classified as continuous or semicontinuous. Filters that are taken off-line for backwashing are considered semicontinuous. Those that operate during the backwash step are classified as continuous. Depth filters are also classified according to depth—shallow, conventional, or deep bed; filter medium—mono (sand only for example), dual media, or multimedia (to include anthracite or a synthetic medium); and flow direction—traditional downflow or upflow. Depth filters can be driven by gravity or pressure.

Surface filtration technologies use the principle of mechanical sieving through a septum. The septum can be made from a number of materials, including cloth fabrics and different synthetic materials. Metcalf & Eddy, AECOM (2007) lists three principal cloth media surface filters: cloth-media filter, disc filter, and diamond cloth-media filter.

Conventional Wastewater Technology

This book deals with biologically treated wastewater after secondary treatment, typically called *secondary effluent*, and in some cases after primary treatment, (called *primary effluent*). The reader is directed to any of the numerous texts and handbooks dealing strictly with municipal and industrial biological wastewater treatment.

Conventional treatment processes used to treat municipal and industrial wastewaters are classified as either suspended growth processes (e.g., activated sludge), attached growth processes (such as trickling filters), or hybrid processes. The most commonly used process for biological treatment of both municipal and industrial wastewaters is

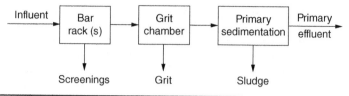

FIGURE **1-3**   Primary treatment.

the activated sludge process. Effluent from the conventional activated sludge process has in most cases no more than 20 mg/L of suspended solids and no more than 20 mg/L of Biochemical Oxygen Demand (BOD).

The activated sludge process includes primary treatment (screening, grit removal, and primary sedimentation), secondary treatment (aeration and secondary clarification), and tertiary treatment (CFS). (See Figs. 1-3 to 1-5.) Primary treatment includes grit removal, also called preliminary treatment, and clarification or sedimentation—used interchangeably. The effluent from the primary clarifier is called primary effluent. Secondary treatment includes aeration—in which the microorganisms use wastewater as a substrate for growth and remove dissolved organic matter—and sedimentation, in which the mass of organisms is separated from the clarified secondary effluent. The effluent from the clarifier following biological treatment is termed secondary effluent. The nutrients, nitrogen and phosphorus, are consumed in the process. When secondary effluent is further treated by some form of conventional filtration, i.e., sand filtration, mixed media filtration, or a cloth media filter, the effluent from the filtration step is termed tertiary treatment.

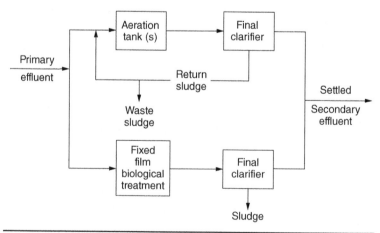

FIGURE **1-4**   Settled secondary treatment.

Figure 1-5 Tertiary treatment.

Membrane Technology

Water reuse and recycling and desalination of seawater and brackish water are two strategies to combat water scarcity. Membrane technologies play a key role in both.

The quality of the treated water is the same 24 hours per day at less than 0.1 ntu (nephlometric turbidity units)—very clean water. Most times it is in the 0.02 to 0.05 ntu range. The term ntu refers to the light scattering property of water and is explained in detail in Chapter 2. This is irrespective of the influent turbidity; if influent turbidity spikes, the permeate remains constant. Recovery—the percentage of treated water with respect to feed water—is especially high for hollow fiber MF (as high as 99.7%, or 98% if a second membrane system is used) and up to 85% for UF. The footprint of membranes is much lower than that of conventional filtration systems, varying between 25% and 50% of the area needed for a conventional plant. The drive for water reuse combined with concerns about the risks and cost of using chemicals and conventional technology has changed attitudes about membranes. Reduced chemical consumption and associated sludge disposal issues are two drivers behind the adoption of membrane technologies. Low pressure membranes generate a slurry, not a sludge, and membranes use less chemicals than conventional systems and can be operated remotely.

Several classes of treatment processes constitute membrane filtration for the purposes of this book. These processes include microfiltration (MF), ultrafiltration (UF), nanofiltration (NF), and reverse osmosis (RO).

MF is a size-exclusion, pressure driven membrane process that operates at ambient temperature. It is usually considered an intermediate option between UF and multimedia granular filtration, with pore sizes ranging from 0.10 to 1.0 μm. Hollow fiber MF used for wastewater reuse falls in the normal range of 0.1 to 0.2 μm and is an effective barrier for particles, bacteria, and protozoan cysts. MF systems operate at pressures between 10 and 30 pounds per square inch (psi; 0.7 and 2.1 bar).

UF membrane systems retain particulates, bacteria, protozoa, viruses, and organic molecules greater than their molecular weight cutoff (MWCO). UF membranes operate at pressures between 10 and 50 psi (0.7 and 3.5 bar) and offer absolute barriers to cysts, bacteria, and most viruses. Operating pressures are usually lower than those for NF and RO and higher than those for MF, although some UF systems are able to run at lower pressure than MF systems. Fluxes in general are lower than MF, and power consumption may be higher depending upon cross-flow operation. UF membrane systems typically receive higher virus removal credit than MF and, depending upon the removal rating, can achieve the required 4-log virus removal without chlorine addition. Most do not tolerate oxidants well.

NF membrane systems retain dissolved organic compounds, essentially all multivalent cations and anions, and a fraction of the monovalent species. NF membranes are often used to soften water. They operate at pressures between 50 and 150 psi (3.5 and 10.3 bar). RO membranes are used to remove over 99% of organic and inorganic species from MF/UF treated municipal water and to treat seawater to drinking water quality.

CHAPTER 2

Water Quality

Introduction

To design a membrane system, we must know not only the required flow rate and water quality of the treated waste stream but also a thorough characterization of the source water, i.e., its physical, chemical, and microbiological constituents. This chapter reviews basic chemistry with emphasis on the concepts and terminology used in membrane applications. Common water and wastewater chemicals are described as the coagulants and polyelectrolytes used in treatment. Source and treated wastewaters are characterized using their physical, chemical, and microbiological parameters. Expected water quality of untreated wastewater; primary effluent, different secondary effluent schemes, tertiary effluent, and quaternary effluent; and standards for water reuse and drinking water are presented.

Basic Chemistry Review

The purpose of this section is to review the fundamental concepts needed to better understand chemistry's role in membrane applications and operations.

Fundamental Concepts

All matter, the material things around us, solids, liquids, or gases, is composed of atoms—the atomic theory of matter which forms the basis of modern chemistry. The reader is directed to any introductory chemistry text for a more detailed treatment.

The purpose of this section is to briefly set definitions that are useful in reviewing chemical principles. The terms substance, element, and compound are defined.

A substance is often defined as a form of matter that cannot be separated into another form of matter by any physical process. Water is a substance. Sodium chloride (salt) is a substance. An element is a substance that cannot be decomposed by a chemical reaction into similar substances. Today there are 111 elements. Hydrogen, carbon, oxygen, and copper are elements. A compound is a substance composed of two or more elements chemically combined. The smallest indivisible unit of an element is an atom. Atoms of an element have a unique structure that provides the element its physical and chemical properties.

Atoms are composed of subatomic particles: protons, electrons, and neutrons. The proton has a positive electrical charge. The neutron has no charge. The electron has a negative electrical charge. Protons and neutrons are concentrated in the nucleus of the atom. A neutron and proton have about the same mass. The nucleus comprises over 99.9% of the atom's mass and carries a positive charge. The atomic number is the number of protons in the nucleus of an atom. Each element is unique in the number of protons, and every atom of that element has the same number. Examples are hydrogen, atomic number 1; carbon, 6; and oxygen, 8.

An electron's mass is about 1/1,840 that of a proton. The number of electrons in an atom varies. Electrons move around the nucleus in regular orbits, also called rings or energy levels. The outer ring contains the valence electrons. Atoms gain or lose electrons to complete the rings. The outer rings of inert elements are complete. *Valence* is a term which describes the "combining power of an element relative to hydrogen." The valence of a hydrogen atom is 1+. An element with a valence of 2+, for example, can replace two hydrogen atoms in a compound. In a neutral atom, the number of electrons equals the number of protons and the magnitude of the negative charge thus equals the magnitude of the positive charge. Nonneutral atoms, called ions, have either more or fewer electrons than protons and thus possess either a negative or a positive charge. Ions retain most of the properties of neutral atoms.

A molecule may have a net positive or negative charge and exist as an ion. An atom that loses an electron becomes a positively charged ion called a cation. An atom or molecule that gains an electron becomes an anion. An ionic compound is composed of cations and anions. The nitrate molecule, for example—chemical symbol NO_3^-— is composed of a single nitrogen atom and three oxygen atoms covalently bonded as one molecule with a net charge of minus one. Other common molecular and atomic ions are sulfate (SO_4^{2-}), a divalent anion with a charge of 2–; phosphate (PO_4^{3-}), a trivalent anion; sodium (Na^+), a monovalent cation; calcium (Ca^{2+}), a divalent cation; and aluminum (Al^{3+}), a trivalent cation.

Sodium and chlorine are good examples of ions. In water, to be more stable, a sodium atom readily donates an electron. A chlorine atom, to be more stable, needs another electron and readily accepts the electron of the sodium ion. In water, sodium is always a positive ion and chloride is always a negative ion.

Bonding

There are three important types of chemical bonding: ionic, covalent, and hydrogen. The two of most importance in ion exchange are ionic and covalent.

Ionic bonding occurs between two ions when one gives up one of its electrons to another; the two are bonded by the attraction of the electrical charges. Ion exchange in the classical sense only applies to ions.

A second type of bonding is covalent bonding, in which a pair or multiple pairs of electrons are shared. A shared pair of electrons is called a covalent bond. A carbon atom needs four electrons to fill in its outer shell and be stable. It can form four covalent bonds with other atoms or molecules.

The last type of bonding is hydrogen bonding, which produces the weakest bonds.

Ionization

The theory of ionization, attributed to Arrhenius, is used to determine the various strengths of acids and bases. Arrhenius attributed the different strengths of acids and bases to a difference in degree of dissociation or ionization, which serves today to explain many of the observed phenomena in aqueous solutions. According to the original theory of Arrhenius, all acids, bases, and salts dissociate into ions when placed in solution in water. Equivalent solutions of different compounds often vary greatly in conductivity. All strong acids and bases are assumed to approach 100% ionization in dilute solutions. The weak acids and bases, however, are so poorly ionized that in most cases it is impractical to express the degree of ionization as a percentage. Tables giving ionization constants of weak acids, bases, and salts are found in the usual handbooks, in many textbooks on quantitative analysis or physical chemistry, and on the Internet.

The equilibrium relationship can be used to describe ionization. For acids, the value of Z is equal to the number of moles of hydrogen ions (H^+) obtainable from one mole of the acid. For HCl, $Z = 1$; for H_2SO_4, $Z = 2$. For acetic acid (CH_3COOH), $Z = 1$, since only one of the hydrogen atoms in the acetic acid molecule will ionize to yield available H^+ ions in solution:

$$CH_3COOH \rightarrow CH_3COO^- + H^+$$

For bases, the value of Z is equal to the number of moles of H^+ with which one mole of the base will react. For NaOH, $Z = 1$; for $Ca(OH)_2$, $Z = 2$.

For example, sulfuric acid (H_2SO_4), has a molecular weight of 98.1 g. When sulfuric acid is added to water, it ionizes, i.e., it dissociates into two H^+ ions and one SO_4^{2-} radical. The net positive valence is $1 \times 2 + 2 = 4$, and the equivalent weight of sulfuric acid is 49 g. Hydrochloric acid

(HCl) dissolved in water ionizes into one H^+ ion and one Cl^- ion. Its molecular weight is 36.1 g and its equivalent weight is also 36.1 g.

Complex Ions—Ligands

Complex ions are soluble species formed through combination of simple species in solution. For example, if Hg^{2+} and Cl^- are present in water, they will combine to form the undissociated but soluble species $HgCl_2(aq)$, where "(aq)" is used to designate that the species is in solution. Chloride can also combine with mercury in other proportions to form a variety of complexes. Soluble molecules or ions which can act as chloride does to form complexes with metals such as mercury are called *ligands*. Among the ligands are H^+, OH^-, NH_3, F^-, CN^-, S_2O^{2-}, and many other inorganic and organic species. NH_3 complexes with metals are common—for example, with silver:

$$Ag(NH_3)_2^+ \rightleftharpoons Ag^+ + 2NH_3$$

All such ions are readily destroyed by creating conditions, physically or chemically, that will remove one of the dissociation products.

Ionic Strength

The solubility of sparingly soluble salts increases with increasing feed total dissolved solids (TDS). To account for this effect in calculating the solubility of a salt (e.g., calcium sulfate, barium sulfate, strontium sulfate, or SDSI), the ionic strength of a water is calculated. The ionic strength of each ion is derived by taking the concentration in parts per million (ppm) of each ion (as calcium carbonate) and multiplying by 1×10^{-5} for each monovalent ion and by 2×10^{-5} for each divalent ion. Summing the ionic strengths of all ions then gives the total ionic strength of the water.

Chelation

The solubility of metal ions is also increased by the presence of chelating agents. These substances have the ability to seize, or sequester, metal ions and hold them in a clawlike grip (the word *chelate* is from the Greek *chele*, meaning "claw"). Like a claw, a chelating molecule forms a ring in which the metal ion is held by a pair of pincers so that it is not free to form an insoluble salt. The pincers of a chelating molecule consist of ligand atoms (usually nitrogen, oxygen, or sulfur), each of which donates two electrons to form a coordinate bond with the ion. There are many natural chelates, such as hemoglobin (containing iron), vitamin B_{12} (containing cobalt), and chlorophyll (containing magnesium). Many well-known substances, such as aspirin, citric acid, adrenaline, and cortisone, can act as chelating agents; EDTA is a chelating agent which has a remarkable affinity for calcium and is used for the determination of hardness. Food-grade citric acid is used to remove inorganic contaminants from low pressure membrane surfaces as part of a clean in place operation to return a membrane to its original flux condition.

Adsorption

Adsorption is the process by which ions or molecules present in one phase tend to condense and concentrate on the surface of another phase. Adsorption of contaminants present in water or wastewater onto activated carbon is frequently used in conjunction with low pressure membranes as an alternative to a coagulation pretreatment step to reduce dissolved organics in water. Activated carbon is used as powdered activated carbon to adsorb dissolved organics from wastewater on its surface for subsequent removal by a low pressure membrane; or in a column as a granular activated carbon unit operation preceding or following the low pressure membrane.

The material being concentrated is the adsorbate, and the adsorbing solid is termed the adsorbent. There are three general types of adsorption: physical, chemical, and exchange adsorption. Physical adsorption is relatively nonspecific and is due to the operation of weak forces of attraction or van der Waals forces between molecules. The adsorbed molecule is not affixed to a particular site on the solid surface, but is free to move about over the surface. In addition, the adsorbed material may condense and form several superimposed layers on the surface of the adsorbent. Physical adsorption is generally quite reversible, i.e., with a decrease in concentration, the material is desorbed to the same extent that it was originally adsorbed.

Chemical adsorption, by contrast, is the result of much stronger forces, comparable with those leading to the formation of chemical compounds. Normally the adsorbed material forms a layer over the surface that is only one molecule thick, and the molecules are not considered free to move from one surface site to another. When the surface of the material is covered by a monomolecular layer, the capacity of the adsorbent is essentially exhausted. Also, chemical adsorption is seldom reversible. The adsorbent must generally be heated to higher temperatures to remove the adsorbed material.

Exchange adsorption is characterized by electrical attraction between the adsorbate and the surface. Ion exchange is included in this class. Here, ions of a substance concentrate at the surface as a result of electrostatic attraction to sites of opposite charge on the surface. In general, ions with greater charge, such as trivalent ions, are attracted more strongly toward a site of opposite charge than are molecules with lesser charge, such as monovalent ions. Also, the smaller the size of the ion (hydrated radius), the greater the attraction. Although there are significant differences among the three types of adsorption, there are instances in which it is difficult to assign a given adsorption to a single type. Ion exchange—strong acid, strong base, weak acid, and weak base—is a unit operation selected in posttreatment of either microfiltration (MF) or reverse osmosis (RO) to produce high quality water for industry. If, for example, silica must be removed to parts per trillion (ppt) levels, electrodeionization (EDI) will follow ion exchange.

Symbols, Formulas, and Equations

Each element is designated by a chemical symbol. The symbol consists of either one or two letters. The first letter is always capitalized, e.g., hydrogen is H, carbon is C, calcium is Ca, and sodium is Na (from its Latin name, *natrium*). Elements combine by transferring electrons. Oxygen, O_2, is a molecule formed when two atoms of the same element combine. Sodium chloride, NaCl, is a molecule formed by two different elements, sodium and chloride. A chemical formula describes each molecule and gives its elemental composition and relative proportions. Water, for example—H_2O—is a compound made of two atoms of hydrogen and one atom of oxygen covalently bonded together.

Atomic weight, the number of protons and neutrons in the nucleus, is the mass of an atom relative to one-twelfth the mass of a carbon 12 atom (which thus has an atomic weight of 12). This is important to membranes because the dalton, a unit of mass equal to one-twelfth the mass of a carbon 12 atom, is typically used as a unit of measure for the molecular weight cutoff of ultrafiltration (UF) membranes. The dalton is also used to measure the rejection of nanofiltration (NF) membranes.

The number of protons in an atom of a particular element is constant, but the number of neutrons may vary. Atoms of the same element with different atomic weights are called isotopes. Carbon has six protons and six neutrons. Its atomic weight is 12. Carbon 14 (^{14}C) has six protons and eight neutrons. It is an isotope.

Gram atomic weight is atomic weight expressed in grams. Gram molecular weight is the atomic weight of a molecule in grams. The gram molecular weight of any compound contains the same number of molecules. Avogadro's number is the number of atoms in 12 grams of pure carbon (CL^2), i.e., 6.02×10^{23}. A mole of any substance contains 6.02×10^{23} entities of that substance; the entities can be atoms, molecules, or ions. Twelve grams of carbon contains 6.02×10^{23} atoms. Sixteen grams of oxygen contains 6.02×10^{23} atoms. Seventeen grams of the hydroxyl radical (OH^-) contains 6.02×10^{23} ions. Equivalent weight is strictly defined as the weight of a compound which contains one gram-atom of available hydrogen, or its chemical equivalent. It is the weight of a substance based on the formula weight and the number of reactive protons associated with the substance. The equivalent weight is determined by the following relationship:

$$\text{Equivalent weight} = \frac{MW}{Z}$$

In this relationship, MW is the molecular weight of the compound and Z is a positive integer whose value depends upon the chemical context: the compound's net positive valence. Net positive valence is the valence of the positive element multiplied by its subscript, e.g., $NaCl = 1 \times 1 = 1$; $CaSO_4 = 2 \times 1 = 2$; and $Fe_2(SO_4)_3 = 3 \times 2 = 6$.

Bar Graphs

Results of a water analysis are normally expressed in milligrams per liter and reported in tabular form. For better visualization of the chemical composition, these data can be expressed in either milliequivalents per liter or in milligrams per liter expressed as $CaCO_3$ to permit graphical presentation.

An order of preference exists in nature for the combining of anions and cations. For cations the order is

Iron
Aluminum
Magnesium
Sodium
Potassium
Hydrogen

For anions the order of preference is

Hydroxide
Bicarbonate
Carbonate
Sulfate
Chloride
Fluoride
Nitrate

To plot a bar chart, two bars are constructed equal to the scale length of total anions or cations. Cations are plotted on the top bar and anions on the bottom in the order just listed.

In most waters, some carbon dioxide (CO_2) is dissolved in the sample but usually not reported in the analysis. The quickest and most accurate way to get CO_2 concentration is to use the total dissolved solids and pH, and calculate carbon dioxide concentration using the chart in any edition of *Standard Methods for the Examination of Water and Wastewater*.

The sum of the positive values of milliequivalents per liter must equal the sum of the negative values for water in equilibrium. Hypothetical combinations of positive and negative ions can be written from a bar graph. These combinations can be used to confirm the accuracy of a chemical analysis. Many times, hollow fiber MF follows a lime-soda softening plant. Bar charts are helpful in calculating the chemicals added for lime-soda ash softening.

A typical problem in membrane operations is—how to dispose of the chemical backwash from a low pressure membrane system? Before a process can be selected, the engineer must know the constituent compounds in the backwash. The first step is to prepare a bar chart that illustrates how to construct a bar chart. Since both phenolphthalein (P) alkalinity and methyl orange (MO) alkalinity are given, we know that the pH is greater than 8.3.

Example 2-1 Bar Chart Example—Chemically Enhanced Backwash Wastewater. Given the following analysis, find the hypothetical combinations of compounds and prepare a bar chart.

Analysis:

$$P\ alk = 25\ mg/L\ as\ CaCO_3$$
$$MO\ alk = 265\ mg/L\ as\ CaCO_3$$
$$Ca^{2+} = 72\ mg/L\ as\ Ca^{2+}$$
$$Mg^{2+} = 50.4\ mg/L\ as\ Mg^{2+}$$
$$Na^+ = 1{,}219\ mg/L\ as\ Na^+$$
$$SO_4^{2-} = 163.2\ mg/L\ as\ SO_4^{2-}$$
$$Cl^- = 1{,}781\ mg/L\ as\ Cl^-$$

Since $P < \frac{1}{2}MO$, $P < \frac{1}{2}MO$, then $CO_3^{2-} = 2P = (2)(25) = 50\ mg/L\ as\ CaCO_3$.

$$HCO_3^- = T - 2P = 260 - (2)(25) = 260 - 50 = 210\ mg/L\ as\ CaCO_3$$

Express constituents as calcium carbonate $(CaCO_3)$:

$$Ca^{2+} = 72 \times 50/20 = 180\ mg/L\ as\ CaCO_3$$
$$Mg^{2+} = 24 \times 50/12 = 210\ mg/L\ as\ CaCO_3$$
$$Na^+ = 1{,}219 \times 50/23 = 2{,}650\ mg/L\ as\ CaCO_3$$
$$SO_4^{2-} = 163.2 \times 50/48 = 170\ mg/L\ as\ CaCO_3$$
$$Cl^- = 1{,}825 \times 50/35.5 = 2{,}570\ mg/L\ as\ CaCO_3$$

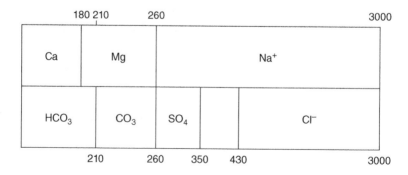

Find hypothetical combinations using order of preference:

Iron	Hydroxide
Aluminum	Bicarbonate
Calcium	Carbonate
Magnesium	Sulfate
Sodium	Chloride
Potassium	Fluoride
Hydrogen	Nitrate

$Ca(HCO_3)_2$ =180 mg/L as $CaCO_3$
$Mg(HCO_3)_2$ = 30 mg/L as $CaCO_3$
$MgCO_3$ = 50 mg/L as $CaCO_3$
$MgSO_4$ = 90 mg/L as $CaCO_3$
Na_2SO_4 = 80 mg/L as $CaCO_3$
$NaCl$ = 2,570 mg/L as $CaCO_3$

The pH is above 8.4, since there are two kinds of alkalinity.

$Ca(HCO_3)_2$ – 200 mg/L as $CaCO_3$ KCl – 10 mg/L as $CaCO_3$
$Mg (HCO_3)$ – 35 mg/L as $CaCO_3$
$MgCO_3$ – 65 mg/L as $CaCO_3$ pH above 8.4, since 2 kinds of alkalinity
Na_2CO_3 – 65 mg/L as $CaCO_3$
K_2CO_3 – 10 mg/L as $CaCO_3$
K_2SO_4 – 15 mg/L as $CaCO_3$

Units of Expression

Parts per billion (ppb): A method for reporting the concentration of an ion or substance in a water. The following conversions apply for dilute waters with a specific gravity of 1.0: One ppb is equal to 1 µg/L. One ppm is equal to 1,000 ppb.

Parts per million (ppm): A method for reporting the concentration of an ion or substance in a water. The following conversions apply for dilute waters with a specific gravity of 1.0: One ppm is equal to 1 mg/L. One grain per U.S. gallon (gpg) is equal to 17.1 ppm. One pound per 1,000 U.S. gallons is equal to 120 ppm. A 1% solution is equal to 10,000 ppm. One ppm is equal to 1,000 ppb.

Parts per million as calcium carbonate ($CaCO_3$): A method for reporting the concentration or equivalent weight of an ion or substance in a given volume of water. Reporting the concentration of ions in ppm as $CaCO_3$ is popular among ion exchange chemists for the calculation of ionic loading of cation or anion resins. It is also popular in determining whether a water analysis is balanced, i.e., whether the sum of the cations equals the sum of the anions when the concentrations of the ions are reported as calcium carbonate equivalents. Water chemists use the concept of equivalency when balancing cation and anion electroneutrality levels, since ions combine in nature based on their valence state and available electrons, not on their actual weight. Calcium carbonate was arbitrarily picked because its molecular weight is 100 and its equivalent weight is 50 (since it is divalent). The formula to convert an ion reported in milligrams per liter to ppm as $CaCO_3$ is to multiply the milligrams per liter by the ratio of the equivalent weight of the ion to the equivalent weight of calcium carbonate.

As an example, a water with sodium at 100 ppm as $CaCO_3$ and chloride at 100 ppm as calcium carbonate is in ionic balance, since every sodium ion has a corresponding chloride ion. However, sodium concentration at 100 ppm as $CaCO_3$ is only 47 mg/L of actual substance (since its equivalent weight is 23.0), and chloride at 100 ppm as CaCO3 is only 71 mg/L of actual substance (since its equivalent weight is 35.5). The calculated TDS of this solution is 118 mg/L.

The concentration of ions or chemical compounds in solution is normally expressed as the weight of the element or compound in milligrams (mg) per liter (L) of water. The density of water is 1.0 mg/mL, depending upon temperature. Although 1 mg per million milligrams is equivalent to 1.0 ppm, and the term *parts per million* is thus acceptable, it is not preferred.

Chemical dosages may also be expressed as pounds of chemical per million gallons. Multiply milligrams per liter by 8.34 to obtain pounds per million gallons:

$$\frac{mg}{L} \times 8.34 = \text{pounds per million gallons}$$

Another way of expressing concentration often used by engineers is grains per gallon. One pound is 7,000 grains. Therefore, 1.0 grain per gallon is 17.1 mg/L. Elemental concentration expressed as milligrams per liter usually means that the solution contains the stated number of milligrams of a substance in 1 L of water; for example, 2.5 mg/L of iron means that each liter of water contains 2.5 mg of iron. Frequently, though, a chemical concentration—most often hardness—is expressed as milligrams per liter or milligrams per liter as calcium carbonate.

Using milliequivalents per liter (mEq/L) is similar to expressing as calcium carbonate, i.e., the term expresses the dissolved constituent in terms of its combining weight. Milliequivalents per liter are computed from milligrams per liter for elemental ions and for radicals or compounds. For elements, mEq/L = mg/L × valence/atomic weight.

For compounds or radicals, mEq/L = mg/L × electrical charge/ molecular weight = mg/L/equivalent weight.

The term *hardness* refers to the concentration of calcium and magnesium combined. When expressed as calcium carbonate, the calcium and magnesium weights are multiplied by the equivalent weight of calcium carbonate (50) divided by the equivalent weight of the cation (calcium = 40/2 = 20 and magnesium = 24/2 = 12). The calcium and magnesium values are added together and the total is used to express the hardness of that water. For example, if the concentration of calcium were 250 mg/L as $CaCO_3$ and the concentration of magnesium were 50 mg/L as $CaCO_3$, the hardness of the water would be expressed as 300 mg/L as $CaCO_3$.

The alkalinity of water may contain any combination of bicarbonate (HCO_3), carbonate (CO_3), or hydroxide (OH) species, but its concentration may be expressed as milligrams per liter as $CaCO_3$ and include all of the species of alkalinity.

Two other compounds are sometimes expressed differently. Nitrogen compounds—ammonia (NH_3), nitrate (NO_3), and nitrite (NO_2)—are sometimes expressed in milligrams per liter as nitrogen (N). Phosphates are often expressed milligrams per liter as phosphorus (P).

Care must be taken to ensure the interpretation is correct.

Solutions

Membranes are periodically cleaned by chemical solutions. An understanding of solution terminology is helpful in understanding the strength of the solutions used and is necessary to compute the quantities of chemical needed to prepare a solution.

Example 2-2 illustrates how to calculate the amount of sodium hydroxide to prepare a 4% solution and how to express it on in pounds per gallon.

Example 2-2 Calculate Percent Solution
How many pounds per gallon (lb/gal) of NaOH are required to prepare a 4% NaOH solution?

Solution What we know:

The specific weight of water is 8.34 lb/gal.

The specific gravity of a 4% NaOH solution is 1.048.

Multiply the specific gravity by the specific weight of water to get the specific weight of a 4% NaOH solution: 8.34 lb/gal × 1.048 = 8.74 lb/gal.

Multiply 8.74 lb/gal × 0.04 = 0.35 lb/gal.

Therefore, a 4% NaOH solution requires 0.35 lb/gal of NaOH.

Another way:

One percent solution is 10,000 mg/L; 4% is 40,000 mg/L.

Multiply 40,000 mg/L × 1.048 = 41,920 mg NaOH/liter of water.

41,920 mg/L × 3.785 L/gal × 1 lb/453,560 mg = 0.35 lb/gal.

The normality of a solution is its relation to a normal solution. The symbol N is used as the abbreviation for normal. A normal solution contains one equivalent weight of a substance per liter of solution. To prepare 1 L of a normal solution of an acid or a base requires sufficient compound to furnish 1.008 g of H^+ or 17 g of OH^- and enough distilled water to make a solution having a volume of 1 L.

The molarity of a solution is its relation to a molar solution. The molarity of a solution is the number of moles of solute per liter of solution, e.g., a solution containing one gram molecular weight per liter of solution is a one molar solution.

A one molar solution is one gram molecular weight in 1,000 g of solution.

The common inorganic chemical compounds used in membrane operations in water are listed in Table 2-1. This table provides the name, formula, atomic or molecular weight, and equivalent weight of chemicals commonly used as inorganic coagulants and in chemical cleaning operations, as well as the multiplication factor used to express their concentrations as $CaCO_3$.

Nomenclature

In dealing with membrane cleaning chemicals, a few basic rules can be helpful. Binary compounds have the ending -ide (e.g., NaOH is sodium hydroxide, NaCl is sodium chloride). Acids containing oxygen can be tricky. The highest oxidation state is indicated by the -ic suffix and is typical of the mineral acids often used with membranes: HCl is hydrochloric acid and H_2SO_4 is sulfuric acid. Oxalic and phosphoric acids are also sometimes used.

The next lower oxidation state uses the -ous suffix, as in sulfurous and phosphorous acids. The lowest oxidation state is indicated by the *hypo-* prefix, e.g., hypochlorous acid, and by both the *hypo-* prefix and the -*ite* suffix, e.g., hypochlorite. The reader is directed to any elementary chemistry book to review other nomenclature issues.

Gas Laws

Air is used alone or in combination with water to periodically remove contaminants from a membrane surface. This process is sometimes called flux maintenance. A basic understanding of the

Substance	Formula	Atomic or Molecular Weight	Equivalent Weight	Multiplication Factor to Express as $CaCO_3$
Aluminum hydrate	$Al(PH)_2$	78.0	26.0	1.92
Aluminum sulfate	$Al_2(SO_4)_3 18H_2O$	666.4	111.1	0.45
Aluminum sulfate	$Al_2(SO_4)_3$ (anhydrous)	342.1	57.0	0.88
Ammonia	NH_3	17.0	17.0	2.94
Ammonium (ion)	NH_4	18.0	18.0	2.78
Barium	Ba	137.4	68.7	0.73
Calcium	Ca	40.1	20.0	2.50
Calcium carbonate	$CaCO_3$	100.08	50.1	1.00
Calcium hydroxide	$Ca(OH)_2$	74.1	37.1	1.35
Calcium hypochlorite	$Ca(ClO)_2$	143.1	35.8	0.70
Calcium sulfate	$CaSO_4$ (anhydrous)	136.1	68.1	0.74
Calcium sulfate	$CaSO_4 2H_2O$ (gypsum)	172.2	86.1	0.58
Chlorine	Cl	35.5	35.5	1.41
Copper sulfate (cupric)	$CuSO_4$	160.0	80.0	0.63
Copper sulfate (cupric)	$CuSO_4 5H_2O$	250.0	125.0	0.40
Iron (ferrous)	Fe^{+2}	55.8	27.9	1.79
Iron (ferric)	Fe^{+3}	55.8	18.6	2.69
Ferrous hydroxide	$Fe(OH)_2$	89.9	44.9	1.11
Ferrous sulfate	$FeSO_4$ (anhydrous)	151.9	76.0	0.66
Ferrous sulfate	$FeSO_4 7H_2O$	278.0	139.0	0.36
Ferrous sulfate	$FeSO_4$ (anhydrous)	151.9	151.9	Oxidation
Ferric chloride	$FeCl_3$	162.0	54.1	0.93

TABLE 2-1   Common Water and Wastewater Treatment Chemicals Used in Membrane Applications

Substance	Formula	Atomic or Molecular Weight	Equivalent Weight	Multiplication Factor to Express as $CaCO_3$
Ferric chloride	$FeCl_2 6H_2O$	270.0	90.1	0.56
Ferric sulfate	$Fe_2(SO_4)_3$	399.9	66.7	0.75
Magnesium	Mg	24.3	12.2	4.10
Magnesium sulfate	$MgSO_4$	120.4	60.2	0.83
Manganese (manganous)	Mn^{+2}	54.9	27.5	1.83
Manganese (manganic)	Mn^{+3}	54.9	18.3	2.73
Manganese chloride	$MnCl_2$	125.8	62.9	0.80
Manganese dioxide	MnO_2	86.9	21.7	2.30
Manganese hydroxide	$Mn(OH)_2$	89.0	44.4	1.13
Nitrogen (valence 3)	N^{+3}	14.0	4.67	10.7
Nitrogen (valence 5)	N^{+5}	14.0	2.80	17.9
Oxygen	O	16.0	8.00	6.25
Phosphorus (valence 3)	P^{+3}	31.0	10.3	4.85
Phosphorus (valence 5)	P^{+5}	31.0	6.20	8.06
Sodium	Na	23.0	28.0	2.18
Sodium bicarbonate	$NaHCO_3$	84.0	84.0	0.60
Sodium bisulfate	$NaHSO_4$	120.0	120.0	0.42
Sodium bisulfite	$NaHSO_3$	104.0	104.0	0.48
Sodium carbonate	$Na_2CO_3 10H_2O$	286.0	143.0	0.35
Sodium chloride	$NaCl$	58.58	58.5	0.85
Trisodium phosphate	$Na_3PO_4 12H_2O$ $(18.7\% \ P_2O_5)$	380.2	126.7	0.40

TABLE 2-1 Common Water and Wastewater Treatment Chemicals Used in Membrane Applications

Substance	Formula	Atomic or Molecular Weight	Equivalent Weight	Multiplication Factor to Express as CaCO$_3$
Trisodium phosphate (anhydrous)	Na$_3$PO$_4$ (43.2% P$_2$O$_5$)	164.0	54.7	0.91
Disodium phosphate	Na$_2$HPO$_4$12H$_2$O (19.8% P$_2$O$_5$)	358.2	119.4	0.42
Disodium phosphate (anhydrous)	Na$_2$HPO$_4$ (50% P$_2$O$_5$)	142.0	47.3	1.06
Monosodium phosphate	NaH$_2$PO$_4$H$_2$O (51.4% P$_2$O$_5$)	138.1	46.0	1.09
Monosodium phosphate (anhydrous)	NaH$_2$PO$_4$ (59.1% P$_2$O$_5$)	120.0	40.0	1.25
Sodium sulfate (anhydrous)	Na$_2$SO$_4$	142.1	71.0	0.70
Sodium sulfate	Na$_2$SO$_4$10H$_2$O	322.1	161.1	0.31
Sodium thiosulfate	Na$_2$S$_2$O$_3$	158.1	158.1	0.63
Sodium sulfite	Na$_2$SO$_3$	126.1	63.0	0.79
Water	H$_2$O	18.0	9.00	5.56
Bicarbonate	HCO$_3$	61.0	61.0	0.82
Carbonate	CO$_3$	60.0	30.0	0.83[a]
Carbon dioxide	CO$_2$	44.0	44.0	1.14

TABLE 2-1   (Continued)

gas laws is required to understand the design and operation of these systems.

- Boyle's law: the volume of a gas varies inversely with the pressure at constant temperature:

$$P_1V_1 = P_2V_2$$

- Charles's law: the volume of a gas at constant pressure varies in direct proportion to the absolute temperature:

$$\frac{V_1}{T_1} = \frac{V_2}{T_2}$$

- Generalized gas law:

$$PV = nRT$$

where n is the number of moles of gas, R is the universal gas constant (0.082 atm/mol), T is the absolute temperature in kelvins, P is the pressure, and V is the volume.

- Dalton's law of partial pressures: the total pressure of a gaseous mixture is equal to the sum of the partial pressures of the component gases. The total pressure of a component of a gas mixture is the pressure that gas would exert if it were the only gas that occupied the entire volume.

- Henry's law (most applicable to membrane operations involving gas and liquids): the mass of a slightly soluble gas dissolved in a liquid at a given temperature is directly proportional to the partial pressure of the gas (CRC Handbook of Chemistry and Physics, 76th ed.):

$$C_{equil} = HP_{gas}$$

where C_{equil} is the concentration of dissolved gas at equilibrium, P_{gas} is the partial pressure of the gas above the liquid, and H is the Henry's law constant for the gas at a given temperature.

Dilutions

Diluting a solution increase the volume while the solute remains constant:

$$V_1(Conc)_1 = V_2(Conc)_2$$

Sampling

Two types of sampling techniques are used to collect wastewater samples for analysis: grab samples and composite samples.

Grab samples are only indicative of water quality at a single point in time. Composite samples are collected at intervals over a specified time, e.g., every hour over a 24 h period, proportioned to flow if appropriate, and then combined to allow characterization of the water. In some cases, the entire volume of water to be measured is collected.

For example, if we wish to characterize the backwash water generated during the operation of an MF system, we may chose to use both grab samples at specific time intervals to characterize water quality with time and a composite sample to indicate the water quality of the entire backwash so appropriate treatment can be specified.

Domestic wastewater remains consistent over time. For most uses, a grab sample is adequate for characterization.

Chemicals Used in Wastewater Reuse

Many chemicals are either added directly to the wastewater or used to clean the membranes. Chemical additives are commonly used in coagulation and flocculation, corrosion control, pretreatment chemical softening, pretreatment for membranes, chemical precipitation, scale inhibition for diffusive membranes, pH adjustment, disinfection and oxidation, and algae and aquatic weed control.

Coagulants

Selection

Coagulants can generally be grouped into four categories:

1. Simple metal salts—aluminum sulfate, ferric sulfate, and ferric chloride. They sold as dry crystallized solids and as concentrated (~2 M) solutions.

2. Prehydrolyzed metal salts—coagulants with base added during manufacturing to reduce the alkalinity consumption from hydrolyzing of metal salts. Polyaluminum chloride (PACL) is the most common prehydrolyzed metal salt. Prehydrolyzed iron is rare as a commercial product. The amount of base added during manufacturing is expressed as basicity—the percent of molar ratio of hydroxide to metal salt. Commercial prehydrolyzed metal salts have basicity raging from 10% to 83%. As basicity increases beyond 75%, it becomes increasingly difficult to keep the metal hydroxide from forming in the solution during shipping and extended storage (Letterman, Amirtharaja, and O'Melia, 1999).

3. Acid-fortified metal salts—coagulants in liquid form supplemented with strong mineral acids (e.g., H_2SO_4 and HCl). Acid-fortified coagulants, when added into water, consume more alkalinity and depress pH to a greater degree than simple metal salts.

4. Additive-supplemented metal salts—metal salt coagulants premixed with additives such as phosphoric acid, sodium silicate, calcium salts, and cationic polymers such as epichlorohydrin polydimethylamine and polydiallyl dimethyl ammonium chloride.

The basic information on the most commonly used coagulants and prehydrolyzed coagulants is presented in Tables 2-2 and 2-3.

TABLE 2-2 Characteristics of Commercial Coagulants

Coagulant	Formula	WM (g/mol)	Form	Concentration (% of Weight)	Specific Gravity (g/mL)	Metal Content (% of Weight)	Alkalinity Consumption[a] (mg alk./mg Coagulant)
Alum	$Al_2(SO_4)_3 14H_2O$	594	Liquid	50	1.34	4.5	0.25
Alum	$Al_2(SO_4)_3 14H_2O$	594	Solid	100	—	9.1	0.51
Ferric chloride	$FeCl_3$	162.3	Liquid	40	1.43	13.8	0.37
Ferric chloride	$FeCl_3$	162.3	Solid	100	—	34.4	0.92
Ferric sulfate	$Fe_2(SO_4)_3 9H_2O$	561.6	Solid	100	—	19.9	0.53
Ferrous sulfate	$FeSO_4$	227.8	Solid	100	—	20.1	0.36

[a]For non-acid-fortified, non-additive-supplemented coagulants only.

Coagulant	Formula	WM (g/mol)	Form	Concentration (% of Weight)	Specific Gravity (g/mL)	Metal Content (% of Weight)	Alkalinity Consumption[a] (mg alk./mg Coagulant)
ProPaC 9700	Unknown	—	Liquid	100	1.34	12.2	—
PAX 18	Unknown	—	Liquid	100	—	9.0	—
PAX XL9	Unknown	—	Liquid	100	1.43	5.5	—
Hyperion 1090	Unknown	—	Liquid	100	—	12.2	—
Samulchlor 50	Unknown	—	Liquid	100	—	12.2	—

[a]For non-acid-fortified, non-additive-supplemented coagulants only.

TABLE 2-3 Characteristics of Commercial Prehydrolyzed Coagulants

There are two different and related criteria for selecting a coagulant for coagulation in MF/UF processes:

- whether the coagulant selected can effectively remove total organic carbon (TOC) and
- whether the coagulant selected can minimize the fouling of membranes.

From the perspective of TOC removal from secondary effluent, there is no consensus on what coagulant works better. Research comparing iron-based coagulants and alum indicated that iron-based coagulants worked better than alum (Grozes, White, and Marshall, 1995). PACL may be more effective at removing natural organic matter (NOM) under certain water quality conditions (Dempsey, Ganho, and O'Melia, 1984). Other researchers have concluded that different coagulants show no substantial difference when compared on the basis of the same metal dose (Howe and Clark et al., 2002). It is apparent that removal efficiency of coagulation is very much affected by the type of NOM and other water quality parameters. High molecular weight, more hydrophobic, and more acidic fractions of NOM are preferably removed by the coagulation process (Randtke, 1988; Owen et al., 1995; Dennett et al., 1996; White et al., 1997; Howe and Clark, 1999). The removal of NOM has been reported between 10% and 90%, with removal between 30% and 50% being more typical. The enhanced coagulation rule under the U.S. Environmental Protection Agency's (USEPA's) Phase II disinfection/disinfection by product (D/DBP) regulations requires removing 15% to 50% of NOM, depending upon raw water TOC and alkalinity, as summarized in Table 2-4.

The enhanced coagulation rule was designed as a compromise between the water qualities desired to minimize DBPs and the practical limitations of removal efficiency of NOM by coagulation. The rule specifies when that specific ultraviolet (UV) absorbency (SUVA; the ratio of UV_{254} to dissolved organic carbon [DOC]) is less than 2, or when DOC reduction is less than 0.3 mg/L for every 10 mg/L of alum, NOM is not amenable by coagulation. One common

TOC (mg/L)	Alkalinity (mg/L CaCO$_3$)		
	< 60	60–120	> 120
4	35%	25%	15%
4–8	45%	35%	25%
> 8	50%	40%	30%

TABLE 2-4   TOC Removal Requirement by Enhanced Coagulation

misunderstanding of the enhanced coagulation rule is that compliance with the rule leads to compliance with the D/DBP rule. This is not necessarily the case for the majority of surface water sources, for which water treated by enhanced coagulation may still contain NOM that generates DBPs when chlorinated. Measures such as altering the disinfection strategy are necessary for utilities with those water sources to be in compliance with the D/DBP rule.

From the perspective of minimizing membrane fouling, even though in general, better removal of NOM typically leads to less fouling, the fouling of membranes may be more complex than as explained by the removal of NOM alone. It has been found that the alkalinity of source water appears to have a significant impact on the selection of coagulants. As a rule of thumb, for low alkalinity water (alkalinity < 60 mg/L CaCO$_3$) prehydrolyzed coagulants (such as PACL) seem to have the least fouling, based on field experiences and measurement of the membrane performance index. A possible explanation is that for low alkalinity water, the formation of ferric hydroxide may be limited thermodynamically and/or kinetically. As a result, metal ions that do not form metal hydroxide can form a metal-NOM complex on the surface of the membrane and foul the membrane. This metal-NOM complex formation and its impact on membrane fouling has been identified at one membrane plant filtering low alkalinity water coagulated with ferric chloride as the reason for membrane fouling.

Dose of Coagulants

Both TOC reduction and solids loading can affect membrane fouling under different coagulant doses. TOC reduction appears to be more crucial for reducing membrane fouling, although the fraction of NOM that causes membrane fouling is likely not more than 10% to 15% of TOC concentration (Howe and Clark, 2002). As a general rule, the coagulant dose required to achieve enhanced coagulation is higher than the dose required for optimal turbidity removal. NOM reduction as measured by TOC has been found typically to be in the range of 0.1 to 0.5 mg/L for every 10 mg/L of alum. In the *Enhanced Coagulation and Enhanced precipitation Guidance Manual*, incremental TOC reduction of 0.3 mg/L for every 10 mg/L of alum is chosen as the criterion for dose determination (USEPA, 1999). That is, if the incremental TOC reduction is less than 0.3 mg/L for every 10 mg/L of alum, increasing the coagulant dose beyond this point is deemed not cost effective.

The relationship between membrane fouling and coagulant dose can be complicated. There is a V curve: at a low coagulant dose, membrane fouling becomes worse than with no coagulant at all. As the dose increases, membrane fouling decreases, as illustrated in Fig. 2-1.

This type of relationship between coagulant dose and membrane fouling was also observed in a pilot study. When there was no alum

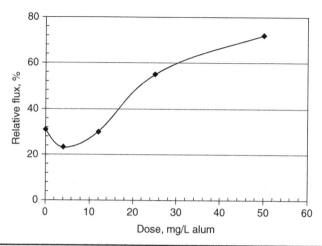

FIGURE 2-1 Effect of alum on MF flux.

addition, the increase in transmembrane pressure (TMP) was about 12 to 13 psid/day. At an alum dose of 9 to 12 mg/L, the increase in TMP reduced to 8 to 9 psid/day. When the alum dose was reduced to a range of 4.5 to 6 mg/L, the TMP increase reached 16 to 25 psid/day. The TMP trend over time at different operating conditions is presented in Fig. 2-2.

The reason for the changes in membrane fouling in response to coagulant dose is not clear yet. However, it is possible that different water chemistry conditions at different coagulant doses might affect

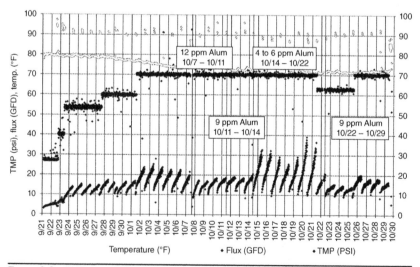

FIGURE 2-2 TMP versus flux for alum coagulant.

membrane fouling. Without coagulation, NOM in natural water typically carries negative charges and is stable. At low coagulant doses, NOM might be destabilized through charge neutralization and become ready to adsorb on the membrane surface. As the coagulant dose increases, NOM begins to adsorb onto the metal hydroxide flocs and fouling is reduced. At a dose far beyond the dose for enhanced coagulation, it is plausible to assume that low pH from the consumption of the added coagulants could cause dissolution of metal hydroxides and NOM could be restabilized. In turn, it would increase membrane fouling and turn the V curve into an S curve. Scanning electron microscope (SEM) images show that coagulation causes distinctive changes to materials captured at the membrane surface. Without coagulant, small grainy particles with diameter about 10 to 30 nm have been observed covering all membrane surfaces. With coagulation, larger nodules (80 to 130 nm) of materials aggregated into clumps, leaving other membrane surfaces clean. An overall reduction of the amount of material on the membrane surface was noted as the alum dose increased (Howe and Clark, 2002).

Direct Coagulation versus Sedimentation

Coagulation reduces membrane fouling by mainly reducing NOM and possibly small inorganic colloids in feed water. At the same time, coagulation generates a large amount of solids from metal hydroxides. As the coagulant dose increases, the amount of solids generated increases. Therefore, at the core, the issue is at what point the impact of increased solids loading would offset the benefits of reduced NOM on membranes.

In many situations, wastewater is treated with coagulants before discharge to a membrane system. Coagulants are also used prior to the membrane system for organics reduction and color removal. Coagulants are used, e.g., to precipitate phosphorus. Oxidants are also used to oxidize and change the state of a contaminant before membranes.

In a simplified explanation, colloidal particles are usually negatively charged, and the coagulants used in water and wastewater treatment normally consist of positively charged ions. The positive charge neutralizes the negative charge and promotes coagulation. Some coagulants contain ions with more positive charges than others. Those consisting of trivalent ions, such as aluminum and iron, are 50 to 60 times as effective as those with bivalent ions, such as calcium. They are 700 to 1,000 times more effective than coagulants with monovalent ions, such as sodium.

Coagulants used in conventional applications are not always the best for membranes. Jar testing combined with bench-scale membrane equipment is the fastest, least expensive way to determine optimum choice of coagulant, coagulant dose, pH, detention time, and expected removal efficiency.

Alum

Alum is the most common coagulant used to treat water and wastewater in conventional plants. Alkalinity is necessary for the reaction to occur. If not enough is naturally present, the alkalinity of the water must be increased. The mechanism of alum and most inorganic coagulants is as follows:

1. Alum (the inorganic coagulant) added to raw water reacts with the alkalinity naturally present to form jellylike floc particles of aluminum hydroxide, $Al(OH)_3$.

2. The positively charged trivalent aluminum ion neutralizes the negatively charged particles of color or turbidity. This occurs within 1 or 2 s after the chemical is added to the water, which is why rapid, thorough mixing is critical to good coagulation.

3. Within a few seconds, the particles begin to attach to each other to form larger particles.

4. The floc that is first formed is made up of microfloc that still has a positive charge from the coagulant; the floc particles continue to neutralize negatively charged particles until they become neutral particles themselves.

5. Finally, the microfloc particles begin to collide and stick together (agglomerate) to form larger, settleable floc particles. For low pressure membranes, the microfloc is large enough in size to be removed.

Membranes require significantly less chemical than conventional treatment. Many physical and chemical factors can affect the success of a coagulant, including mixing conditions; pH, alkalinity, and turbidity levels; and water temperature. Alum works best in a pH range of about 5.8 to 8.5. If it is used outside this range, the floc either will not form completely or may form and then dissolve back into the water.

Iron Salts

Iron salts such as ferric chloride and ferrous sulfate can operate effectively over a wider range of pH values than alum. However, they are quite corrosive and require special facilities for storage and handling. Both alum and ferrous sulfate are affected by the alkalinity of the raw water. If the alkalinity is not high enough, an effective floc will not form. If floc is not completely formed because of insufficient alkalinity or a pH value outside the optimal range, but the alkalinity or pH is later changed during treatment or in the system, the floc can re-form in the distribution system. This will, of course, cause customer complaints and problems due to a buildup of sediment in the system piping.

Other Coagulants

Aluminum chlorohydrate and polyaluminum chloride (PACL), in this author's experience, are the least problematic for low pressure membranes. These work well in waters with low alkalinity. However, most of the work has been done with NOM reduction. Chapter 6 discusses membrane-coupled bioprocesses. One option beside membrane bioreactors (MBRs) is to follow secondary or tertiary treatment with a low pressure membrane. The applications utilize coagulants for both flux enhancement and organic reduction prior to the NF or RO.

Coagulant Aids

A coagulant aid is a chemical added during coagulation in conventional treatment plants to improve coagulation, build a stronger floc, overcome temperature drops, reduce the amount of coagulant needed, and reduce the amount of sludge generated. Coagulant aid chemistry is important in membrane operations because carryover impacts both the low pressure and diffusive membranes downstream.

The three general types of coagulant aids are activated silica, weighting agents, and polyelectrolytes. Most aids are used with alum but can be used alone. Knowing about coagulant aids is important because often both low pressure and diffusive membranes follow a clarifier, which may be using one or more coagulant aids. The carryover can impact membrane operation; for example, all concentrations of cationic polymer will have some effect on low pressure membranes.

Activated Silica

Activated silica—sodium silicate (Na_2SiO_3), activated by the addition of a strong acid, e.g., hypochlorous acid—is still used to strengthen floc and improve color removal, especially at low temperatures. Activated silica is not typically used with low pressure membranes, as only a pin floc is needed. Little work, though, has been done in evaluating it with low pressure membranes for color removal.

A major disadvantage of using activated silica in conventional treatment, and one magnified in use with membranes, is that too much silica causes the formation of a gel, which clogs the membrane pores.

Adding sodium and silica to a wastewater further treated by membranes can have a negative effect in the membranes. Silica must be reduced to the ppb range for high pressure boilers. Silica precipitates on an RO membrane at concentrations greater than 120 mg/L; the solubility of silica is 120 mg/L at 25°C. Silica is further discussed in Chapter 5.

Weighting Agents

Weighting agents are described in *Water Treatment*, 3rd ed., as natural materials such as bentonite clay, limestone, and silica used in conventional water treatment at dosages in the 10 to 50 mg/L range to enhance the flocculation process by increasing the probability of particle collisions. In a proprietary process, sand is used as a weighting agent.

Polyelectrolytes

Polyelectrolytes, aka polymers, are organic molecules that, when dissolved in water, produce highly charged ions. They are classified as

- Cationic polyelectrolytes
- Anionic polyelectrolytes
- Nonionic polyelectrolytes

The impact of polyelectrolytes on membrane design is discussed in Chapter 7.

Cationic Polyelectrolytes   Cationic polyelectrolytes produce positively charged ions when dissolved in water. They are used—either alone or together with an inorganic coagulant, such as ferric sulfate or ferric chloride—to coagulate the colloids typically found in surface waters, which are negatively charged. Jar testing is typically used to determine the concentration, detention time, and best combination of cationic polymer and inorganic coagulant for a given water.

Cationic polymers are not recommended for use with membrane technology.

Anionic Polyelectrolytes   Anionic polyelectrolytes form negatively charged ions and often are complimentary with aluminum and iron coagulants. They are used in applications where the particles are positively charged. Anionic polymers are not materially affected by pH, alkalinity, hardness, or turbidity (Wachinski, 2003).

Nonionic Polyelectrolytes   Nonionic polyelectrolytes, which are neutral in charge, release both positively and negatively charged ions. They must be added in larger doses than other polymers but are less expensive (Wachinski, 2003). The normal dosage range of cationic and anionic polymers is 0.1 to 1.0 mg/L; for nonionic polymers, the dosage range is 1 to 10 mg/L. Compared with other coagulant aids, the required dosages of polyelectrolytes are extremely small.

Chemicals Used to Raise Alkalinity

Lime ($CaCO_3$)—used as either quicklime or hydrated lime—soda ash, caustic soda, and sodium bicarbonate are used to raise a water's alkalinity. Lime is the least expensive, but the others are easier to handle and feed.

Water Reuse Standards

Current water reuse standards can be found in the Appendices. The following summary is from *Water Treatment*, 3rd ed.:

NSF International Standards and Approval

As more wastewater is reused to higher reuse standards, the time is approaching when water reuse standards will be stricter than water standards. We see this now in the discharge standards for copper and soon for mercury. With that in mind, a cursory treatment of NSF International and American Water Works Association (AWWA) standards are presented.

The grades of chemical used in membrane applications for drinking water are governed by NSF/ANSI Standard 60 (NSF 60). NSF 60 regulates the quantity of trace contaminants—the USEPA primary drinking water standards—these standards are presented later in the text—allowed in any chemical used in the treatment of drinking water. For reuse applications, final water reuse application govern chemical grades. In many cases, though, wastewater discharge regulations may be stricter than drinking water standards.

For many years, the water industry relied on AWWA standards and on approval by the USEPA of individual products to ensure that harmful chemicals were not unknowingly added to potable water. However, there was no actual testing of the products, and the water treatment industry had to rely on the manufacturers' word that their products did not contain toxic materials.

Recent toxicological research has revealed potential adverse health effects due to rather low levels of continuous exposure to many chemicals and substances previously considered safe.

In the early 1980s, it became evident that more exacting standards were needed, as well as more definite assurance that products positively met safety standards. In view of the growing complexity of testing and approving water treatment chemicals and components, the USEPA awarded a grant in 1985 for the development of private-sector standards and a certification program. The grant was awarded to a consortium of partners. NSF International was designated the responsible lead. Other cooperating organizations were the Association of State Drinking Water Administrators, AWWA, the AWWA Research Foundation, and the Conference of State Health and Environmental Managers.

Two standards were developed by these organizations, along with the help of many volunteers from the water supply and manufacturing industries who served on development committees. These standards have now been adopted by the American National Standards Institute (ANSI), so they bear an ANSI designation in addition to the NSF reference.

NSF/ANSI Standard 60 essentially covers treatment chemicals for drinking water. The standard sets up testing procedures for each type of chemical, provides limits on the percentage of the chemical that can safely be added to potable water, and places limitations on any harmful substances that might be present as impurities in the chemical. NSF/ANSI Standard 61 covers materials that are in contact with water, such as coatings, construction materials, and components used in processing and distributing potable water. The standard sets up testing procedures for each type of product to ensure that they do not unduly contribute to microbiological growth, leach harmful chemicals into the water, or otherwise cause problems or adverse effects on public health.

The reader is directed to www.nsf.org for detailed information on the standards.

NSF Standards Nomenclature

Today NSF International is accredited by the American National Standards Institute (ANSI) to develop US national standards. These standards, formerly known as ANSI/NSF standards, are now called NSF/ANSI standards.

NSF Certification

The NSF certification program involves two fundamental documents. One is the standard to which the product is evaluated, including test procedures and pass/fail criteria. All NSF drinking water treatment unit standards are NSF/ANSI standards.

The second document is the NSF Certification Policies for Drinking Water Treatment Systems and Components. This document details the procedures and criteria required to achieve and maintain NSF certification of a company's products.

Manufacturers of chemicals and other products that are sold for the purpose of being added to water, or that will be in contact with potable water, must submit samples of their product to NSF International or another qualifying laboratory for testing based on standards 60 and 61. If a product qualifies, it is then "listed." There are also provisions for periodic retesting and inspection of the manufacturer's processes by the testing laboratory.

The listings of certified products provided by NSF International are used in particular by three groups in the water supply industry.

- In the design of water treatment facilities, engineers must specify that pipes, paints, caulks, liners, and other products that will be used in construction, as well as the chemicals to be used in the treatment process, are listed under one of the standards. This ensures in a very simple manner that only appropriate materials will be used. It also provides contractors with specific information on what materials qualify for use, without limiting competition.

- Individual states and local agencies have the right to impose more stringent requirements or to allow the use of products based on other criteria, but most states have basically agreed to accept NSF International standards. When state authorities approve plans and specifications for the construction of new water systems or improvements to older systems, they will generally specify that all additives, coatings, and components must be listed as having been tested for compliance with the standards.

- Water system operators can best protect themselves and their water system from customer complaints, or possibly even lawsuits, by insisting that only listed products be used for everything that is added to, or in contact with, potable water. Whether it concerns purchasing paint for plant maintenance or taking bids for supplies of chemicals, the manufacturer or representative should be asked to provide proof that the exact product has been tested and is listed.

Copies of the current listing of products approved based on standards 60 and 61 should be available at state drinking water program offices. A copy of the current listing can also be obtained on the NSF website.

AWWA Standards

The reader is directed to the AWWA website (www.awwa.org). A very good summary of the AWWA standards, from *Water Treatment*, 3rd ed. (498–499), is given here:

Since 1908, AWWA has developed and maintained a series of voluntary consensus standards for products and procedures used in the water supply community. These standards provide minimum requirements for most aspects of drinking water systems and cover products such as pipe, valves, meters, filtration media, and water treatment chemicals. They also cover procedures such as disinfection of storage tanks and design of pipe. As of 2002, AWWA had approximately 120 standards in existence. A list of standards currently available may be obtained from AWWA at any time.

AWWA offers and encourages use of its standards by anyone on a voluntary basis. AWWA has no authority to require the use of its standards by any water utility, manufacturer, or other person. Many individuals in the water supply community, however, choose to use AWWA standards. Manufacturers often produce products complying with the provisions of AWWA standards. Water utilities and consulting engineers frequently include the provisions of AWWA standards in their specifications for projects or purchase of products. Regulatory agencies require compliance with AWWA standards as part of their public water supply regulations. All of these uses of AWWA standards can establish a mandatory relationship, for example, between a buyer and a seller, but AWWA is not part of that relationship.

AWWA recognizes that others use its standards extensively in mandatory relationships and takes great care to avoid provisions in the standards that could give one party a disadvantage relative to another. Proprietary products are avoided whenever possible in favor of generic descriptions of functionality or construction. AWWA standards are not intended to describe the highest level of quality available, but rather describe minimum levels of quality and performance expected to provide long and useful service in the water supply community.

While maintaining product and procedural standards, AWWA does not endorse, test, approve, or certify any product. No product is or ever has been AWWA approved. Compliance with AWWA standards is encouraged, and demonstration of such compliance is entirely between the buyer and seller, with no involvement by AWWA.

AWWA standards are developed by balanced committees of persons from the water supply community who serve on a voluntary basis. Product users and producers, as well as those with general interest, are all involved in AWWA committees. Persons from water utilities, manufacturing companies, consulting engineering firms, regulatory agencies, universities, and others gather to provide their expertise in developing the content of the standards. Agreement of such a group is intended to provide standards, and thereby products, that serve the water supply community well.

Wastewater Reuse Source Waters

Candidate source waters for wastewater reuse include various degrees of treated wastewater:

- Secondary effluent: primary-effluent—screened wastewater with grit removed—followed by primary settling and secondary biological treatment (usually activated sludge). See Fig. 2-3.

- Tertiary effluent: secondary effluent followed by either sand filtration, mixed media filtration, granular media filtration, or cloth media disc filtration. See Fig. 2-4.

- Quaternary effluent: secondary or tertiary effluent further treated by low pressure membranes (MF or UF) alone or further with diffusive membranes (NF or RO). See Fig. 2-5.

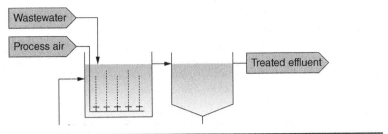

FIGURE 2-3   Conventional activated sludge schematic.

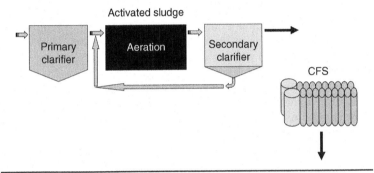

FIGURE 2-4   Tertiary treatment.

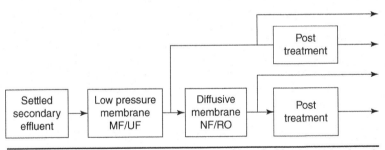

FIGURE 2-5   Quaternary membrane treatment, with and without post treatment.

Important Characteristics of Raw and Treated Wastewaters

The wastewaters discharged from municipal wastewater treatment plants include a wide range of biological, inorganic, and organic constituents. Some of these constituents can be harmful to people and/or ecosystems, depending on concentration and duration of exposure. Some are essential nutrients at low concentrations (e.g., certain trace elements) but may become hazardous at higher concentrations.

This section describes briefly the key constituents of concern when municipal wastewater is reused.

Membrane technology is rarely used to further treat secondary effluent for discharge to a watercourse except for nutrient removal. Treatment of surface water for drinking water is a topic covered in drinking water treatment texts. Water reuse can involve multiple potential applications, many of which do not expose humans to microbiological constituents that may affect human health, e.g., landscape irrigation or industrial applications where risk to human health from incidental consumption is negligible. Other constituents may have an adverse impact on aquatic species but no adverse impact on human health at the same concentration.

Water is a primary design consideration for membrane filtration systems. Poor water quality may require pretreatment to avoid lower fluxes, which in turn increases the necessary membrane area and required number of modules, increasing both the cost and the size of the system.

Contaminants of concern in biological wastewater treatment are presented in the Tables 2-5 and 2-6. Table 2-5 shows characteristics of nontreated municipal wastewater. Tables 2-6a to 2-6f summarize the contaminants regulated by the Safe Drinking Water Act.

Standard Methods (1995) classifies waters according to color, turbidity, odor, taste, acidity, alkalinity, calcium carbonate saturation, hardness, oxidant demand, conductivity, salinity, floatables (fats, oils, and grease—FOG), solids, temperature, metals (metals of concern in reuse are aluminum, arsenic, barium, and heavy metals—calcium, magnesium, iron, manganese, and strontium), inorganic nonmetallic constituents (e.g., boron, carbon dioxide, chlorine residual, chloride, pH, nitrogen—as ammonia, nitrates, nitrites, and organic nitrogen—dissolved oxygen, phosphorus, silicon, sulfide, and sulfate) aggregate organic constituents (BOD, COD, TOC, ThOD, FOG, and UV-absorbing compounds), and microbiological organisms (bacteria, viruses, protozoa, and algae).

Color

Color is used to assess the age or condition of a water. Color in a water or untreated or treated wastewater can be caused by the presence of inorganic substances such as iron and/or manganese, humic materials, certain weeds, plankton, industrial wastes, or other sources such as landfill leachate. Untreated municipal wastewater is typically gray or dark gray. If a wastewater is black, it has very low dissolved oxygen and is anaerobic. Color is either apparent color or true color. Apparent color can be caused by dissolved substances and by particulate matter. True color is dissolved color. Color is measured by visual comparison to a standard or by measurement using a spectrophotometer. It is expressed in color units or by its characteristics, i.e., pH 7.6 and

Contaminant	Unit[a]	Concentration[b]		
		Low Strength	Medium Strength	High Strength
Solids, total	mg/L	390	720	1,230
Dissolved, total	mg/L	270	500	860
Fixed	mg/L	160	300	520
Volatile	mg/L	110	300	340
Suspended solids, total	mg/L	120	210	400
Fixed	mg/L	25	50	85
Volatile	mg/L	95	160	315
Biochemical oxygen demand 5-d 20°C ($BOD_5$20°C)	mg/L	110	190	350
Total organic carbon	mg/L	80	140	260
Chemical oxygen demand	mg/L	250	430	800
Nitrogen (total as N)	mg/L	20	40	70
Organic	mg/L	8	15	25
Free ammonia	mg/L	12	25	45
Nitrites	mg/L	0	0	0
Nitrates	mg/L	0	0	0
Phosphorus (total as P)	mg/L	4	7	12
Organic	mg/L	1	2	4
Inorganic	mg/L	3	5	10
Chlorides[c]	mg/L	30	50	90
Sulfate[c]	mg/L	20	30	50
Oil and grease	mg/L	50	90	100
Volatile organic compounds	mg/L	< 100	100–400	> 400
Total coliform	no./100 mL	10^6–10^8	10^7–10^9	10^7–10^{10}
Fecal coliform	no./100 mL	10^3–10^5	10^4–10^6	10^5–10^8
Cryptosporidium oocysts	no./100 mL	10^{-1}–10^0	10^{-1}–10^1	10^{-1}–10^2
Giardia lamblia cysts	no./100 mL	10^{-1}–10^1	10^{-1}–10^2	10^{-1}–10^3

[a]mg/L = g/m^3
[b]Low strength is based on an approximate wastewater flow rate of 750 L/cap-d (200 gpd/cap); medium strength is based on an approximate wastewater flow rate of 460 L/cap-d (120 gpd/cap); and high strength is based on an approximate wastewater flow rate of 240 L/cap-d (60 gpd/cap).
[c]Values should be increased by the amount of constituent present in the domestic water supply.
Adapted from Metcalf and Eddy (2003).

TABLE 2-5 Typical Raw Wastewater Characteristics

Contaminant	Maximum Contaminant Level (mg/L) or Treatment Technique	Public Health Goal (mg/L)
Atrazine	0.003	0.003
Benzene	0.005	0
Benzo[a]pyrene (PAHs)	0.0002	0
Carbofuran	0.04	0.04
Carbon tetrachloride	0.005	0
Chlordane	0.002	0
Chlorobenzene	0.1	0.1
2,4-D	0.07	0.07
Dalapon	0.2	0.2
1,2-Dibromo-3-chloropropane (DBCP)	0.0002	zero
o-Dichlorobenzene	0.6	0.6
p-Dichlorobenzene	0.075	0.075
1,2-Dichloroethane	0.005	0
1,1-Dichloroethylene	0.007	0.007
cis-1,2-Dichloroethylene	0.07	0.07
trans-1,2-Dichloroethylene	0.1	0.1
Dichloromethane	0.005	0
1,2-Dichloropropane	0.005	0
Di(2-ethylhexyl) adipate	0.4	0.4
Di(2-ethylhexyl) phthalate	0.006	0
Dinoseb	0.007	0.007
Dioxin (2,3,7,8-TCDD)	0.00000003	0
Diquat	0.02	0.02
Endothall	0.1	0.1
Endrin	0.002	0.002
Epichlorohydrin	TT	0
Ethylbenzene	0.7	0.7
Ethylene dibromide	0.00005	0
Glycophosphate	0.7	0.7
Heptachlor	0.0004	0
Heptachlor epoxide	0.0002	0
Hexachlorobenzene	0.001	0
Hexachlorocyclopentadiene	0.05	0.05
Lindane	0.0002	0.0002
Methoxychlor	0.04	0.04
Oxamyl (Vydate)	0.2	0.2

TABLE 2-6a   National Primary Drinking Water Regulations: Organic Contaminants

Contaminant	Maximum Contaminant Level (mg/L) or Treatment Technique	Public Health Goal (mg/L)
Pentachlorophenol	0.001	0
Picloram	0.5	0.5
Polychlorinated biphenyls (PCBs)	0.0005	0
Simazine	0.004	0.004
Styrene	0.1	0.1
Tetrachloroethylene	0.005	0
Toluene	1	1
Toxaphene	0.003	0
2,4,5-TP (Silvex)	0.05	0.05
1,2,4-Trichlorobenzene	0.07	0.07
1,1,1-Trichloroethane	0.2	0.2
1,1,2-Trichloroethane	0.005	0.003
Trichloroethane	0.005	0
Vinyl chloride	0.002	0
Xylenes (total)	10	10

TABLE 2-6a   (Continued)

Contaminant	Maximum Contaminant Level (mg/L) or Treatment Technique	Public Health Goal (mg/L)
Antimony	0.006	0.006
Arsenic	0.010	0
Asbestos (fiber > 10μm)	7 million fibers/L (MFL)	7 MFL
Barium	2	2
Beryllium	0.004	0.004
Cadmium	0.005	0.005
Chromium (total)	0.1	0.1
Copper	TT action level 1.3	1.3
Cyanide (as free cyanide)	0.2	0.2
Fluoride	4.0	4.0
Lead	TT action level 0.015	0
Mercury (inorganic)	0.002	0.002
Nitrate (measured as nitrogen)	10	10
Nitrite (as nitrogen)	1	1
Selenium	0.05	0.05
Thallium	0.0005	0.0005

TABLE 2-6b   National Primary Drinking Water Regulations: Inorganic Contaminants

Contaminant	Maximum Contaminant Level (mg/L) or Treatment Technique	Public Health Goal (mg/L)
Chloramines (as Cl$_2$)	4	4
Chlorine (as Cl$_2$)	4	4
Chlorine dioxide (as ClO$_2$)	0.8	0.8

TABLE 2-6c   National Primary Drinking Water Regulations: Disinfection By-Products

Contaminant	Maximum Contaminant Level Treatment Technique	Public Health Goal (pCi/L)
Alpha/photon emitters	15 pCi/L	0
Beta photon emitters	4 mrem/y	0
Radium 226 and radium 228 (combined)	5 pCi/L	0
Uranium	30 µg/L	0

TABLE 2-6d   National Primary Drinking Water Regulations: Radionuclides

Contaminant	Secondary Maximum Contaminant Level
Aluminum	0.05 to 0.2 mg/L
Chloride	250 mg/L
Color	15 color units
Copper	1.0 mg/L
Corrosivity	No corrosivity
Fluoride	2.0 mg/L
Foaming agents	0.5 mg/L
Iron	0.3 mg/L
Manganese	0.05 mg/L
Odor	3 threshold odor number
pH	6.5 to 8.5
Silver	0.1 mg/L
Sulfate	250 mg/L
Total dissolved solids	500 mg/L
Zinc	5 mg/L

TABLE 2-6e   National Primary Drinking Water Regulations: Secondary Maximum Contaminants

Contaminant	Maximum Contaminant Level Treatment Technique	Public Health Goal (mg/L)
Cryptosporidium	TT Safe Drinking Water Act	0
Giardia lamblia	TT Safe Drinking Water Act	0
Heterotrophic plate count	TT	Measures a range of bacteria naturally present in the environment
Total coliforms (including fecal coliform and *Escherichia coli*)	5.0%[a]	0
Turbidity	TT	N/A
Viruses (enteric)	TT	0

[a]See the relevant discussion in this chapter.

TABLE 2-6f   National Primary Drinking Water Regulations: Microbial Contaminants

original pH, and in terms of dominant wavelength. The reader is directed to any edition of *Standard Methods for Examination of Water and Wastewater* for a thorough treatment. Low pressure membranes—MF and UF—with or without coagulant(s) can remove apparent color. True color is only removed by diffusive membranes.

Turbidity

Turbidity as defined by the USEPA is

> a measure of the cloudiness of water. It is used to indicate water quality and filtration effectiveness, e.g., whether disease-causing organisms are present. Higher turbidity levels are often associated with higher levels of disease-causing microorganisms such as viruses, parasites, and some bacteria.

In a more technical sense, it is the measure of the scatter of incident light caused by particulate matter in water, and is typically used instead of suspended solids to characterize drinking water sources. Turbidity is expressed in nephelometric turbidity units (ntus). In wastewater reuse applications, it is used together with suspended solids to measure both raw water quality and filtered water quality.

In the Safe Drinking Water Act, turbidity is used in lieu of specific microorganism measurement. Higher turbidity levels in water are associated with a higher concentration of pathogenic microorganisms.

Taste and Odor

As of the writing of this book, "toilet to tap" wastewater reuse in the United States has not yet been accepted. As such, taste and odor detection in recycled wastewater is not an issue. Title 22 regulations in California do not address odor on an analytical scale. It is an issue for any irrigation project close to population.

Acidity

Acidity is base-alkalinity-neutralizing power, the quantitative capacity to react with a strong base to a designated pH. The value varies with the end-point pH. Acidity measures aggregate capacity. In general, strong mineral acids, weak acids, and hydrolyzing salts all contribute to acidity. It is usually determined by titration.

Alkalinity

The alkalinity of water is its acid-neutralizing capacity, a measure of the water's capacity to absorb hydrogen ions without significant pH change, i.e., a measure of the capacity of the water to neutralize strong acids. Just like acidity, alkalinity is an aggregate capacity and varies with the end-point pH. Alkalinity consists of three forms: bicarbonate, carbonate, and hydroxyl or hydroxide alkalinity. In natural waters this capacity is usually attributable to bases such as bicarbonate (HCO_3^-) and carbonate (CO_3^{2-}). Other contributors of alkalinity are phosphates, silicates, and other bases.

In water analysis and certain membrane operations, the concentrations of the various forms of alkalinity present in the water must be known. This information is needed, e.g., when designing NF and RO processes to soften water or remove targeted contaminants, when designing backwash systems for MF and UF, and when determining inorganic coagulant doses. The alkalinity of the water used to perform chemically enhanced backwashes (CEBs)—the Pall Corporation uses the term enhanced flux maintenance minicleans—using sodium hypochlorite or citric acid every 1 to 3 days must be known to avoid precipitation on the hollow fiber membranes during the cleaning.

Bicarbonate (HCO_3)   The bicarbonate ion plays a significant role in RO scaling when it combined with calcium. The solubility of calcium bicarbonate is low; it can cause a scaling problem in the back end of an RO system. Calcium bicarbonate solubility is measured using the LSI (Langelier Saturation Index) for brackish waters or the Stiff-Davis Solubility Index (SDSI) for seawaters and is lower with increasing temperature and increasing pH. Bicarbonate is one component of alkalinity. Its concentration is in balance with carbon dioxide in the pH range of 4.4 to 8.2 and in balance with carbonate in the pH range of 8.2 to 9.6. Bicarbonate alkalinity is the most common and the most frequently used because bicarbonates are formed from the reactions between carbon dioxide and water.

Alkalinity is determined by titration using standard sulfuric acid solutions and is often expressed as phenolphthalein (P) alkalinity and methyl orange (MO) or total (T) alkalinity. P alkalinity is determined at

Result of Titration	Hydroxide Alkalinity as CaCO$_3$	Carbonate Alkalinity as CaCO$_3$	Bicarbonate Alkalinity as CaCO$_3$
P = 0	0	0	T
P < ½T	0	2P	T−2P
P = ½T	0	2P	0
P > ½T	2P−T	2(T−P)	0
P = T	T	0	0

Key: P = phenolphthalein alkalinity; T = total alkalinity.

TABLE 2-7 Alkalinity Relationships

a titration to an end point of pH 8.3 (reported range is 8.2 to 8.5). MO alkalinity, or T alkalinity, is determined at titration to an end point of pH 4.3 (reported range is 4.2 to 4.5).

Carbonate (CO$_3$) Carbonate is a divalent anion. The solubility of calcium carbonate is low; it can cause a scaling problem in the back end of an RO system. Calcium carbonate solubility is measured using the LSI for brackish waters or SDSI for seawaters and is lower with increasing temperature and increasing pH. Carbonate is one component of alkalinity. Its concentration is in balance with bicarbonate in the pH range of 8.2 to 9.6. At a pH of 9.6 or higher, there is no carbon dioxide or bicarbonate—all alkalinity is in the carbonate form.

Five situations are possible, as shown in Table 2-7.

Hydroxide only. Samples containing only hydroxide alkalinity have a high pH, usually well above 10. Titration is essentially complete at the phenolphthalein end point. In this case, hydroxide alkalinity is equal to the phenolphthalein alkalinity.

Carbonate only. Samples containing only carbonate alkalinity have a pH of 8.5 or higher. The titration to the phenolphthalein end point is equal to exactly one-half of the total titration. In this case, carbonate alkalinity is equal to the total alkalinity.

Hydroxide-carbonate. Samples containing hydroxide and carbonate alkalinity have a high pH, usually well above 10. The titration from the phenolphthalein to the methyl orange end point represents one-half of the carbonate alkalinity. Therefore, carbonate alkalinity may be calculated as follows:

$$\text{Carbonate alk.} = 2\,(\text{titration from phenol. to methyl orange}) \times \frac{10,000}{\text{ml sample}}$$

and

$$\text{Hydroxide alk.} = \text{total alk.} - \text{carbonate alk.}$$

Carbonate-bicarbonate. Sample containing carbonate and bicarbonate alkalinity have a pH above 8.3 and usually less than 11. The titration to the phenolphthalein end point represents one-half of the carbonate concentration. Carbonate alkalinity may be calculated as follows:

$$\text{Carbonate alk.} = 2(\text{titration from phenol. end point}) \times \frac{1,000}{\text{ml sample}}$$

and

$$\text{Bicarbonate alk.} = \text{total alk.} - \text{carbonate alk.}$$

Bicarbonate only. Samples containing only bicarbonate alkalinity have a pH of 8.3 or less, usually less. In this case, bicarbonate alkalinity is equal to the total alkalinity. Buffers are substances that offer resistance to changes in pH when acids or bases are added to water.

No alkalinity form exists below pH 4.3, as dissolved carbon dioxide is in equilibrium with carbonic acid in solution. Between pH 4.3 and 8.3, the equilibrium shifts to the right, reducing carbon dioxide and creating bicarbonate (HCO_3^-) ions. At pH greater than 8.3, bicarbonate ions are converted to carbonate ions. Hydroxide appears at pH greater than 9.5 and reacts with carbon dioxide to yield both bicarbonate and carbonate. The maximum carbonate concentration occurs between pH values of 10 and 11.

Calcium Carbonate Saturation

Calcium carbonate saturation indices are used to determine whether a wastewater is scale forming, neutral, or scale dissolving. Waters oversaturated with $CaCO_3$ tend to precipitate $CaCO_3$. Waters undersaturated with $CaCO_3$ tend to dissolve $CaCO_3$. This information is especially useful in NF/RO posttreatment and in assessing water used in flux maintenance, chemical enhanced backwash (CEB), and clean in place (CIP) solution makeup.

Langelier Saturation Index (LSI)   LSI is a method of reporting the scaling or corrosive potential of low TDS brackish water based on the level of saturation of calcium carbonate. LSI is important to boiler water and municipal plant chemists in determining whether a water is corrosive (has a negative LSI) or will tend to scale calcium carbonate (has a positive LSI). LSI is important to RO chemists as a measurement of the scaling potential for calcium carbonate. The LSI value is calculated by subtracting the calculated pH of saturation of calcium carbonate from the actual feed pH. Calcium carbonate solubility decreases with increasing temperature (as evidenced by the liming of a teakettle), higher pH, higher calcium concentration, and higher alkalinity levels. The LSI value can be lowered by reducing pH through the injection of an acid (typically sulfuric or hydrochloric)

into the RO feed water. A recommended target LSI in the RO concentrate is −0.2 (which indicates that the concentrate is 0.2 pH units
below the point of calcium carbonate saturation). A −0.2 LSI allows
for pH excursions in actual plant operation. A polymer-based antiscalant can also be used to inhibit the precipitation of calcium carbonate. Some antiscalant suppliers have claimed the efficacy of their
product up to a positive LSI value of 2.5 in the RO concentrate
(though a more conservative design LSI level is +1.8). Sodium
hexametaphosphate, an inorganic antiscalant, was used in the early
days of RO but its maximum concentrate LSI was +0.5 and it had to
be made in short-lived batches, as the air easily oxidized it.

Hardness

Hard water impacts membrane operations in many ways. Lime soda
softening is often used to remove hardness, to soften the water that
contacts the membrane. NF is a viable option for removing hardness.
NF is explained in detail in Chapter 5. Hard water cannot be used for
membrane cleaning operations or to prepare chemical backwash or CIP
solutions, as the scale can cause severe inorganic membrane fouling.

Hardness is a measure of the capacity of water to precipitate soap.
Hard waters are generally considered those that require considerable
amounts of soap to produce a foam or lather and that also produce
scale in hot-water pipes, heaters, boilers, and other units in which the
temperature of water is increased materially.

Hardness is due to the presence of calcium and magnesium salts:
bicarbonates, carbonates, sulfates, chlorides, and nitrates. Strontium,
iron, manganese, and aluminum also cause water to be hard but are
not usually present in appreciable quantities. Hardness is classified
as permanent or temporary. Temporary hardness, from carbonate or
bicarbonate, is partially reduced by boiling. Permanent hardness is
not reduced by boiling. Hardness is expressed in milligrams per liter
as $CaCO_3$. Waters are commonly classified in terms of the degree of
hardness as follows:

> 0 to 75 mg/L: soft
>
> 75 to 150 mg/L: moderately hard
>
> 150 to 300 mg/L: hard
>
> 300 mg/L and up: very hard

Hard water is caused by soluble, divalent, metallic cations (positive ions having a valence of 2). The principal elements that cause
water hardness are calcium (Ca) and magnesium (Mg). The degrees
of hardness table uses the combined values of calcium and magnesium expressed as calcium carbonate. Strontium, aluminum, barium,
iron, manganese, and zinc can also cause hardness in water, but they
are not usually present in large enough concentrations to contribute
significantly to the total hardness.

Water hardness varies considerably in different geographic areas of the contiguous 48 states. This is due to different geologic formations and is also a function of the contact time between water and these formations. Calcium is dissolved as water passes over and through limestone deposits. Magnesium is dissolved as water passes over and through dolomite and other magnesium-bearing minerals. Because groundwater is in contact with these formations for a longer period of time than surface water, groundwater is normally harder than surface water.

Hardness can be categorized by either of two methods: calcium versus magnesium hardness and carbonate versus noncarbonate hardness. The calcium-magnesium distinction is based on the minerals involved. Hardness caused by calcium is called calcium hardness, regardless of the salts associated with it, which include calcium sulfate ($CaSO_4$), calcium chloride ($CaCl_2$), and others. Likewise, hardness caused by magnesium is called magnesium hardness. Calcium and magnesium are normally the only significant minerals that cause hardness, so it is generally assumed that

Total hardness = calcium hardness + magnesium hardness

The carbonate-noncarbonate distinction, however, is based on hardness from either the bicarbonate salts of calcium or the normal salts of calcium and magnesium involved in causing water hardness. Carbonate hardness is caused primarily by the bicarbonate salts of calcium and magnesium, which are calcium bicarbonate, $Ca(HCO_3)_2$, and magnesium bicarbonate, $Mg(HCO_3)_2$. Calcium and magnesium combined with carbonate (CO_3) also contribute to carbonate hardness. Noncarbonate hardness is a measure of calcium and magnesium salts other than carbonate and bicarbonate salts. These salts are calcium sulfate, calcium chloride, magnesium sulfate ($MgSO_4$), and magnesium chloride ($MgCl_2$). Calcium and magnesium combined with nitrate (NO_3) may also contribute to noncarbonate hardness, although it is a very rare condition. For carbonate and noncarbonate hardness,

Total hardness = carbonate hardness + noncarbonate hardness

When hard water is boiled, carbon dioxide (CO_2) is given off. Bicarbonate salts of calcium and magnesium then settle out of the water to form calcium and magnesium carbonate precipitates. These precipitates form the familiar chalky deposits on teapots. Because it can be removed by heating, carbonate hardness is sometimes called temporary hardness. Because noncarbonate hardness cannot be removed or precipitated by prolonged boiling, it is sometimes called permanent hardness.

Hard water forms scale, usually calcium carbonate, which can cause a variety of problems. If it is left to dry on the surface of MF/UF membranes, white scale forms on the inside of water pipes, which will eventually reduce the flow capacity or possibly block the pipes entirely.

When hard water is heated (some chemical backwash schemes utilize heated water) scale forms much faster. In particular, when the magnesium hardness is more than about 40 mg/L (as $CaCO_3$), magnesium hydroxide scale will deposit in hot-water heaters (used to heat the CEB solutions that are operated at normal temperatures of 140°F to 150°F [60°C to 66°C]). A coating of only 0.04 in. (1 mm) of scale on the heating surfaces of a hot-water heater creates an insulation effect that will increase heating costs by about 10%—and many solutions to clean membranes are heated.

Oxidant Demand

Oxidants are added to a wastewater for disinfection and for oxidation of insoluble contaminants like ferrous iron, reduced manganese, sulfides, and others. Oxidants used in wastewater reuse applications include chlorine, hydrogen peroxide, ozone, and potassium permanganate. Oxidant demand is the difference between the added dose and the residual oxidant concentration, usually measured right before the membrane(s). Oxidant demand is important to protect any membrane from oxidation if that membrane cannot withstand it. The demand is reported as sample temperature, pH, contact time, oxidant dose, and the method used to measure the oxidant. The fate of oxidants in wastewater is complex; if destruction or damage to the membrane may occur, the demand should be evaluated over a range of conditions and proper safeguards should be installed in the system.

Conductivity

Conductivity is a measure of the ability of water to transmit electricity due to the presence of dissolved ions. It depends on the concentration of ions in the water, mobility, valence, and temperature at the time of measurement. Solutions of inorganics are good conductors. Solutions of organic compounds, which do ionize to any degree, are poor conductors. Absolutely pure water with no ions will not conduct an electrical current. Conductivity is measured by a conductivity meter and is reported as microhms per centimeter ($\mu\Omega$/cm) or microsiemens per centimeter (μS/cm). Conductivity is a convenient method of determining the level of ions in water but is nonspecific as to what the ions are. The electrical conductivity, i.e., specific conductance of ions will vary by ion and will decrease as the concentration of ions increase. TDS meters utilize conductivity measurements with a conversion factor applied. Conductivity can also be estimated using individual conversion factors from the reported ion concentrations of a water analysis or by using a single conversion factor based on the sum of the ions (TDS). Carbon dioxide conductivity can be estimated by taking the square root of the ppm concentration and then multiplying by 0.6. The silica ion does not contribute to conductivity. The most accurate conductivity readings for high quality RO permeate are obtained on-site, since levels of carbon dioxide levels—a gas—can vary when exposed to the atmosphere.

Salinity

Salinity is the saltiness or dissolved salt content of a body of water—originally the mass of dissolved salts in a given mass of solution (*Standard Methods*, 1995), it has been expanded as a general term to describe the levels of different salts, such as sodium chloride, magnesium and calcium sulfates, and bicarbonates. It is a dimensionless number usually given as a percentage—note that 1% = 10,000 mg/L.

Solids

The term *solids* and particulate are interchangeably used in wastewater. The term *turbidity* is most used in drinking water applications. Both are used in water recycling and reuse. *Particulate* is a term to describe all solids, bacteria, viruses, and protozoa. Particulate matter can interfere with disinfection.

Solids can be classified as suspended or dissolved, and further divided into volatile (fixed or inert) or nonvolatile. The reader is referred to *Standard Methods for the Examination of Water and Wastewater* for a detailed explanation of analytical techniques to measure the solids content of a water sample. What constitutes dissolved solids is important when membranes are involved.

Total Solids   *Total solids* is the term applied to the residue remaining after a water sample has been evaporated and dried at 103°C. Total solids include total suspended solids—the mass of solids retained after filtration—and total dissolved solids—those solids which pass through the filter. *Standard Methods* defines the normal pore size of the filter through which the water is filtered and defined as total dissolved solids to be 2.0 μm. Fixed solids are the residue remaining after burning the sample at 550°C. The solids that are evaporated are called volatile solids.

The following are useful relationships involving solids:

- Total solids = total volatile solids + total fixed solids.
- Total suspended solids = total volatile suspended solids + total fixed suspended solids.
- Total dissolved solids = total volatile dissolved solids + total fixed dissolved solids.
- Total solids = total suspended solids + total dissolved solids.

Total volatile solids can be used to estimate the chemical oxygen demand (COD) of a water by simply multiplying by 1.1:

$$\text{Total volatile solids} \times 1.1 = COD$$

Total Dissolved Solids (TDS) TDS, in water treatment, is the inorganic residue left after the filtration of colloidal and suspended solids and then the evaporation of a known volume of water. TDS is reported as parts per million or milligrams per liter. TDS, in RO design projections, is determined by calculation using the sum of the cations, anions, and silica ions (with the ions reported as such, not as calcium carbonate). Feed or permeate TDS, in RO design projections, can also be estimated by applying a conversion factor to the conductivity of the solution. TDS can also be determined in the field by use of a TDS meter. TDS meters measure the conductivity of the water and then apply a conversion factor that reports TDS to a known reference solution (e.g., ppm sodium chloride or ppm potassium chloride). The user is cautioned that TDS levels for waters with a mixture of ions and determined from conductivity measurements may not agree with TDS calculated as a sum of the ions. As a rough rule of thumb, 1 ppm of TDS (when referenced to a NaCl solution) correlates to a conductivity of $2\ \mu\Omega/cm$ ($\mu S/cm$).

Dissolved inorganics such as dissolved metals are determined by filtering a wastewater sample through a 0.45 μm nominal filter and measuring the metals captured on the filter using atomic absorption. MF nominal pore size is typically in the 0.1 μm range (see Chapters 4 and 6). UF membranes have even finer pores. Since dissolved solids are measured using a 2 μm filter, it is possible for MF or UF to retain dissolved salts.

For this reason, expressing the pore size of the filtration media is required to ensure proper design of filtration systems. The TDS and the particular species of dissolved solids present in the membrane feed are both critical considerations for NF and RO systems. Species such as silica, calcium, barium, and strontium, which can precipitate on the membrane, can cause scaling and a consequent rapid decline in flux under certain conditions. Scaling is typically controlled using pretreatment chemicals such as an acid to lower the pH and/or a proprietary scale inhibitor. However, the total quantity of dissolved solids of any species also influences system operation, as the net driving pressure required to achieve a target flux is related to the osmotic pressure of the system, which is directly proportional to the TDS. Thus, as the TDS increases, so does the required feed pressure.

TDS is generally not a significant consideration for MF and UF systems, since these processes do not remove dissolved solids. In some cases, however, the use of upstream oxidants may cause the precipitation of iron or manganese salts (either unintentionally or by design as a pretreatment process), which could accelerate membrane fouling.

Colloids   Colloids are usually defined as the solids filtered through the 2 μm nominal filter and retained on a 0.45 nominal filter.

Temperature

Temperature is a critical design parameter. It has significant effects on feed pump pressure requirements, hydraulic flux balance between stages, permeate quality, and solubility of sparingly soluble salts. As a rough rule of thumb, every 10°F decrease in feed temperature increases the feed pump pressure requirement by 15%. The hydraulic flux balance between stages (in other words, the amount of permeate produced by each stage) is impacted by temperature. When water temperature increases, the elements located in the front end of the system produce more permeate, which results in reduced permeate flow by the elements located at the rear of the system. A better hydraulic flux balance between stages occurs at colder temperatures. At warmer temperatures, salt passage increases due to the increased mobility of the ions through the membrane. Warmer temperatures decrease the solubility of calcium carbonate. Colder temperatures decrease the solubility of calcium sulfate, barium sulfate, strontium sulfate, and silica.

The temperature of municipal wastewater is on average higher than that of surface water or groundwater. Temperature is important in designing wastewater treatment facilities. It plays a major role in designing low pressure and diffusive membrane systems. Temperature is typically taken during sample collection. The impact on membrane design and operation is discussed in Chapters 3, 4, 6, and 7.

Transmittance

Measured directly as a percentage, transmittance is the clarity of the water to transmit UV rays. It is important any time membranes—MF, UF, NF, or RO—are used to pretreat for UV.

Metals

Aluminum   Aluminum, based on its low solubility, is typically not found in any significant concentration in well or surface waters. When present in an RO feed water, it is typically colloidal in nature (not ionic) and is the result of alum carryover by an on-site or municipal clarifier or lime softener. Alum (aluminum sulfate) is a popular coagulant that is effective in the absorption and precipitation of naturally occurring negatively charged colloidal material (e.g., clay and silt) from surface waters and wastewaters. When introduced into water, alum dissociates into trivalent aluminum and sulfate. The hydrated aluminum ion reacts with the water to form a number of complex hydrated aluminum hydroxides, which then polymerize and start adsorbing the negatively charged colloids in water. Fouling by aluminum-based colloid carryover can occur, with alert levels for the RO designer ranging from 0.1 to 1.0 ppm aluminum in the feed water. Aluminum chemistry is complicated by the fact that aluminum is amphoteric. At low pH values, it can exist as a positively charged trivalent cation or as an aluminum hydroxide compound. Aluminum at high pH values can exist as a negatively charged anionic compound. Typically, the pH range of least solubility for aluminum compounds is 5.5 to 7.5.

Barium (Ba)   Barium is a divalent cation. The solubility of barium sulfate ($BaSO_4$) is low, which can cause scaling problems in the back end of an RO system. Barium sulfate solubility is lower with increasing sulfate levels and decreasing temperatures. Typically, barium can be found in some well waters, with typical concentrations less than 0.05 to 0.2 ppm. It is important that barium be measured with instruments capable of 0.01 ppm (10 ppb) minimum detection levels. With saturation at 100%, supersaturation up to 6000% is typical with an antiscalant.

Heavy Metals
The heavy metals cadmium, chromium, lead, mercury, nickel, and zinc are not present in secondary effluent usually, unless there is an industrial discharge. Most times, low concentrations are removed in the biological process and concentrated in the sludge. Any time a waste stream is recycled to the head of a treatment plant, that stream should be screened for heavy metals.

Calcium (Ca)   Calcium is a divalent cation. Along with magnesium, it is a major component of hardness in brackish water. The solubility of calcium sulfate ($CaSO_4$; gypsum) is typically limited to 230% with the use of an antiscalant. The solubility of calcium carbonate is typically limited to an LSI value of positive 1.8 to 2.5.

Iron (Fe)   Iron is a water contaminant that takes two major forms. The water-soluble form is known as the ferrous state and has a 2+ valence. In nonaerated well waters, ferrous iron behaves much like calcium or magnesium hardness, in that it can be removed by softeners and its precipitation in the back end of an RO system can be controlled by the use of a dispersant chemical in the RO feed water. The water-insoluble form is known as the ferric state and has a 3+ valence. Typically, RO manufacturers will recommend that combined iron levels be less than 0.05 ppm in the RO feed. If all iron is in the soluble ferrous form, iron levels up to 0.5 ppm in the feed can be tolerated if the pH is less than 7.0 (though an iron dispersant is recommended). The introduction of air into water with soluble ferrous iron will result in oxidation to insoluble ferric iron. Soluble iron can be found in deep wells, but can be converted into the more troublesome insoluble iron by the introduction of air through placement in tanks or through leaky pump seals. Soluble iron can be treated with dispersants or can be removed by iron filters, softeners, or lime softening. Insoluble ferric iron oxides or ferric hydroxides, being colloidal in nature, will foul the front end of the RO system. Sources of insoluble iron are aerated well waters, surface sources, and iron scale from unlined pipe and tanks. Insoluble iron can be removed by iron filters, lime softening, softeners (with limits), MF and UF (with limits) to less than 0.05 mg/L, and multimedia filtration with polyelectrolyte feed (with limits).

Precautions are required with the use of potassium permanganate in manganese greensand iron filters, in that potassium permanganate is an oxidant that could damage any polyamide membrane. Precautions are also required with cationic polyelectrolytes, in that they can irreversibly foul a negatively charged polyamide membrane. Corrosion proof vessels and piping (e.g., fiber reinforced plastic [FRP], polyvinyl chloride [PVC], or stainless steel) are recommended for all RO systems, RO pretreatment, and distribution piping coming to the RO system.

Iron as a foulant will quickly increase RO feed pressure requirements and permeate TDS. In some cases, the presence of iron can create a biofouling problem by being the energy source for iron-reducing bacteria. Iron-reducing bacteria can cause the formation of a slimy biofilm that can plug the RO feed path.

Magnesium (Mg)   Magnesium, a divalent cation, can account for about one-third of the hardness in a brackish water, but can have a concentration 5 times higher than calcium in seawater. The solubility of magnesium salts is high and typically does not cause a scaling problem in RO systems.

Manganese (Mn)   Manganese is a water contaminant present in both well and surface waters in levels up to 3 ppm. Like iron, it can be found in organic complexes in surface waters. In oxygen-free water, it is soluble. In the oxidized state, it is insoluble and usually in the form of black manganese dioxide (MnO_2) precipitate. An alert level for potential manganese fouling in aerated RO feed waters is 0.05 ppm. Drinking water regulations limit manganese to 0.05 ppm due to its ability to cause black stains. Dispersants used to control iron fouling can be used to help control manganese fouling.

Sodium (Na)   Sodium is a monovalent cation. The solubility of sodium salts is high and does not cause an RO scaling problem. Sodium is the prevalent cation in seawater, and is the cation used to automatically balance an RO feed water analysis. Dietary sodium levels can range from 2,000 mg/L for low-sodium diets to 3,500 mg/L for average consumption levels. The USEPA has set a drinking water equivalent limit of 20 mg/L for potable water but is reevaluating that limit as too low. Daily consumption of 2 L (0.53 gallons) of water with 100 mg/L of sodium would be only 200 mg. A relatively hard water with 10 gpg (171.2 mg/L) of hardness (as calcium carbonate) results in only an additional 79 mg/L of sodium when softened.

Strontium (Sr)   Strontium is a divalent cation. The solubility of strontium sulfate is low and can cause a scaling problem in the back end of an RO system. Strontium sulfate solubility is lower with increasing sulfate levels and decreasing temperatures. Typically, strontium can be found in some well waters where lead ores are also present, with typical concentrations less than 15 ppm. With saturation at 100%, supersaturation up to 800% is typical with an antiscalant.

Inorganic Nonmetallic Constituents

Boron (B)   Boron can be found in seawater at levels up to 5 ppm and at lower levels in brackish waters where inland seas once existed. Boron is not a foulant. The removal of boron to ppb levels is an important issue in the electronics industry, as it adversely affects the process in some applications. The removal of boron is important in the production of potable/irrigation water in seawater desalination, with suggested limits of 0.5 ppm boron. The element boron exists in equilibrium as the borate monovalent anion $B(OH)_4^-$ at higher pH levels and as nonionized boric acid $B(OH)_3$ at lower pH levels. The relative concentrations of borate and boric acid are dependent on pH, temperature, and salinity. The borate ion becomes more prevalent at higher pH levels, higher salinity, and higher temperature. The rejection of boron by RO is better with the borate ion, due to its charge. The rejection of nonionized boric acid is low, due to its smaller size and lack of electric charge.

Chloride (Cl)   Chloride is a monovalent anion. Chlorides are found in municipal wastewaters from human excreta, about 6 g per person per day. Infiltration by brackish water or seawater can significantly increase chloride concentration. The solubility of chloride salts is high and does not create an RO scaling problem. Chloride is the prevalent anion in seawater and is the anion used to automatically balance an RO feed water analysis. The upper limit recommended by the USEPA and World Health Organization for chloride in potable water is 250 ppm, based on taste issues.

pH   Hydrogen activity in water, expressed by the term *pH*, is a measure of the acid or alkaline condition. It is expressed mathematically as

$$pH = \log \frac{1}{H^+}$$

Water dissociates, yielding hydrogen ions equal to 10^{-7} mol/L. Pure water has a pH of 7. Hydroxyl ions are equally produced at 10^{-7} mol/L, and water is neutral. The ion product of the hydrogen ions and the hydroxyl ions is a constant.

$$H_2O = H^+ + OH^-$$

Adding acid to water increases the hydrogen ion concentration and lowers the water's pH. Adding an alkali to water adds hydroxyl ions and increases the pH. The chemical equilibrium of water is changed when the hydrogen ion concentration is increased. In accordance with the chemical equilibrium in the dissociation equation, the hydrogen ion concentration must decrease so that the ion product equals the constant 10^{-14}.

To water chemists, pH is important in defining the alkalinity equilibrium levels of carbon dioxide, bicarbonate, carbonate, and hydroxide ions. The pH of the concentrate is typically higher than that of the feed due to the higher concentration of bicarbonate/ carbonate ions relative to the concentration of carbon dioxide. The RODESIGN program allows the user to adjust the pH of the feed water using hydrochloric and sulfuric acid. Lowering the feed pH with acid results in a lower LSI value, which reduces the scaling potential for calcium carbonate. Feed and concentrate (reject) pH can also affect the solubility and fouling potential of silica, aluminum, organics, and oil. Variations in feed pH can also affect the rejection of ions; for example, fluoride, boron, and silica rejection are lower when the pH becomes lower.

Nitrogen The forms of nitrogen of greatest interest in water and wastewater (in order of decreasing oxidation state) are nitrate (NO_3), nitrite (NO_2), ammonia (NH_3), and organic nitrogen (see Table 2-8). Organic nitrogen—organically bound nitrogen in the trivalent (trinegative) oxidation state, includes proteins, peptides, nucleic acids, and urea. Cells are 7% to 9% nitrogen.

Ammonium (NH_4) Ammonium is a monovalent cation. Ammonium salts are very soluble and do not cause an RO scaling problem. The ammonium ion is the result of very soluble gaseous ammonia (NH_3) being dissolved in water. Nonionized ammonia ionizes in water to form the ammonium ion and hydroxide ion. The degree of ionization of ammonia to ammonium is dependent on pH, temperature, and the ionic strength of the solution. At higher pH levels the ammonia gas is prevalent and, being a gas, will not be rejected by RO (similar to carbon dioxide gas). At lower pH levels the ammonium ion is prevalent and is rejected by RO. Ammonia and ammonium exist in equilibrium at varying relative concentrations in the general pH

Compound	Abbreviation	Form	Definition
Ammonia nitrogen	NH_3-N	Soluble	NH_3-N
Ammonium nitrogen	NH_4^+-N	Soluble	NH_4-N
Total ammonia nitrogen	TAN	Soluble	NH_3-N + NH_4N$^+$
Total Kjeldahl nitrogen	TKN	Particulate soluble	Organic N + NH_3-N + NH_4-N
Organic nitrogen	Organic N	Particulate soluble	TKN − NH_3-N + NH_4-N
Total nitrogen	TN	Particulate soluble	Organic N + NH_3-N + NH_4-N + NO_3-N

Adapted from Metcalf and Eddy (2003).

TABLE 2-8 Forms of Nitrogen and Their Definitions

range of 7.2 to 11.5. Ammonium is typically not found in well water sources, having been converted by bacterial action in soils to the transitory nitrite (NO_2) ion and then oxidized into the more prevalent nitrate ion. Ammonium is found in surface water sources at low levels (up to 1 ppm as the ion), the result of biological activity and the breakdown of organic nitrogen compounds. Surface sources can be contaminated with ammonium from septic systems, animal feed lot runoff, or agricultural runoff from fields fertilized with ammonia or urea. Ammonium is prevalent in municipal waste facilities with levels up to 20 ppm as the ion in the effluent, the result of high levels of organic nitrogen compounds and biological activity. Another source of ammonium is the addition of ammonia to chlorine to form biocidal chloramines.

Ammonia Ammonia nitrogen is toxic in the 2 to 4 mg/L range. It exerts a nitrogenous oxygen demand, reacts with chlorine, and will deplete the chlorine. It can be used with chlorine to form chloramines, or as a disinfectant. The range in sewage is 10 µg/L to 30 mg/L. Ammonia must be added if it is not present in sufficient concentrations; the ratio of biochemical oxygen demand (BOD) to ammonia nitrogen to phosphorus is 100:20:1. Ammonia can cause bulking sludge in activated sludge if not present in sufficient quantities.

Carbon Dioxide (CO_2) Carbon dioxide is a gas that when dissolved in water reacts with the water to form weak carbonic acid (H_2CO_3). If a pure water were completely saturated with carbon dioxide, its concentration would be about 1,600 ppm and its pH would be about 4.0.

A typical source for carbon dioxide in natural waters is the result of a balance with bicarbonate alkalinity based on the pH of the water. The concentration of carbon dioxide in water is typically indirectly determined by graphical comparison to the bicarbonate concentration and pH. Carbon dioxide and the bicarbonate ion are in balance in the pH range of 4.4 to 8.2. (The alkalinity is all carbon dioxide at pH 4.4 and is all bicarbonate at pH 8.4.) The *RODESIGN program* calculates the carbon dioxide level based on the bicarbonate level and pH of the water. Carbon dioxide, being a gas, is not rejected or concentrated by an RO membrane; therefore, its concentration will the same in the feed, permeate, and concentrate.

Acidifying the RO feed water will lower pH by converting bicarbonate to carbon dioxide.

Nitrate (NO_3)   Nitrate is a monovalent anion. Nitrate salts are highly soluble and do not cause an RO scaling problem. Nitrate, along with ammonia gas and ammonium, is a nitrogen-based ion whose presence is tied to nature's nitrogen cycle. The primary sources of nitrogen introduction in a feed water are decomposing animal and plant waste, septic systems, animal feed lot runoff, and agricultural field runoff from fields fertilized with ammonia. In well water sources, ammonia and ammonium are not found, having been converted to the transitory nitrite ion by certain types of bacteria in soils and then oxidized into the more prevalent nitrate ion. Frequently, nitrate concentrations are reported as ppm as nitrogen in water analysis rather than as ppm as nitrate, as required for RO projections. To convert ppm as nitrogen to ppm as nitrate, multiply by 4.43. The USEPA has set a maximum recommended limit of nitrate at 10 ppm as nitrogen (44.3 ppm as nitrate) for potable drinking water. Nitrates are harmful in that they compete with oxygen for carrying sites in blood hemoglobin. The reduced oxygen content can result in the "blue-baby syndrome," which is why babies and pregnant women are at higher risk from the effects of nitrates.

Dissolved Oxygen (DO)   The dissolved oxygen (DO) of a treated wastewater is important to manage taste and odor issues. Many compounds, including ammonia, extract an oxygen demand, which can make water septic. Water under anaerobic conditions—i.e., low to zero DO—when combined with manganese can exhibit a high fouling potential, and in some cases irreversible fouling. The fouling potential of some supernatants and all clarifier sludges should be considered when determining whether or not to recycle a stream to the head of a plant.

Phosphorus   The concentration of phosphorus in raw wastewater typically ranges from 4 to 8 mg/L, depending on a number of factors including the contribution by industrial dischargers and the nature of the drinking water supply. In general, phosphorus levels decreased after the detergent ban was imposed, but there are still some cleaning agents that contain phosphorus. Some drinking water plants chelate iron and manganese using metabisulfate, a form of phosphorus as a corrosion inhibitor, and thus will contribute phosphorus to the raw wastewater.

The total phosphorus (TP) is comprised of both inorganic and organic forms. The inorganic forms, which are soluble, include ortho-phosphate and polyphosphates. The orthophosphate form $(PO_4)^{-3}$ is the simplest form of phosphorus and accounts for 70% to 90% of the TP. It is the form that is available for biological metabolism without further breakdown. It is also the form that is precipitated by metal salts in a chemical phosphorus removal system. Polyphosphates consist of more complex forms of inorganic orthophosphates that are generally synthetic in nature. The polyphosphates are broken down to orthophosphates during the treatment process.

Organically bound phosphorus can be in both a soluble and a particulate form. Organically bound phosphorus includes a wide variety of more complex forms of phosphorus that derive from proteins, amino acids, and nucleic acids that, to some extent, are degraded and are present as waste products. Organic phosphorus is contributed by a variety of industrial and commercial sources.

The organically bound phosphorus can further subdivided into biodegradable and nonbiodegradable fractions. The soluble nonbiodegradable fraction will pass through the treatment plant and be discharged in the effluent at a concentration equal to its concentration in the influent. The particulate nonbiodegradable form, if not settled, will be removed with the sludge.

The ratio of BOD to TP is significant, particularly to a plant that has incorporated some form of enhanced biological phosphorus removal. Ratios less than 20:1 can indicate potential problems achieving effective enhanced biological phosphorus removal, because the organisms responsible for the uptake of phosphorus require adequate amounts of carbon; more specifically, they require carbon in the form of volatile formic acids (VFAs). Phosphorus is an essential nutrient for biological growth; if it is not present in sufficient quantities, growth can be inhibited and the efficiency of a biological treatment process reduced. This can be an issue in treatment processes that incorporate enhanced primary settling, because the additional coagulation and settling involved from the addition of chemical salts to the primary clarifiers to improve solids removal will also reduce the TP available for the downstream biological process. Phosphorus inhibitation may also continue to be an issue as treatment plants update for low levels of effluent nitrogen and phosphorus, particularly with plants that use a denitrification filter downstream of a final clarifier or filter for phosphorus removal. Phosphorous is removed to less than 0.5 mg/L levels by precipitation with alum or a ferric—any tri-valent salt.

Silica (SiO$_2$)   Silica (silicon dioxide), in some cases, is an anion. The chemistry of silica is a complex and somewhat unpredictable subject. In a similar fashion as TOC reports the total concentration of organics (as carbon) without detailing what the organic compounds are, silica reports the total concentration of silicon (as silica) without detailing what the silicon compounds are. The total silica content of a water is composed of reactive silica and unreactive silica. Reactive silica (e.g., silicates [SiO$_4$]) is dissolved silica that is slightly ionized and has not been polymerized into a long chain. Reactive silica is the form that RO and ion exchange chemists hope for. It is the form of silica to be used in RO projection programs. Reactive silica, though it has anionic characteristics, is not counted as an anion in terms of balancing a water analysis but it is counted as a part of total TDS. Unreactive silica is polymerized or colloidal silica, acting more like a solid than a dissolved ion. Silica in the colloidal form can be removed by RO but can cause colloidal fouling of the front end of a RO system. Colloidal silica—the size of which can be as small as 0.008 μm—can be measured empirically by the Silt Density Index, but only that portion that is 0.45 μm or larger. Particulate silica compounds (e.g., clays, silts, and sand) are usually 1 μm or larger and can be measured using the Silt Density Index. Polymerized silica, which uses silicon dioxide as the building block, exists in nature (e.g., quartzes and agates). It also results when the reactive silica saturation level is exceeded. The solubility of reactive silica is typically limited to 200% to 300% with the use of a silica dispersant. Reactive silica solubility increases with increasing temperature and with pH levels below 7.0 or above 7.8, and decreases in the presence of iron, which acts as a catalyst in the polymerization of silica. Silica rejection is pH sensitive, with increasing rejection at a more basic pH as the reactive silica exists more in the salt form than in the acidic form.

Hydrogen Sulfide (H$_2$S)   Hydrogen sulfide is a gas that causes the noticeable "rotten egg" smell in feed waters, with a threshold odor level of 0.1 ppm and a noticeable offensive odor at 3 to 5 ppm. Hydrogen sulfide is readily oxidized to elemental sulfur by oxidants (e.g., air, chlorine, or potassium permanganate). Sulfur acts as a colloidal foulant and has a history of not being removed well by conventional multimedia filtration. The preferred RO system design suggests leaving the hydrogen sulfide in its gaseous form, letting it pass through the RO into the permeate, and then treating the permeate for its removal.

Sulfate (SO$_4$)   Sulfate is a divalent anion. The solubility of calcium, barium, and strontium sulfate is low and can cause a scaling problem in the back end of an RO system. The solubility of these sparingly soluble salts is lower with decreasing temperature. The recommended upper limit for sulfate in potable water is 250 ppm, based on taste issues.

Sulfates, like chlorides, are found in most water supplies and in wastewaters from the presence of proteins. Sulfate is reduced under anaerobic conditions to hydrogen sulfide. High sulfates are also found in brackish water. Another contribution (along with chlorides) to TDS, sulfates are removed by NF and RO.

Aggregate Organic Constituents

Biochemical Oxygen Demand (BOD)   BOD is used to measure the quantity of oxygen required for oxidation of biodegradable organic matter present in water (wastewater) by aerobic biochemical action. Oxygen demand is exerted by three classes of materials:

1. carbonaceous organic materials usable as a food source by aerobic organisms;

2. oxidizable nitrogen from nitrate, ammonia, (NH_3), and organic nitrogen—food for specific bacteria, e.g., *Nitrosomonas* and *Nitrobacter*; and

3. chemical reducing compounds, e.g., Fe^{2+}, sulfates (SO_3), and sulfide (S), which are oxidized by dissolved oxygen.

In domestic sewage, all oxygen demand is due to carbonaceous and oxidizable nitrogen. Industrial waste though, can contain all three in addition to toxics and nonbiodegradable substances (refractory organics). The units of BOD are milligrams of dissolved oxygen utilized per liter of organic waste oxidized.

Chemical Oxygen Demand (COD)   COD is the amount of oxygen required to oxidize the organic fraction of a sample which is susceptible to permanganate or dichromate oxidation in acid solution. Standard COD tests correspond to 80% to 85% of theoretical oxygen demand (ThOD); rapid COD tests 70% of ThOD.

Following are the relationships between total volatile solids (TVS) and COD:

Carbohydrates:	(1.1 to 1.2) × TVS = COD
Proteins:	1.8 × TVS = COD
Lipids (fats):	3.0 × TVS = COD

Organic Carbon   Another water quality constituent that influences membrane flux is the organic carbon content, which is typically expressed in terms of either total (TOC) or dissolved organic carbon (DOC). Organic carbon in the feed water can contribute to membrane fouling, by either adsorption of the dissolved fraction onto the membrane material or obstruction by the particulate fraction. Thus, lower fluxes may be necessary if membrane filtration is applied to treat a water with significant organic carbon content.

Theoretical Oxygen Demand (ThOD)   We can write a ThOD equation for a compound—e.g., 100 mg/L of glucose—by assuming complete oxidation to CO_2 and H_2O:

$$C_6H_{12}O_6 + 6O_2 \rightarrow 6CO_2 + 6H_2O$$

$$MW \text{ of glucose} = 180$$

$$MW \text{ of } O_2 = 32$$

Set up the proportion:

$$\frac{100 \text{ mg/L}}{180} = \frac{COD}{6 \times 32}$$

$$COD = \frac{100 \times 6 \times 32}{180}$$

$$COD = 106 \text{ mg/L}$$

This method of choice for measuring the normally low concentrations of organic matter in drinking water sources is now more common in municipal wastewater reuse applications. Organic carbon is oxidized to carbon dioxide and water in a high temperature furnace.

Total Organic Carbon (TOC)   TOC, an acronym for total organic carbon or total oxidizable carbon, is a nonspecific test that measures the amount of carbon bound in organic material and is reported in ppm as carbon. Since TOC only measures the amount of carbon in organic matter, the actual weight of the organic mass can be up to 3 times higher in natural surface waters. Organics are compounds that contain carbon (with the exception of carbon dioxide, bicarbonate, and carbonate). In water treatment, organics can be classified as naturally occurring or man-made. Naturally occurring organic matter is typically negatively charged colloids or suspended solids, comprised of tannins, lignins, water-soluble humic acid compounds resulting from the decay of certain vegetative matter, and fulvic acid compounds resulting from the decay of certain vegetative matter. Naturally occurring organic material can be a foulant to RO membranes, particularly the negatively charged composite polyamides. Neutrally charged RO membranes (e.g., neutrally charged composite polyamides and cellulose acetate) are more resistant to organic fouling. An RO membrane will reject organic compounds. Generally, organic compounds

Parameter Projected Average mg/L (Except Where Noted)	Secondary Effluent	MF Filtrate/Permeate (Design Goal)
BOD5	20	< 5
COD	37 (max 42)	< 10
TOC	12	< 4
TSS	< 10	< 0.1
TKN	4	< 4
NH$_3$-N	0.1 (max 1.0)	< 0.1
P (total)	< 3 (max 10)	< 0.10
Ortho-P	< 2	< 0.10
Turbidity	< 10	< 0.1
Fecal coliform (CFU/100 mL)	—	0

TABLE 2-9 MF/UF Effect on Secondary Effluent

with a molecular weight greater than 200 Da are rejected at levels greater than 99%. The rejection of compounds with molecular weights less than 200 Da will vary based on molecular weight, shape, and ionic charge. As a rough rule of thumb, alert levels for potential organic fouling in natural water sources are TOC at 3 ppm, BOD at 5 ppm, and COD at 8 ppm. Table 2-9 shows the TOC values of secondary effluent and MF-treated secondary effluent. For a given wastewater, a correlation curve can be drawn between TOC and BOD; TOC and COD; or TOC and TVS. TOC can be reduced through direct coagulation and MF/UF or through MF and RO.

Fats, Oils, and Grease (FOG)   The term *grease* includes fats, oils, waxes, and other related constituents found in wastewater. Grease content has been determined in the past by extraction with hexane. At this time, freon is used.

Significant fats are stable organic compounds not easily decomposed by bacteria. If you combine a fat with a strong alkali, you form a soap—soluble in water but high in hardness, and reacts with calcium and magnesium to form insoluble precipitates. Fats, oils, and grease (FOG) can cause problems in sewers and wastewater plants; they coat membrane surfaces, interfere with biological processes, and cause maintenance problems. If discharged into streams, they interfere with biological life streams and create unsightly flowing material and films. FOG levels higher than 20 ppm may prove problematic for membranes. It is an important characteristic to measure in the design of a system.

UV-Absorbing Compounds   The tendency for a membrane to be affected by TOC is partially influenced by the nature of the organic matter in the water. TOC can be characterized as either hydrophilic or hydrophobic in composition, and studies suggest that the hydrophobic fraction contributes more significantly to membrane fouling. The character of the organic carbon content can be roughly quantified by measuring the SUVA of the water, as calculated using the following equation:

$$SUVA = \frac{UV_{254}}{DOC}$$

where SUVA = specific ultraviolet (light) absorbance (L/mg-m)
    UV_{254} = ultraviolet (light) absorbance at 254 nm (L/m)
    DOC = dissolved organic carbon (mg/L)

Because TOC is more commonly measured than DOC in drinking water treatment, SUVA is sometimes estimated using values for TOC in place of those for DOC.

Higher SUVA values tend to indicate a greater fraction of hydrophobic organic material, thus suggesting a greater potential for membrane fouling. Generally, SUVA values exceeding 4 L/mg-m are considered somewhat more difficult to treat. However, organic carbon (and turbidity) can often be removed effectively via coagulation and presettling, particularly if the organic material is more hydrophobic in character, thus minimizing the potential for membrane fouling and facilitating operation at higher fluxes. Coagulation can also be conducted in-line (i.e., without presettling) with MF/UF systems. Pretreatment using the injection of powdered activated carbon may also reduce DOC in the membrane feed; however, because spiral-wound membrane modules cannot be backwashed, powdered activated carbon should not be used in conjunction with NF/RO systems unless provisions are made to remove the particles upstream.

Microbiological Pathogens

Wastewater contains many microorganisms but only a subportion of the organisms—notably enteric pathogens—are potential human health hazards. Classes of microbes that can cause infection in humans include helminths (parasitic worms), parasitic protozoa, bacteria, and viruses. Some microorganisms are obligate pathogens (i.e., they must cause disease to be transferred from host to host), whereas others are opportunistic pathogens, which may or may not cause disease. In the United States, the most frequently documented waterborne enteric pathogens are the protozoa *Cryptosporidium* and *Giardia*; the bacteria *Salmonella*, *Shigella*, and toxigenic *Escherichia coli* O157:H7; and the enteroviruses and norovirus (Craun et al., 2006). They cause acute gastrointestinal illness and have the potential to create large-scale epidemics. The occurrence and concentrations of microbial pathogens in reclaimed water depend on the health of the tributary population and the applied wastewater treatment processes.

Primary treatment does little to reduce pathogens. Secondary treatment does not eliminate microbes. Tertiary treatment reduces coliforms to less than 100 per milliliter. Integral MF systems are considered disinfection membranes for protozoa and bacteria but not viruses, which are killed by chlorine. Viruses are reduced up to 6 log, typically 1 to 4 log, by UF membranes.

Helminths   Often known as parasitic worms, helminths pose significant health problems in developing countries where wastewater reuse is practiced in agriculture using raw sewage or primary effluents (Shuval et al., 1986). *For wastewater reuse in the* Western Hemisphere using secondary treatment membranes, helminths are not a consideration, as they are removed in secondary treatment.

Protozoa   Protozoa are single-celled eukaryotes that are heterotrophic and generally larger in size than bacteria. Some protozoa are mobile, using flagella, cilia, or pseudopods, whereas others are essentially immobile. Malaria, probably the best-known disease caused by protozoa, is caused by the genus *Plasmodium*. In U.S. water systems, *Giardia lamblia*, *Cryptosporidium parvum*, and *C. hominis* have been associated with gastrointestinal disease outbreaks through contaminated water. In 1993, an outbreak of cryptosporidiosis caused an estimated 400,000 illnesses and more than 50 deaths through contaminated drinking water in Milwaukee, Wisconsin (Mac Kenzie et al., 1994; Hoxie et al., 1997). Part of the protozoan life cycle often involves spores, cysts, or oocysts, which can be highly resistant to chlorine. *Cryptosporidium* oocysts and *Giardia* cysts of human origin are frequently detected in secondary wastewater effluent (Bitton, 2005), and these may still persist in disinfected effluent after granular media or membrane filtration (see, e.g., Rose et al., 1996). Thus, in potable reuse applications, additional treatment processes are needed to reduce the risk of infection from *Cryptosporidium* and *Giardia*.

Protozoa are especially important in reuse applications, especially *Cryptosporidium* and *G. lamblia*. These pathogenic organisms are the basis for drinking water regulations in the United States. Both are chlorine resistant. No cure exists for either one. Both are easily removed by MF and UF membranes.

Bacteria   Bacteria are single-celled prokaryotes and are ubiquitous in the environment. Domestic wastewaters contain many pathogenic bacteria that are shed by the human population in the sewershed. Particularly important are pathogenic bacteria that cause gastroenteritis and are transmitted by the fecal-oral route (enteric bacterial pathogens). Enteric bacteria were estimated to account for 14% of all waterborne disease outbreaks in the United States between 1971 and 1990 (Craun, 1991) and 32% between 1991 and 2002 (Craun et al., 2006). Based on hospitalization records (Gerba et al., 1994), the most severe bacterial infections result from *E. coli* (14%), *Shigella* (5.4%), and *Salmonella* (4.1%).

Because of the public health significance of bacterial pathogens, monitoring systems and water quality standards have been established based on fecal coliforms (a classification that includes *E. coli*) and *Enterococcus* in the United States and in many nations around the world (NRC, 2004). It is important to note that most *E. coli* and *Enterococcus* organisms are not pathogenic. Rather, they are part of the normal microflora in the human digestive tract and are necessary for proper digestion and nutrient uptake. *E. coli* and *Enterococcus* are employed as indicators of the presence of human waste (also called fecal indicator bacteria) in water quality monitoring and protection because they are present in high concentrations in human feces and sewage and they are more persistent than most bacterial pathogens. They are, therefore, used to indicate inadequate treatment of sewage to remove bacterial pathogens (NRC, 2004). Fecal indicator bacteria in undisinfected secondary effluent range from 102 to 105 per 100 mL depending on the quality of the influent water (Bitton, 2005). However, the concentration of fecal indicator bacteria (i.e., total coliform, fecal coliform, *Enterococcus*, and *E. coli*) in filtered, disinfected secondary effluent can be brought below the nominal detection limit of 2.2 organisms per 100 mL, and with advanced treatment it can be brought even lower.

Viruses   Viruses are extremely small infectious agents that require a host cell to replicate. They are of special interest in potable reuse applications because of their small size, resistance to disinfection, and low infectious dose. There are many different viruses, and they infect nearly all types of organisms, including animals, plants, and even bacteria. Aquatic viruses can occur at concentrations of 108 to 109 per 100 mL of water in the ocean (Suttle, 2007) and 109 to 110 per 100 mL in sewage (Wu and Liu, 2009); however, most of these are bacteriophages—viruses that infect bacteria. The viruses of concern in water reuse or in the discharge of treated wastewater to drinking water sources are human enteroviruses (e.g., poliovirus, hepatitis A), noroviruses (i.e., Norwalk virus), rotaviruses, and adenoviruses. Human viruses are usually present in undisinfected secondary effluent and may still persist in effluents after some advanced treatment (see, e.g., Blatchley et al., 2007; Simmons and Xagoraraki, 2011). Fecal indicator bacteria that are currently used for water quality monitoring are not an adequate indication of the presence or absence of viruses, because bacteria are more efficiently removed or inactivated by some wastewater treatment processes than are enteric viruses

(Berg, 1973; Harwood et al., 2005). Thus, viruses need to be carefully addressed whenever treated municipal wastewater is discharged or reused in a context where there may be human contact, particularly when it makes up all or part of a drinking water supply.

Viruses are a concern in wastewater reuse. This strand deoxyribo-nucleic acid (DNA) or ribonucleic acid (RNA) are easily killed by chlorine. UF can remove viruses up to logs. MF removal varies between 0.5 and 2.5 log.

Terminology Relevant to Basic Chemistry Review

18-Megohm Water   A high purity water that conducts electrical current poorly because of the lack of ionized impurities (electrolytes). It has an electrical resistivity of approximately 18 MΩ-cm (180,000 Ω-m) and a conductivity of 0.0556 mΩ/cm (0.00000556 S/m) at a specified temperature, typically 77°F (25°C). This type of water is also called ultrapure water.

Anion   A negatively charged atom or molecule that migrates toward the anode.

Anionic   Having a negative ion charge.

Atomic Weight   The relative weight of an element compared with the carbon 12 standard.

Cation   A positively charged ion or radical (as NH_4^+) that migrates toward the cathode.

Calcium (Ca)   The presence of calcium in water is a factor contributing to the formation of scale and insoluble soup curds that are a means of clearly identifying hard water.

Calcium Carbonate ($CaCO_3$)   A colorless or white crystalline compound. $CaCO_3$ is a sparingly soluble salt, the solubility of which decreases with increasing temperature. It has the potential to cause scaling if it is concentrated to supersaturation.

Calcium Carbonate ($CaCO_3$) Equivalent   An expression of the concentration of specified constituents in water in terms of the equivalent value to calcium carbonate ($CaCO_3$). For example, the hardness in water that is caused by calcium, magnesium, and other ions is usually described in terms of the calcium carbonate equivalent. For example, the concentration of calcium ions (i.e., $[Ca^{2+}]$) can be multiplied by 100/40 (the ratio of the molecular weight of $CaCO_3$ to the atomic weight of Ca^{2+}) to give $[Ca^{2+}]$ as equivalent $CaCO_3$.

Calcium Hardness   The portion of total hardness caused by calcium compounds such as calcium carbonate ($CaCO_3$) and calcium sulfate ($CaSO_4$).

Calcium Sulfate ($CaSO_4$)   A sparingly soluble salt, the solubility of which decreases with increasing temperature. It is called gypsum in its hydrated form. It is a potential source of scaling in desalting systems if it is concentrated to supersaturation.

Cationic   Having a positive ionic charge.

Caustic   Caustic soda (NaOH) or any compound chemically similar to caustic soda. The term is usually applied to strong bases.

Caustic Soda (NaOH)   Sodium hydroxide, a strongly alkaline chemical used for pH adjustment, water softening, anion exchange demineralizer regeneration, and other purposes. It is sometimes called caustic.

Chelating Agent   A chemical reagent; typically a water-soluble organic molecule such as citric acid or EDTA, that reacts with metal ions to keep them in a aqueous solution, therefore increasing the solubility of the metal in water.

Conductivity   A measure of the ability of an aqueous solution to conduct an electric charge; related to the amount of total dissolved solids (TDS).

Dalton (Da)   A unit of mass equal to one-twelfth the mass of a carbon 12 atom (i.e., one atomic mass unit [amu]); typically used as a unit of measure for the molecular weight cutoff of a UF, NF, or RO membrane.

Dissolved Carbon Dioxide   The carbon dioxide (CO_2) that is dissolved in a liquid medium, typically expressed in milligrams per liter. The saturation concentration is dependent on several factors, including partial pressure, temperature, and pH.

Dissolved Concentration   The amount per unit volume of a constituent of a water sample filtered through a 0.45 μm pore-diameter membrane filter before analysis.

Dissolved Gases   The sum of gaseous components that are dissolved in a liquid medium. Typical dissolved gases found in water include oxygen (O_2), nitrogen (N_2), carbon dioxide (CO_2), methane (CH_4), and hydrogen sulfide (H_2S), among others. High concentrations of dissolved gases can result in filter air binding and pump cavitation.

Dissolved Organic Matter (DOM)   The carbon portion of the organic matter in water that passes through a 0.45 μm pore-diameter filter. For aquatic humic substances, the carbon level typically represents approximately 50% of the organic matter (the rest being hydrogen, oxygen, nitrogen, and sulfur).

Dissolved Solids   In operational terms, the constituents in water that can pass through a 0.45 μm pore-diameter filter. See also *total dissolved solids*.

Divalent Ion   An ion with two positive or negative electrical charges, such as ferrous (Fe^{2+}) or sulfate (SO_4^{2-}).

Equivalent Weight   The weight of a compound that contains one gram-atom of available hydrogen or its chemical equivalent; the molecular weight of the element or compound divided by the net positive valence.

Equivalent Weight   The weight of a substance based on the formula weight and number of reactive protons associated with the substance. For an element, the equivalent weight is the atomic weight divided by the number of reactive protons. For example, calcium (Ca) has an atomic weight of 40 and has two reactive protons, so its equivalent weight is $40/2 = 20$. For a compound, the equivalent weight is the molecular weight divided by the number of reactive protons. For example, sodium chloride (NaCl) has a molecular weight of 58.5 and one reactive proton, so the equivalent weight is $58.5/1 = 58.5$. Calcium carbonate ($CaCO_3$) has a molecular weight of 100 and two reactive protons (in the Ca^{2+} ion), so the equivalent weight is $100/2 = 50$. Equivalent weight is also known as combining weight.

Gram atomic weight   The quantity of an element in grams corresponding to its atomic weight (e.g., 40 grams of Calcium, Ca^{40}, or 16 grams of oxygen, O^{16}).

Gram Molecular Weight   The molecular weight in grams of any particular compound. For example, oxygen (O) = 16 g, sodium hydroxide (NaOH) = $23 + 16 + 1 = 40$ g, and sodium hypochlorite (NaOCl) = 74.5.

Hydrophilic   Attracting water.

Hydrophobic   Repelling water.

Inorganic Chemicals   Arsenic, asbestos, cadmium, chromium, copper, cyanide, fluoride, lead, mercury, nickel, nitrate, and nitrite.

Ionic Bond   A type of chemical bond in which electrons are transferred.

Ionization   The splitting or dissociation (separation) of molecules into negatively and positively charged ions.

Langelier Saturation Index (LSI)   A method to predict the precipitation of calcium carbonate under certain conditions of temperature, pH, hardness, alkalinity, and total dissolved solids.

Lime (CaO)   A calcined chemical material, calcium oxide. Lime is used in lime softening and in lime-soda ash water treatment, but first it must be slaked to calcium hydroxide, $Ca(OH)_2$. Lime is also called burnt lime, calyx, fluxing lime, quicklime, or unslaked lime.

Metals   Elements that easily lose electrons (oxidize) to form positively formed ions called cations; e.g., copper (Cu^{2+}) and ammonium (NH^{4+}).

Micron   A metric measurement equivalent to 10^{-6} m (4.0×10^{-5} in.). It is represented by the symbol μm. Used to classify the pore size in a membrane.

Mineral Acid   An inorganic acid, especially hydrochloric acid (HCl), nitric acid (HNO_2), or sulfuric acid (H_2SO_4).

Mineral Acidity   Acidity in water caused by the presence of strong inorganic acids—such as hydrochloric acid (HCl), nitric acid (HNO_3), or sulfuric acid (H_2SO_4)—as opposed to weak acidity caused by such acids such as carbonic acid (H_2CO_3) or acetic acid (CH_2COOH). Mineral acidity is usually expressed in water analysis as free mineral acidity.

Mineral Salt   A chemical compound formed by the combination of a mineral acid and a base. Minerals from dissolved rock exist in water in the form of dissolved mineral salts. An excess of mineral salts can give water a disagreeable taste or even be harmful to human health.

Mineral-Free Water   Water produced by either distillation or deionization. The term is sometimes found on labels of bottled water as a substitute term for *distilled water* or *deionized water*.

Molecular Weight (MW)   The sum of the atomic weights of the atoms in a molecule.

Molecule   The smallest particle of an element or compound that retains all of the characteristics of the element or compound. A molecule is made up of one or more atoms. The helium molecule, for example, has only one atom per molecule. Oxygen molecules (O_2) have two atoms; ozone molecules (O_3) have three atoms. Molecules found in chemical compounds often have many atoms of various kinds.

Neutralization   (1) A chemical reaction in which water is formed by mutual destruction of the ions that characterize acids and bases when both are present in an aqueous solution. Typically, the hydrogen and hydroxide ions react to form water ($H^+ + OH^- \rightarrow H_2O$) and the remaining product is a salt. Neutralization occurs with both inorganic and organic compounds. Neutralization can also occur without water being formed, as in the reaction of calcium oxide and carbon dioxide to form calcium carbonate ($CaO + CO_2 \rightarrow CaCO_3$). Neutralization does not mean that a pH of 7.0 has been attained; rather, it means that the equivalence point for an acid-base reaction has been reached. (2) The process by which the cell-attachment protein of a virus is bound by an antibody, thereby inhibiting infection by the virus.

pH   A measure of the acidity or alkalinity of a solution, such that a value of 7 is neutral; lower numbers represent acidic solutions and higher numbers, alkaline solutions. Strictly speaking, pH is the negative logarithm of the hydrogen ion concentration (in moles per liter). For example, if the concentration of hydrogen ions is 10^{-7} mol/L, the pH will be 7.0. As a measure of the intensity of a solution's acidic or basic nature, pH is operationally defined relative to standard conditions that were developed so that most can agree on the meaning of a particular measurement. The pH of an aqueous solution is an important characteristic that affects many features of water treatment and analysis.

Potable Water   Water that is safe and satisfactory for drinking and cooking.

Salinity   The amount of salt in a solution; usually used in association with salt solutions in excess of 1,000 mg/L and synonymously with the term *total dissolved solids* (TDS).

Sequester   To keep a substance (e.g., iron or manganese) in solution through the addition of a chemical agent (e.g., sodium hexametaphosphate) that forms chemical complexes with the substance. In the sequestered form, the substance cannot be oxidized into a particular form that will deposit on or stain fixtures. Sequestering chemicals are aggressive compounds with respect to metals, and they may dissolve precipitated metals or corrode metallic pipe materials.

Softening   The removal of hardness (i.e., divalent metal ions, primarily calcium and magnesium) from water.

Solutes   The materials (such as chemicals) contained in a solution.

Solvent   Water (or any liquid) that contains dissolved matters or total dissolved solids. A solution is made up of the solvent and the solute.

Synthetic Organic Chemicals   Pesticides, solvents, monomers, pharmaceuticals, endocrine disruptors, and hormones.

Total Dissolved Solids   In salinity, sulfates and chlorides.

CHAPTER 3

Basic Concepts

Introduction

This chapter provides an overview of the basic concepts of the membrane filtration processes: low pressure membranes—microfiltration (MF) and ultrafiltration (UF); and diffusive membranes—nanofiltration (NF) and reverse osmosis (RO). Major topics covered include definitions and terminology as well as descriptions of the various membrane classifications, membrane materials, geometry, modules, construction, driving forces, operation, and hydraulic configurations.

Membrane technology used for water recycling and reuse is based upon the concept that a thin layer of material called a semipermeable membrane can be used to separate contaminants from a treated wastewater source when a driving force is applied across the membrane.

Membranes can be classified in many ways. One is according to separation mechanism, i.e., removal based on size of the contaminant, also known as the sieve effect; and separation based upon diffusivity, the osmotic pressure of the solutes. Membranes are also classified as porous (with fixed pores) or nonporous dense membranes, also called diffusion membranes.

Porous membranes have fixed pores. The International Union of Pure and Applied Chemistry (1985) classifies porous membranes according to pore size:

- Macropores—pores larger than 50 nm
- Mesopores—pores ranging from 2 to 50 nm
- Micropores—pores smaller than 2 nm

Porous membranes include MF and UF membranes. Nonporous, or dense, membranes include NF and RO membranes. In some cases, NF membranes can be classified as both.

Characterized by pore size, membranes are ranked (in decreasing order) as MF, UF, NF, RO. Low pressure membrane technology is a pressure driven (positive or vacuum) separation process which uses a layer of semi-permeable material (membrane) polymeric or ceramic.

Low pressure membranes—MF and UF—are the two most often associated with the term *membrane filtration*. MF and UF are characterized by their ability to remove suspended or colloidal particles via a sieving mechanism based on the size of the membrane pores relative to that of the particulate matter.

Characterized by operating pressure, membranes are ranked (in decreasing order) as RO, NF, UF, MF. High pressure or diffusion-driven membrane technology is also pressure driven (no vacuum here); it uses the diffusivity of soluble constituents. Constituents removed include divalent cations and anions, e.g., calcium and magnesium removal (membrane softening), nitrate removal, and arsenic removal. Dissolved organics such as natural organic matter, dissolved organic compounds, and synthetic organic compounds are also removed.

The physical and chemical properties of the contaminants in the wastewater, the driving force required, and size exclusion requirements are used to select the membrane process. Separation of contaminants from water using diffusive membranes—i.e., NF and RO—occurs when there is a significant difference in the transport coefficients (i.e., osmotic pressure) through the membranes for the contaminants.

The ability of each type of membrane filtration system to remove water pathogens of interest on the basis of size is illustrated in Fig. 3-1, the filtration spectrum. The figure shows the approximate size range of viruses, bacteria, *Cryptosporidium* oocysts, and *Giardia* cysts, as well as the ability of MF, UF, NF, and RO to remove each of these pathogens on the basis of size exclusion. Overlap between the range covered by a membrane filtration process with a given pathogen size range indicates the ability of that process to remove the pathogen. Note that the molecular weights listed do not correspond precisely to the indicated pathogen size range, but are rough generalizations depicted as a result of the fact that NF, RO, and some UF processes are rated according to a molecular weight cutoff (MWCO) on the basis of their ability to remove dissolved solids and larger macromolecules. Figure 3-1 also illustrates contaminants of interest, size range, and typical processes required for removal. Figure 3-2 illustrates particle size relative to MF pore size. It shows the relative sizes of a pencil dot, a large siliceous particle (grain of sand), a *Cryptosporidium* oocyst, and a *Giardia* cyst, in relation to a 0.1 µm MF pore.

Although each of the classes of membrane filtration functions as a filter for various sizes of particulate matter, the basic principles of operation vary between MF and UF on the one hand and NF and RO on the other. Each type of system is described in the following sections.

The typical range of MWCO levels is less than 200 Da for RO membranes and between 200 and 1,000 Da for NF membranes. Diffusive membranes are not "filters" of dissolved solids; NF and RO utilize semipermeable membranes that do not have definable pores. NF and RO processes achieve removal of dissolved contaminants

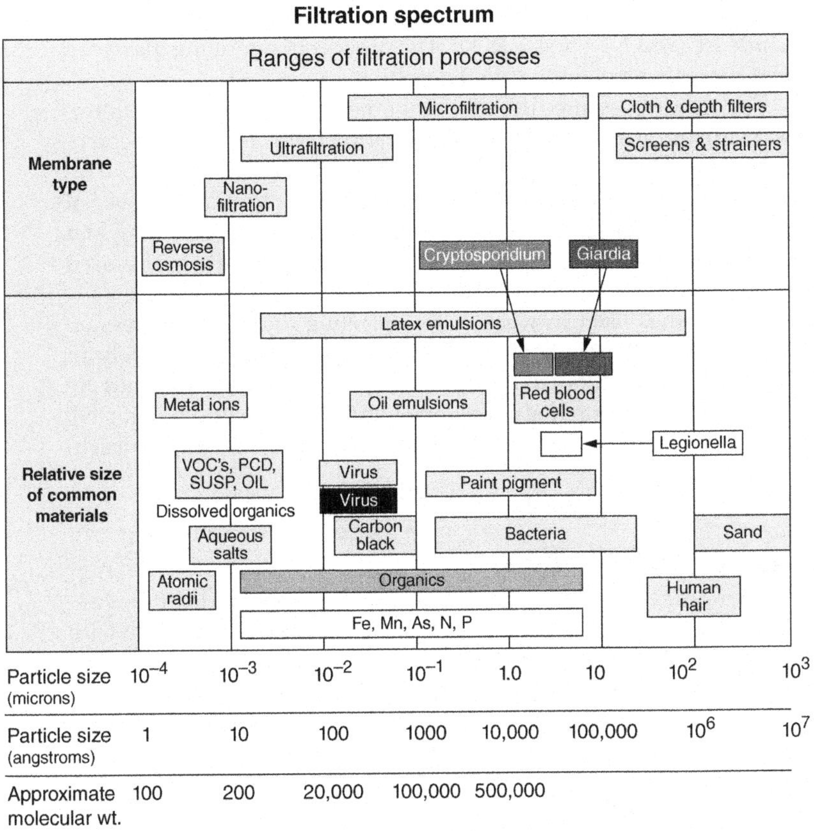

Filtration spectrum

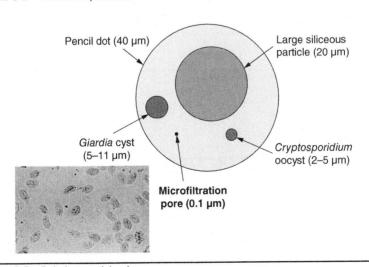

FIGURE 3-1 Filtration spectrum.

FIGURE 3-2 Relative particle sizes.

through the process of reverse osmosis. High pressure membranes include RO and NF ("loose RO"). The process of removing dissolved constituents is sometimes called desalting.

The major pressure driven membrane processes are microfiltration (MF), ultrafiltration (UF), nanofiltration (NF), and reverse osmosis (RO). In addition, important membrane processes used to treat wastewaters include electrodialysis (ED) and electrodialysis reversal (EDR). The membrane processes most used in water recycling and reuse applications are MF, UF, NF, and RO. ED and EDR are also used but are not discussed in this book.

Membranes used in wastewater recycling are, for low pressure MF and UF, either hollow fiber, tubular, flat-plate (exclusively in membrane bioreactors); in NF and RO, spiral wound membranes are used. Disk tube technology uses plate and frame technology for NF and RO applications requiring very high pressures, like landfill leachate treatment.

Terminology and Definitions

All membrane processes—i.e., both low pressure and diffusive—are represented in a Typical membrane Schematic, a rectangle with a diagonal line, as shown in Fig. 3-3. They are also presented in block form as shown in Fig. 3-4. The influent or source water (secondary effluent in the case of wastewater reuse), enters the membrane.

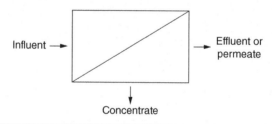

Figure 3-3 Typical membrane schematic.

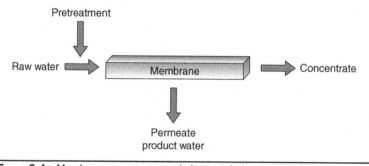

Figure 3-4 Membrane processes and characteristics.

The treated water is designated *permeate*. The concentrate, reject, or—in the case of RO membranes treating sea or brackish water—brine, contains all of the contaminants removed by the membrane. For low pressure membranes the concentrate or reject is all of the particulate and biological matter removed. For diffusive membranes, the concentrate contains all rejected dissolved species.

Low Pressure Membranes—Microfiltration and Ultrafiltration

All membranes have a distribution of pore sizes, and this distribution will vary according to the membrane material and manufacturing process. When a pore size is stated, it can be presented as either nominal (i.e., the average pore size) or absolute (i.e., the maximum pore size), in terms of microns (μm). The nominal pore sizes of MF and UF membranes used in wastewater reuse applications typically range from 0.001 to 0.2 μm—author's experience. Both membranes are designed to remove particular matter, including turbidity. Turbidity is widely used as a performance gauge for conventional media filters, and among the various types of membrane filtration systems. Turbidity, and not suspended solids, is most often used as an assessment tool for low pressure membrane performance, since these systems are specifically designed to remove particulate matter. Higher turbidity measurements are also indicative of higher concentrations of suspended solids, and thus the potential to cause membrane fouling.

MF membranes used in wastewater recycling and reuse are generally considered to have a pore size range of 0.1 to 0.2 μm (nominally 0.1 μm), although there are exceptions, as MF membranes with pores sizes of up to 10 μm are available. Although MWCO is widely accepted as the means to classify UF membranes, pore sizes generally range from 0.001 to 0.05 μm (nominally 0.01 μm) or less, decreasing to an extent at which the concept of a discernible pore becomes inappropriate—a point at which some discrete macromolecules can be retained by the membrane material. In terms of pore size, the lower cutoff for a UF membrane is approximately 0.005 μm.

Overlap exists as to the designation of whether a membrane is MF or UF. In some cases a particular membrane can be considered either a UF or an MF membrane, based upon pore size. Table 3-1, general characteristics of wastewater reuse membrane processes, and Table 3-2, constituent removal by membrane illustrate the overlap. Although the concept of using the nominal or absolute pore size is sometimes used in reference to the filtration capabilities of membrane material, this concept, in the author's opinion, is overly simplistic and does not fully characterize the removal efficiency of a membrane. For example, the mechanisms of filtering particles close in size to the pore distribution of a membrane become more complex than sieving, as particles smaller than most pores may be removed through probabilistic interception through the depth of the filter media. In addition, for some

Membrane Process	Membrane Driving Force	Separation Mechanism	Typical Pore Size	Typical Reuse Pore Size	
Microfiltration	Pressure vacuum difference or vacuum in open vessels	Sieve	Macropores (> 50 nm)	0.08–0.2	Ceramic (various materials) polypropylene, polysulfone (pvdf) polyethan sulfane
Ultrafiltration	Hydrostatic pressure difference or vacuum in open vessels	Sieve	Mesopores (2–50 nm)	0.005–0.2	Aromatic, polyamides, ceramic (various materials), cellulose acetate, polypropylene, polysulfone, PES, polyvinylidene fluoride (PVDF)
Nanofiltration	Hydrostatic pressure difference	Sieve Solution/ diffusion + exclusion	Micropores (< 2 nm)	0.001–0.01	Cellulosic, aromatic polyamide, polysulfone, polyvinyl fluoride (PVDF), thin film composite
Reverse osmosis	Hydrostatic pressure difference	Solution/diffusion + exclusion	Dense (< 2 nm)	0.0001–0.001	Cellulosic aromatic polyamide, thin film composite

(Adapted from Metalf & Eddy AECOM (2007).)

TABLE 3-1 General Characteristics of Wastewater Reuse Membrane Processes

Contaminant	MF	UF	NF	RO
Suspended solids	Yes[1]	Yes[1]	Yes	Yes
Dissolved solids	No	No	Some	Yes
Bacteria and Cysts	Yes	Yes	Yes	Yes
Viruses	No	Yes	Yes	Yes
Dissolved Organic Matter	No[2]	No[2]	Yes	Yes
Iron and Manganese	Yes, if oxidized	Yes, if oxidized	Yes	Yes
Hardness	No	No	Yes	Yes

[1]High levels will foul these membranes.
[2]Could remove some with appropriate pretreatment.

TABLE **3-2**   Constituent Removal by Membrane

membrane materials, particles may be rejected through electrostatic repulsion and adsorption to the membrane material. The filtration properties of the membrane may also depend on the formation of a cake layer during operation, as the deposition of particles can obscure the pores over the course of a filter run, thus increasing the removal efficiency.

Currently, no established standard method exists for characterizing and reporting pore sizes for the various membrane filtration processes. Manufacturers have developed and used their own methods. As a result, the meaning of this information varies among membrane manufacturers, limiting its value. The concept of pore size does not address the integrity of the manufactured membrane module assembly over time, which could potentially pass particles larger than the indicated pore size after months and years of operation. UF membranes, for example, are awarded log removal credit for viruses based upon challenge testing of a new membrane. Little to no data exist on the virus removal capability of a UF membrane in operation for many years—warranties can extend to 10 years. This is significant, as UF membranes are often specified in water reuse applications because of their virus removal capability—UF membranes are capable of virus log removals ranging from 1 to 5 log.

Table 3-3 compares challenge testing of MF and UF performance in removing *Giardia* cysts, *Cryptosporidium* oocysts, MS-2 bacteriophage, particle counts, and turbidity. Except for virus removal (MS-2), no difference exists between the permeate quality of the two membranes.

Because some UF membranes have the ability to retain larger organic macromolecules, they have been historically characterized by MWCO rather than by a particular pore size. The concept of the MWCO (expressed in daltons, a unit of mass) is a measure of the removal characteristic of a membrane in terms of atomic weight (or mass) rather

	MF	UF
Giardia cysts	4.5 to 7 log	5 to 7 log
Cryptosporidium oocysts	4.5 to 7 log	5 to 7 log
MS-2 bacteriophage virus	0.5 to 3.0 log	4.5 to 6 log
Particles < 2 μm	<10/mL	<10/mL
Particles 2 to 5 μm	<10/mL	<10/mL
Particles 5 to 15 μm	<1/mL	<1/mL
Average turbidity	0.01 to 0.03 ntu	0.01 to 0.03 ntu

Pall Corporation (2004).

TABLE 3-3 Typical Summary of Removals, MF vs. UF

than size. Thus, UF membranes with a specified MWCO are presumed to act as a barrier to compounds or molecules with a molecular weight exceeding the MWCO. Because such organic macromolecules are morphologically difficult to define and are typically found in solution rather than as suspended solids, it may be convenient in conceptual terms to use MWCO rather than a particular pore size to define UF membranes when discussed in reference to these types of compounds. Typical MWCO levels for UF membranes range from 10,000 to 500,000 Da, with most membranes used for wastewater, recycling, and reuse treatment at approximately 100,000 Da. However, UF membranes remove particulate contaminants via a size exclusion mechanism and not on the basis of weight or mass; thus, UF membranes used for wastewater, recycling, and reuse are also characterized according to pore size with respect to microbial and particulate removal capabilities.

Transmembrane Pressure

The driving force necessary to force water through a low pressure membrane—the pressure drop across the membrane—is termed *transmembrane pressure,* or TMP. It is the difference between the average feed pressure and the permeate pressure; it is the driving force, or hydraulic head loss, associated with any given flux. The TMP of the membrane system is an overall indication of the feed pressure requirement. It is used, with flux, to assess membrane fouling and the need for chemical cleaning. Units of TMP measure are either pounds per square inch (psi) or bars. A bar is an international unit of measure and is equivalent to 14.5038 psi. Pressure is also given in kilopascals, 100 kilopascals = 1 bar.

The TMP for low pressure membranes is calculated as

$$TMP = \frac{(\text{feed pressure} + \text{retentate pressure})}{2} - \text{permeate pressure}$$

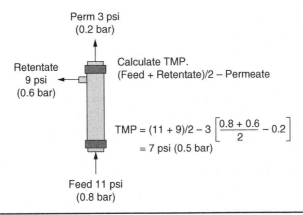

The figure shows a membrane module with labels:

Perm 3 psi
(0.2 bar)

Retentate
9 psi
(0.6 bar)

Calculate TMP.
(Feed + Retentate)/2 − Permeate

$$TMP = (11 + 9)/2 - 3 \left[\frac{0.8 + 0.6}{2} - 0.2 \right]$$
$$= 7 \text{ psi } (0.5 \text{ bar})$$

Feed 11 psi
(0.8 bar)

Figure 3-5   TMP computation.

In Fig. 3-5, TMP computation, the feed pressure is measured as 11 psi (0.8 bar), the retentate pressure is 9 psi (0.6 bar), and the permeate pressure is 3 psi (0.2 bar); thus, the TMP is calculated as $[(11 + 9)/2]$ $-3 = 7$ psi (0.5 bar).

Flux

A number of underlying basic principles are applicable to the design and operation of all types of pressure driven membrane filtration systems. One concept is flux, the volumetric flow rate per unit area through the membrane. Membrane systems operating at high flux are more efficient than ones operating at low flux. Low pressure membrane flux affects capital and operating cost, membrane replacement (membrane materials), and chemical usage. Flux is calculated as the permeate flow divided by the membrane area. It is expressed either as gallons per day per square foot of membrane area (gpd/ft², usually abbreviated *gfd*), as liters per square meter per hour (L/m² h, abbreviated *lmh*), or sometimes as cubic meters per square meter per hour (m³/m² h, abbreviated *m/h*).

Typical fluxes for low pressure polymeric membranes are 5 to 15 gfd (8.5 to 25.5 lmh) for membrane bioreactors, 10 gfd (17 lmh) for diffusive membranes, and as high as 120 gfd (204 lmh) for low pressure membranes treating clean (low turbidity) groundwater. Fluxes for low pressure membranes (LPMs) treating secondary effluent vary between 15 and 80 gfd (25.5 and 135.8 lmh) depending upon the source water, membrane manufacturer, the type of membrane (MF or UF), and the pretreatment selected. Ceramic membranes claim greater than 40 gfd/psi (984 lmh/bar) specific flux treating secondary effluent with coagulant addition and ozone treatment (Lehmann et al., 2009). Specific flux or permeability of a membrane is the flow per unit area per unit of TMP. It is usually expressed as gfd/psi or lmh/bar.

Turbidity Effects

Turbidity is widely used as a performance gauge for conventional media filters; among the various types of membrane filtration systems, it is most often used as an assessment tool for the performance of low pressure membranes—i.e., MF and UF—since these systems are specifically designed to remove particulate matter. Turbidity is the light-scattering property of particles and should not be confused with suspended solids, as there is no direct mathematical correlation. Protozoa such as *Cryptosporidium* and *Giardia*, bacteria such as the coliform group of bacteria (including *E. coli* and the organisms that cause Legionnaires' disease), and suspended materials such as iron and manganese all contribute to turbidity.

Water with higher turbidity may or may not be filtered at lower fluxes to minimize fouling and the consequent backwash and chemical cleaning frequency. This design decision is based upon the membrane particulars, the economic trade-offs, and the customer. Higher turbidity measurements are also indicative of higher concentrations of suspended solids and thus the potential for membrane fouling.

Integrity Testing

Membrane treatment is based upon the premise that an integral membrane will remove all contaminants larger than its pore size. An integrity test is a means of determining whether or a membrane is integral, i.e., free of any defects, breaches, or holes that will allow species in the feed water to pass through the membrane. In low pressure membrane applications for wastewater treatment and reuse, the objectives are either to remove microbial pathogens—protozoa, bacteria, and viruses—or to pretreat for other treatment processes for wastewater reuse (i.e., NF or RO). If intact, a membrane offers essentially 100% removal efficiency for targeted microbial pathogens. If the membrane is not integral—and depending upon the number of fibers compromised—microbial contaminants will pass through the membrane. Therefore, it is important to know if a compromise to the membrane has occurred. Several direct and indirect integrity testing methods are currently available, including the air pressure hold test, bubble point determination, online particle monitoring, and online turbidity monitoring.

Although the objectives and performance criteria for membrane integrity testing could be different from those for drinking water applications, the basic principles outlined in the United States Environmental Protection Agency's (USEPA) guidance manual (2005) are applicable.

As indicated in Fig. 3-6, the required pressure differentials across the membrane to meet the criteria for the removal of *Cryptosporidium*, bacteria, and viruses are approximately 14.5, 97, and 1,750 psid (1.0, 6.7, and 120.6 bar) respectively, at 5°C. Obviously, it is not feasible

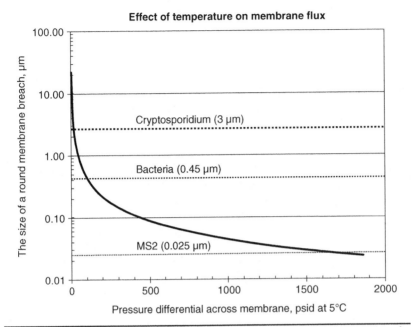

Effect of temperature on membrane flux

FIGURE 3-6 Relationship of pressure differential across membrane and the size of membrane breach. (*From Liu, 2012.*)

to use the pressure decay method for integrity testing if the goal is to completely remove viruses or even bacteria, as the pressure is too high to be practical. Therefore, developing a membrane integrity testing method that is capable of meeting the resolution criterion for bacteria and virus removal would be very beneficial.

Another issue is the sensitivity of the test. As all methods that are able to monitor membrane integrity continuously are generally of low sensitivity, and methods of relatively high sensitivity are unable to be performed on a continuous basis, the current membrane plants have to shut down periodically to perform integrity testing that is sufficiently sensitive. This operation mode is the result of balancing the risk to water quality due to membrane integrity breach and plant productivity.

Membrane Fouling

Membrane fouling describes the phenomenon in which a membrane loses flux. It is caused by organic and inorganic foulants "clogging" the membrane's pores. Fouling is a result of (a) interactions between fouling materials and the membrane medium, (b) interactions between fouling materials, and (c) interactions between fouling materials and aquatic environments.

Natural organic material and total organic carbon (TOC) have been identified as major fouling components for MF and UF membranes, especially for applications of surface water filtration. The dissolved organics (DOC) and organics associated with suspended solids and colloids in secondary effluent are the fouling species of interest in water reuse applications. TOC and DOC can be either hydrophilic or hydrophobic in nature, as determined by the specific ultraviolet absorbance (SUVA) of the wastewater (see Chapter 2). Higher SUVA values tend to indicate more hydrophobic species. The suspended solids and biochemical oxygen demand in primary effluent have delayed the development of membrane technology to treat primary effluent directly. Fouling of diffusive membranes is different, as pretreatment is designed to reduce the fouling, and is discussed later in the chapter. As analytical techniques and knowledge of the structural details of natural organic matter and the organics in secondary effluent progress, the mechanisms of membrane fouling will be better understood and approaches to prevent and control fouling will be developed.

The accumulation of microbial (biofouling), organic (organic fouling), and inorganic foulants on a membrane is called reversible when it can be removed by mechanical membrane cleaning—using air, product water, and chemicals—and irreversible when it cannot be removed from the membrane surface. Figure 3-7 shows the effect of membrane fouling with time. The longer cycle with flux maintenance curve shows a gradual increase of TMP with time even with flux maintenance. This gradual increase is caused by membrane fouling not removed by flux maintenance. When this same phenomenon still occurs after clean-in-place (cip), the fouling is irreversible. In the industry, mechanical removal of foulants is termed *flux maintenance*. Flux maintenance describes the periodic use of air, water, and chemicals—such as sodium hypochlorite, sodium hydroxide, mineral acids such as sulfuric and hydrochloric, and citric acid—to lessen the rate of fouling.

Temperature Effects

The temperature of municipal wastewater is, on average, higher than that of surface water or groundwater. Temperature is important in

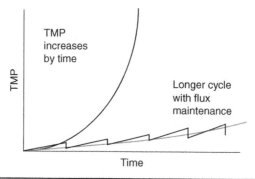

FIGURE 3-7 Effects of flux maintenance on TMP.

designing wastewater treatment facilities. It plays a major role in designing low pressure and diffusive membrane systems. Temperature is typically taken at the time of sample collection.

Like other water quality parameters such as turbidity and total dissolved solids (for NF and RO systems), the feed water temperature also affects the flux of a membrane filtration system. At lower temperatures, water becomes increasingly viscous; thus, lower temperatures reduce the flux across the membrane at constant TMP or alternatively require an increase in pressure to maintain constant flux. The means of compensating for this phenomenon varies with the type of membrane filtration system used. General viscosity-based means of compensating for temperature fluctuations for both MF or UF and NF or RO systems are available. Membrane manufacturers may have a preferred product-specific approach.

MF and UF membrane systems can operate within a relatively narrow range of TMP, limiting TMP to maintain constant flux as the water temperature decreases. This is especially true of vacuum systems. Because the membrane modules can be damaged if the TMP exceeds an upper limit, as specified by the manufacturer, it may not be possible to operate the system at a TMP that is sufficient to meet the required treated water production during colder months. In vacuum systems, the TMP cannot exceed the design value. As a result, additional treatment capacity (i.e., increased membrane area or number of membrane modules) is incorporated into the design of the system, such that the water treatment production requirements can be satisfied throughout the year. Figure 3-8 graphs flux versus temperatures of 8°C and 16°C. It illustrates the impact of temperature

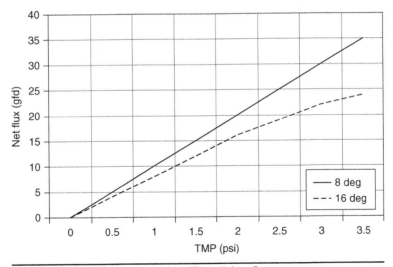

Figure 3-8 Effects of temperature differential on flux.

and why temperature must be considered in the design of membranes operating in changing temperatures.

For the microporous MF and UF membranes, the relationship between flux, TMP, and water viscosity is given in the Membrane Filtration Guidance Manual (2005) by the following equation:

$$J = \frac{TMP}{R_t \times \mu_w} \tag{3.1}$$

where J = flux (gfd)
        TMP = transmembrane pressure (psi)
        R_t = total membrane resistance (psi/gfd-cP)
        μ_w = viscosity of water (cP)

If the system is operated at constant flux, increases in viscosity require proportional increases in operating TMP (assuming constant membrane resistance). However, once the TMP approaches the rated maximum for the membrane, further increases in viscosity necessitate a reduction in flux. Thus, in order to maintain the required filtered water production flow (so as to satisfy customer demand), the membrane area must increase in proportion to the flux decrease, as shown in Equation 3.2:

$$J = \frac{Q_p}{A_m} \tag{3.2}$$

where J = flux (gfd)
        Q_p = filtrate flow (gpd)
        A_m = membrane surface area (ft²)

Combining Equations 3.1 and 3.2 demonstrates that the additional membrane area required is directly proportional to the increase in water viscosity (for constant flow, TMP, and membrane resistance), as shown in Equation 3.3:

$$\frac{Q_p}{A_m} = \frac{TMP}{R_t \times \mu_w} \tag{3.3}$$

where Q_p = filtrate flow (gpd)
        A_m = membrane surface area (ft²)
        TMP = transmembrane pressure (psi)
        R_t = total membrane resistance (psi/gfd-cP)
        μ_w = viscosity of water (cP)

Membrane filtration systems are commonly designed to operate at a particular flux (e.g., as determined via pilot testing or mandated by the state) to produce a specific flow (i.e., rated system capacity) at a given reference temperature. Thus, the required membrane area at the reference temperature can be calculated using Equation 3.3. The increased membrane area required to compensate for cold weather flow can be determined by multiplying this area by the ratio of the viscosity at the coldest anticipated temperature (e.g., the coldest average monthly temperature) to that at the reference temperature. Values for water viscosity can be found in the literature.

After the appropriate values for water viscosity have been determined for both the reference temperature (commonly 20°C for MF and UF systems) and the coldest anticipated temperature, then the design membrane area, as compensated for seasonal temperature variation, can be calculated as shown in Equation 3.4:

$$A_d = A_{20} \times \frac{\mu_T}{\mu_{20}}$$
(3.4)

where A_d = design membrane area (adjusted to temperature T; ft^2)
A_{20} = membrane area required at 20°C reference temperature (ft^2)
μ_T = viscosity of water at temperature T (cP)
μ_{20} = viscosity of water at 20°C (cP)

Membrane Materials

The membrane material is the substance from which the membrane itself is made. Synthetic membranes have been developing since the 1960s. Membranes are now available in a vast range of materials, from cellulosics to thermoplastics to ceramics to sintered metals. Each material has advantages and disadvantages that make it suitable to various applications. When selecting a membrane material, many factors must be considered, with each application having its own unique set of requirements. These factors are discussed in Chapter 4, Low Pressure Membrane Technology—Microfiltration and Ultrafiltration, and Chapter 5, Diffusive Membrane Technology—Nanofiltration and Reverse Osmosis.

Membranes can be solid as well as immobilized liquid phases, can be composed of organic or inorganic material, and can have catalytic and nonseparating properties as well as separating functions.

Traditional materials used in the fabrications of low pressure membranes are cellulose acetate, polyamide, polysulfone, and polyethersulfone. Other polymers used in the manufacturing of membrane materials include polypropylene, nylon, polyacrylonitrile (PAN), polycarbonate, polyvinyl alcohol (PVA), and polyvinylidene fluoride (PVDF). Ceramic and metallic membranes are used in MF and UF applications.

The material properties of the membrane impact the design and operation of the filtration system. Normally, the membrane material is manufactured from a synthetic polymer, although other forms, including ceramic and metallic membranes, may be available. Until just recently, almost all membranes manufactured for wastewater recycling and reuse application were made of polymeric material, since they were considered significantly less expensive than membranes constructed of other materials.

A new generation of ceramic membranes introduced into the market in the last few years could offer unique advantages over polymeric membranes. Ceramic membranes are more resistant to severe chemical and thermal environments and claim to operate more efficiently than polymeric membranes. Pilot ceramic membranes are able to operate at higher fluxes and higher feed water recoveries and have more efficient backwash intervals and low chemical cleaning requirements, with extended life and warranties. Ceramic membranes are able to operate after ozone and with very high specific fluxes of 40 gfd/psi (998 lmh/bar; Lehmann et al., 2009).

Membrane media are manufactured as flat sheets or as hollow fibers and then configured into membrane modules. Today, the most common membrane module configurations are hollow fiber, tubular, and cassette. Spiral wound elements, which consist of a flat sheet of membrane material wrapped around a central collection tube, are rare in MF or UF membrane reuse applications. The term *module*—defined as the smallest component of a membrane unit in which a specific membrane surface area is housed in a device with a filtrate outlet structure—is most often used to refer to all of the various membrane module configurations. Nomenclature varies between membrane manufacturers.

Membrane Modules

There are a number of different types of membrane materials, modules, and associated systems that are utilized by the various classes of membrane filtration. While several different types of membrane modules may be employed for any single membrane filtration technology, each class of membrane technology is typically associated with only one type of membrane module in wastewater recycling, reuse, and water treatment applications. In general, MF and UF use hollow fiber membranes and NF and RO use spiral wound membranes. The terms *hollow fiber* and *spiral wound* refer to the module in which the membrane media are manufactured.

Each of these technologies utilizes a membrane barrier that allows the passage of water but removes contaminants to various degrees. In addition to the various module configurations, there are a number of different types of membrane materials, hydraulic modes of operation, and operational driving forces (i.e., pressure or vacuum) that can vary among the different classes of membrane filtration (MF, UF, NF, and RO).

Membrane filtration media are usually manufactured as a flat sheet or as hollow fibers and then configured into one of several different types of membrane module. A membrane module is the smallest discrete filtration unit in a membrane system. Module construction typically involves potting or sealing the membrane material into a corresponding assembly, which may incorporate an integral containment structure, as with hollow fiber modules. These types of modules are designed for long-term use over the course of a number of years. Spiral wound modules are also manufactured for long-term use, although the design of the membrane filtration systems that utilize spiral wound modules requires that the modules be encased in a separate pressure vessel that is independent of the module itself.

In addition to hollow fiber and spiral wound modules, there are several other types of less common configurations that may be used in membrane filtration systems. These configurations include hollow fine fiber (HFF), tubular, and plate and frame modules. These configurations are sometimes used in membrane filtration systems if coupled with bioprocesses such as membrane bioreactors.

Semipermeable HFF membranes were the original hollow fiber membranes and were developed for desalting (i.e., RO) applications. With the development of widely used porous hollow fiber MF and UF membranes for particulate filtrations with much larger fiber diameters, the semipermeable variety gradually became known as hollow fine fiber membranes. HFF membranes are bundled lengthwise and shaped into a U (called a *U-tube*), which is potted in a cylindrical pressure vessel. Feed water enters an HFF module via a perforated tube in the middle of the vessel and flows radially outward through the membrane bundle. The water that permeates the membrane is collected in the fiber lumen and exits the element at the open end of the U-tube. The remaining water that does not permeate into the fiber lumen carries concentrated salts and suspended solids out of the pressure vessel through the concentrated port. Typically, hollow fine fibers are only about 40 μm in diameter (inside), allowing a very large number of fibers to be contained in a single pressure vessel and maximizing the available membrane surface area per unit volume in the pressure vessel. However, the high packing density also significantly increases the potential for fouling and reduces the flux to levels well below those possible using spiral wound membranes, more than offsetting the advantage in increased surface area. HFF membranes are most commonly used today in seawater desalting applications, particularly in the Middle East.

Tubular membranes are essentially a larger, more rigid version of hollow fiber membranes. With diameters as large as 1 or 2 in. (25.4 to 50.8 mm), the tubes are not prone to clogging and the membrane material (i.e., the tube wall) is comparatively easy to clean. However, the large tubes also result in a very inefficient amount of membrane surface area per unit volume in the pressure vessel. Both porous (for MF and UF) and semipermeable (for NF and RO) membranes have been manufactured for tubular systems in wastewater reuse applications.

A plate and frame configuration, which is one of the earliest membrane modules developed, is simply a series of flat sheet membranes separated by alternating filtrated spacers and feed or concentrate spacers. Because of the very low ratio of surface area to volume, the plate and frame configuration is considered inefficient; but it has special application in treating the leachate from sanitary landfills. Figure 3-9 shows a disk tube module which uses plate and frame construction to attain higher pressures than are capable with spiral wound elements.

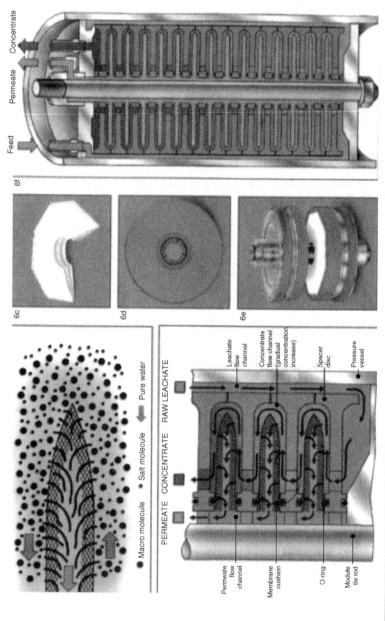

FIGURE 3-9 Disk tube plate and frame. (*Courtesy Pall Corporation*)

High-Performance Low Pressure Membranes—Theoretical Considerations*

Introduction

Enhanced process efficiency is the result of the use of high-performance membranes that can sustain a high flux without sacrificing the energy penalty and service life of membranes. There are many attributes for high-performance membranes. Some of those attributes are intrinsic, as they are inherently related to membrane permeability, whereas others are enabling, as they allow membranes to sustain membrane permeability. Pore size and pore size distribution, thickness, and surface porosity are examples of intrinsic attributes.

Given limitations in pore size by removal requirements and the membrane thickness required for structural stability, surface porosity has a significant effect on membrane permeability. Enabling attributes include chemical composition, molecular architecture, chain orientation with respect to the direction of stress, and morphology; these are directly associated with either membrane surface properties or mechanical strength and chemical stability of membranes. Within membrane media of the same chemical composition and molecular architecture, morphology of polymers (crystalline vs. amorphous) becomes the most crucial attribute to maintaining a membrane's mechanical strength and chemical stability. A membrane with high crystallinity would have high mechanical strength, as it is able to transfer and distribute stress more evenly, and to retard the growth of crack fracture. In addition, crystallinity enhances a membrane's ability to sustain chemical attacks, as it reduces diffusion within the polymer and slows chemical reaction between chemical reagent and polymer. Some attributes may have the opposite effects. For example, a thin and porous membrane may not have the required structural stability and mechanical strength to endure the vigorous cleaning regime, which in turn limits its ability to sustain performance. Therefore, a high-performance membrane is one that has the right combination of attributes.

Attributes Contributing to the Enhancement of Flux

There are many factors contributing to the flux of a membrane. Generally, there are three categories of factors:

- Intrinsic properties. These are the inherent membrane characteristics that directly relate to the permeability of a membrane. This category includes the physical properties of a membrane such as surface porosity, pore size and pore size distribution, and membrane thickness.

---

*This section is from Liu, C. and A. M. Wachinski (2011).

- Membrane surface properties related to fouling, which is the major factor contributing to flux reduction. These properties include surface charge, roughness, and hydrophobicity.

- Properties contributing to mechanical strength and chemical stability of a membrane. As membranes are fouled, they need to be cleaned through both mechanical (hydraulic) and chemical means. To maintain the permeability of a membrane, cleaning measures can be frequent and rigorous. Therefore, a high-performance membrane has to be of mechanical strength and chemical stability to withstand the frequent and rigorous cleaning regime.

It should be noted that some membrane properties can affect both permeability and mechanical strength and chemical stability, in contradictory ways. Sometimes a factor can have opposite effects on permeability and on mechanical strength and chemical stability. This will become apparent in the following discussions. Therefore, it is essential to balance the different aspects of membrane properties to optimize globally, in order to have a membrane with the highest performance.

Intrinsic Properties Affecting the Permeability of Membranes

Intrinsic properties are a membrane's physical and chemical properties that are relevant to flux. Relevant physical properties include membrane configuration, porosity, thickness, pore diameter, and surface roughness; chemical properties include chemical makeup, molecular weight, distribution of polymers, structural morphology, surface charge, and hydrophobicity.

According to the Hagen–Poiseuille equation, the flux of a membrane can be expressed as

$$J = \frac{\varepsilon d_p^2 \Delta P}{32 \delta \mu} \tag{3.5}$$

where J = flux
   ε = surface porosity of the membrane (ratio of the area of membrane pores to the total membrane area)
   d_p = mean pore diameter of the membrane
   ΔP = transmembrane pressure
   δ = effective thickness of the membrane
   μ = viscosity of fluid

The Hagen–Poiseuille equation is a special form of the Darcy equation where (1) flow within membrane pores is in a laminar (uncompressible) flow regime and (2) the pores are round in shape.

Equation 3.5 can be rewritten in the following way:

$$J = K \frac{\Delta P}{\mu} \qquad (3.6)$$

where $K = \frac{\varepsilon d_p^2}{32\delta}$ is defined as permeability

As indicated in Equation 3.6, the permeability of a membrane is proportional to the surface porosity of the membrane (ε) and the square of the diameter of the flow channel in pores, also the mean pore diameter of the membrane (d_p^2), and inversely proportional to the effective thickness of the membrane (δ). As a general rule, surface porosity increases with increasing pore size. Many UF membranes have a surface porosity less than 10%. On the other hand, MF can have a surface porosity ranging from 30% to 80% (Cheryan, 1998). Figure 3-10 shows scanning electron microscope (SEM) images of an MF membrane (left) and a UF membrane (right). The difference in porosity of the two membranes is striking. As a general rule, MF membranes have higher permeability than UF membranes, as they have both a larger pore size and a high porosity.

Because of the need to exclude microbial pathogens such as protozoa and bacteria, the upper boundary of nominal pore size for low pressure membranes used in drinking water filtration is typically in the range of 0.1 to 0.2 μm. In addition, membranes with large pore sizes also tend to have rough surfaces, which in turn facilitate the deposition of fouling materials on the surface. With the limitation on pore size, an ideal membrane would be very porous and very thin if high flux were

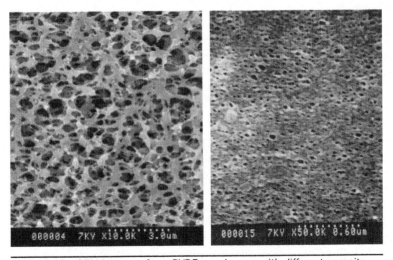

FIGURE **3-10**   SEM images of two PVDF membranes with different porosity.

the only consideration. In reality this would not work, because such a membrane would have poor structural stability and low mechanical strength, and could not withstand the stresses from vigorous hydraulic regeneration. Therefore, the key to a high-performance membrane is the right combination of physical properties.

One idea to enhance membrane flux is to have a composite membrane—a thin, functional membrane layer cast on more porous support media. This method has been widely used for making tighter high-pressure membranes such as NF and RO. Even though some UF membranes also use the composite configuration, there are inherent problem when such a membrane is subjected to vigorous hydraulic stresses, as during backwash. In the composite membrane, the thin membrane layer is "glued" to the supporting medium through interfacial polymerization. Due to the difference in material elasticity of the membrane and supporting medium, the tension upon the axial direction of the hollow fiber membranes (the most common configuration used for drinking water filtration) is converted into shearing on the interfaces of membrane and supporting medium. In addition, backwash operation also applies tension that pushes the two layers apart. This can result in delaminating—the peeling off the functional membrane layer from the underlying support layer. A detailed analysis of mechanical stresses on composite membranes is given later, in the section on the effects of membrane geometry and construction.

Surface Chemical Properties Affecting Fouling

The physical properties of membranes discussed previously are intrinsic for membrane flux. However, in filtering natural water, many constituents can adhere to the membrane and reduce its permeability—a phenomenon called membrane fouling. When a membrane is fouled, surface porosity decreases, hydraulic diameter decreases, and effective thickness increases. The influence of membrane fouling on flux can be expressed by the flux ratio of the fouled membrane to the clean membrane:

$$\frac{J}{J_0} = \frac{(\varepsilon/\varepsilon_0)(D_H/D_{H0})^2}{(\delta/\delta_0)} = \left(\frac{\varepsilon}{\varepsilon_0}\right)\left(\frac{D_H}{D_{H0}}\right)^2\left(\frac{\delta}{\delta_0}\right)^{-1} \tag{3.7}$$

where J, ε, D_H, and δ represent the flux, surface porosity, hydraulic diameter, and effective thickness (respectively) of the membrane after fouling and J_0, ε_0, D_{H0}, and δ_0 represent flux, surface porosity, hydraulic diameter, and effective thickness (respectively) of a clean membrane.

The three terms on the right-hand side of Equation 3.7 correspond to three fouling mechanisms: pore blocking, internal pore plugging, and cake filtration, respectively (Cheryan, 1998). The impacts of last two terms on flux reduction are graphically presented in Fig. 3-11, in which $\delta/\delta_0 = 1$. Since the effect of δ/δ_0 on J/J_0 is linear, the response surface in Fig. 3-11 will move downward to $1 - (\delta/\delta_0)$ when $\delta/\delta_0 < 1.0$.

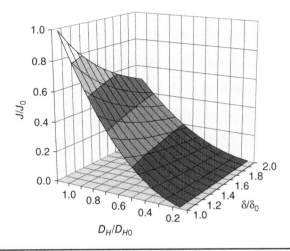

Figure 3-11 J/J_0 as a function of D_H/D_{H0} and δ/δ_0 ($\delta/\delta_0 = 1.0$).

Membrane fouling is the most difficult problem for maintaining membrane permeability, and a very complex phenomenon that involves many factors and is poorly understood. Nevertheless, chemical properties of a membrane medium would have a significant bearing on fouling.

The chemical properties of membranes include chemical makeup and structure and morphology of the polymers, which in turn determine the properties such as surface charge, hydrophobicity, and roughness of a membrane.

Surface charge may somewhat inversely correlate to hydrophobicity of a membrane medium. This should not be surprising, considering that water is a polar solvent. As a result, a membrane medium with high surface charge would tend to absorb water molecules through hydrogen bonds. However, this correlation is not universal, since surface charge is a necessary, but not a sufficient condition for hydrophilicity. Membranes with negative charges are thought to be less prone to fouling, as fouling materials in natural water, such as colloids and natural organic matter, are also usually negatively charged. However, the relationship is less straightforward; the phenomenon of membrane fouling is very complicated and may be affected by many factors, as indicated in Table 3-4. The role of hydrophobicity is equally ambiguous, even though the general thought is that hydrophilic membranes are less prone to fouling. This is again because the complexity of membrane fouling is far from fully understood. Recent research on membrane fouling implies that the components of NOM that have the most impact on fouling appear to be the hydrophilic, high MW fraction.

Membrane fouling is the result of the sum of all the interactions between fouling materials and membranes. If the individual

Interactions	Nature	Energy Level (KJ/mol.)	Range of Influence
Chemical bonds	Bonding	> 150	Intramolecular
Electrostatic	Generally repulsive	≤ 30	1 nm to 1 μm, depending on electrolyte concentration
Hydrogen bonds	Generally repulsive	5 to 30	Intra- and intermolecular, < 4 nm
Steric	Repulsive	Structure specific	Intermolecular, up to 10s of nm
Van der Waals	Attractive	≤ 1	Intermolecular, $\propto r^{-6}$
Hydrophobic	Attractive	~ 2.5 /mol. $-CH_2$	Up to 10s of nm

TABLE **3-4**   Interactions Affecting Membrane Fouling

interactions can be quantified, then the overall fouling potential can be assessed via models such as the Deryagin–Landau–Verwey–Overbeek (DLVO) theory (Gregory, 1993). However, current understanding is still far from the quantitative assessment of fouling potential and thus the effect of individual surface properties of membranes.

Factors Affecting the Mechanical Strength and Chemical Stability of Membranes

As discussed previously, the key to maintaining membrane flux is overcoming membrane fouling. In actual applications, membrane fouling is controlled through hydraulic and chemical means to restore permeability. The hydraulic means include backwash with or without air scrubbing; chemical means are various forms of chemical cleaning. During chemical cleaning, fouling materials either are dissolved or become less adherent to membranes as a result of chemical reactions between the cleaning chemicals and materials causing membrane fouling. Because low pressure membranes are expected to have years of service life—up to a decade—they have to withstand repeated and vigorous cleaning regimes (both hydraulic and chemical). Therefore, the mechanical strength and chemical stability, even though not directly contributing to membrane permeability, are of profound impact on the long-term sustainability of maintaining high flux.

The major factors affecting mechanical and chemical properties of a polymer include the following (Alfrey, 1985; Kumar and Gupta, 1996):

- Chemical composition, including structure of monomers and the ways by which the monomers make up the chains

- Molecular architecture, referring to molecular structure at large, including average molecular weight (AMW) and its distribution, branching, cross-linking, etc.

- Chain orientation with respect to the direction of major stress
- Morphology of polymers (crystalline vs. amorphous)
- Geometry and structure of membranes

The polymers for manufacturing low pressure membranes include polyvinylidene fluoride (PVDF), polyethersulfone (PES), polysulfone (PS), polypropylene (PP), and cellulose acetate (CA). According to a survey by the American Water Works Association Research Foundation (AWWARF) in 2003, over one half of the installed capacity in millions of gallons per day (mgd) of different MF and UF membrane materials is PVDF. Other membrane materials (PES, PS, PP, and CA) range from 8% to 15% of the installed capacity (Adham et al., 2005). The dominance of PVDF membranes may relate to the material's general characteristics of high chemical resistance and mechanical strength.

However, material alone is not the sole determinant of the mechanical strength of a membrane. The manufacturing of hollow fiber membranes is typically carried out through a melt-extrusion process. In this process, polymers are melted and mixed with solvent first, injected through a spinning nozzle, and cooled. The membrane pores are formed as a result of chemically or thermally induced phase separation. Each of those individual steps can play an important role in forming the molecular architecture, chain orientation, and morphology of the final product, which will in turn exhibit different characteristics of mechanical strength and chemical stability.

One of the most critical determinants for a polymer's mechanical strength and chemical stability is the crystallinity of the polymer. Unlike metals, polymers cannot form a perfect crystalline structure. The crystallinity of a polymer is related to the mechanical and chemical properties of the polymer in the following ways:

- Higher crystallinity generally enhances mechanical strength by transferring and distributing the stress more evenly.

- Higher crystallinity can retard crack propagation (a prelude to failure) in a polymer tremendously. The crack growth rate of an amorphous polymer can be an order of magnitude higher than that of a semicrystalline polymer (Hertzberg and Manson, 1980).

- Crystalline polymers not only can dissipate energy when they deform, but also can re-form a crystalline structure that is exceedingly strong.

- Higher crystallinity can considerably limit the diffusion of solvents and nonsolvents (e.g., water) into and through a polymer (Vieth, 1991). This is an important aspect as far as chemical stability of a polymer is concerned. Practically, chemical degradation of polymers is most likely diffusion limited.

- Higher crystallinity is likely to greatly slow down the reaction kinetics of chemical degradation of a polymer. The attack of a chemical reagent on an amorphous structure generally occurs more readily than on a crystalline structure (Reich and Stivala, 1971).

The effect of crystallinity on chemical stability can be illustrated by the example of PVDF membranes. In general, PVDF is highly resistant to oxidants. However, it becomes less refractory in the presence of caustic (NaOH). This is because PVDF reacts with caustic via dehydrofluorination:

$$CF_2-CH-CF-CH_2+NaOH \rightarrow CF_2-CH=CF-CH_2+NaF+H_2O$$
$$\quad\quad\;|\quad\;\;|$$
$$\quad\quad\;H\quad F$$

The reaction forms double bonds between two neighboring carbon atoms by stripping a hydrogen atom and a fluorine atom from them; the product shows a distinct brown color, as in samples B and C (two low-crystallinity PVDF membranes) in Fig. 3-12.

The $C=C$ double bonds in the polymer chains restrict the rotation motion that existed on the original $C-C$ bond. As a result, the polymer chain becomes much more rigid and brittle. Furthermore, the sites of double carbon bonds are vulnerable to the attack of oxidants, and chain scission can occur on those sites. One example of oxidative cleavage of a $C=C$ double bond is

$$R_1-HC=CH-R_2+NaOCl \rightarrow R_1-HC=O+O=CH-R_2$$

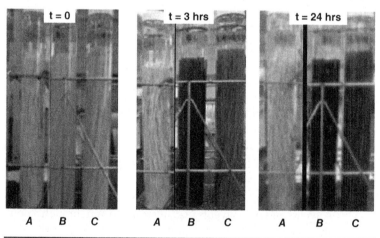

FIGURE 3-12 Color changes of three PVDF hollow fiber samples after soaking in a solution of 0.5% wt NaOCl & 1% wt NaOH at ambient temperature.

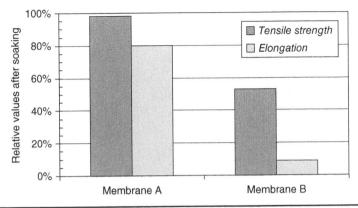

FIGURE 3-13   Changes in tensile strength and elongation after 14-day soaking in a solution of 0.5% wt NaOCl and 4% wt NaOH at 25°C.

As the dehydrofluorination and oxidative cleavage of PVDF occur, the membrane becomes increasingly brittle and fragile, eventually fracturing under the mechanical stresses. Figure 3-13 depicts the relative tensile strength and elongation of two PVDF hollow fiber samples (A and B) after soaking in a solution containing 0.5% wt NaOCl and 4% wt NaOH at 25°C for 24 hours.

As indicated in Figs. 3-12 and 3-13, dehydrofluorination did not occur to a significant degree for the membrane with high crystallinity (A), but occurred for the membranes with low crystallinity (B and C). As the result of dehydrofluorination and subsequent oxidation, the mechanical strength of membranes B and C was reduced considerably. Tensile strength was reduced by almost one-half of the original, value and elongation was less than one-tenth of the original value. On the other hand, the membrane with high crystallinity (A) retained 98% of its original tensile strength and 80% of elongation. Using the actual chemical cleaning conditions (solution concentrations and time), and assuming monthly cleaning, the equivalent exposure from the soaking test above would be equal to 14 years of membrane service time. The high chemical stability of membrane A would allow a more aggressive cleaning regime to be used for controlling fouling whenever necessary so that high flux could be maintained.

Geometry and Structure of Membranes   The majority of low pressure membranes used for water filtration are of hollow fiber configuration, which in general is structurally more stable under mechanical stress. Within this configuration, there is a distinction between monocloth and multicloth membranes: the former refers to membranes made from a single material and the latter to those made from more than one material (multicloth is mostly used for UF membranes). The multicloth membranes consist of a thin layer of

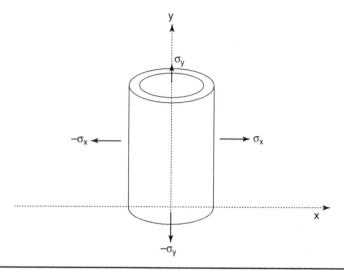

Figure 3-14   Major stress applied upon hollow fiber membranes.

functional membrane cast on a more porous supporting substrate. This configuration is structurally similar to thin-film composite (TFC) membranes that are popular for NF and RO. The
functional layer and underlying support layer are "glued"
together through interfacial polymerization. Under the typical
hydraulic cleaning regimes of backwash and air scrubbing, the
hollow fibers are subjected to two types of major stresses: the
stress in the radius direction (σ_x) and the stress in the longitude
direction (σ_y), as illustrated in Fig. 3-14.

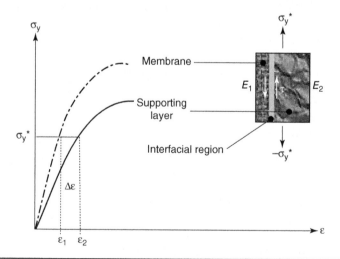

Figure 3-15   Plot of stress versus elasticity of a supported membrane.

The stress in the radius direction exerts a pressure that pushes the functional membrane layer and supporting layer apart. On the other hand, the stress in the longitude direction exerts a tension on the hollow fiber. Due to the difference in elasticity of the membrane and supporting materials, this tension creates a shearing stress on the interfacial region, as illustrated in Fig. 3-15.

When the hollow fiber is subjected to a longitudinal tension σ_y^*, the strains for the membrane medium and the supporting layer are given by

$$\varepsilon_1 = \sigma_y^*/E_1 \tag{3.8}$$

and

$$\varepsilon_2 = \sigma_y^*/E_2 \tag{3.9}$$

where E_1 and E_2 are the modulus of elasticity of the membrane and the supporting layer, respectively.

Due to the difference in elasticity of the membrane medium and supporting layer, the strains caused by the tension would be different for the two materials:

$$\Delta\varepsilon = \varepsilon_2 - \varepsilon_1 = \sigma_y^* \left(\frac{1}{E_2} - \frac{1}{E_1} \right) \tag{3.10}$$

The differential expansion of neighboring materials due to the tension then creates a shearing force on the interfacial region:

$$\tau = G\Delta\varepsilon = G\sigma_y^* \left(\frac{1}{E_2} - \frac{1}{E_1} \right) = \frac{G\sigma_y^*(E_1 - E_2)}{E_1 E_2} \tag{3.11}$$

Equation 3.11 indicates that the shearing force is proportional to the amplitude of the longitudinal tension, the difference in moduli of elasticity, and the shear modulus G of the interfacial region, and inversely proportional to the product of the moduli of elasticity of the membrane and supporting material.

One consequence of cyclic stresses on multicloth membranes is the delaminating of membranes from the supporting layer—that is, membranes are peeled off from the supporting layer as stresses weaken the bond in the interfacial region. Delaminating renders the membranes useless as they lose integrity and allow contaminants to pass through, even though the membranes are not structurally broken.

Diffusive Membranes—NF and RO

NF and RO membranes are designed to remove dissolved solids through the process of reverse osmosis. Osmosis is the natural flow of a solvent, such as water, through a semipermeable membrane (acting as a barrier to dissolved solids) from a less concentrated solution to a

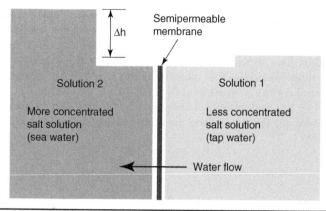

FIGURE 3-16   Osmosis.

more concentrated solution, i.e., the total dissolved solids concentration on opposite sides of the membrane. Osmosis occurs when a plant draws water from the ground via its root system to its leaves. This flow will continue until the chemical potentials (or, concentrations for practical purposes) on both sides of the membrane are equal. The amount of pressure that must be applied to the more concentrated solution to stop this flow is called the osmotic pressure. A rule of thumb for the osmotic pressure of fresh or brackish water is approximately 1 psi (0.07 bar) for every 100 mg/L difference in total dissolved solids.

Osmosis and reverse osmosis are illustrated in Figs. 3-16 and 3-17. Reverse osmosis is the reversal of the natural osmotic process, accomplished by applying pressure in excess of the osmotic pressure to the more concentrated solution. This pressure forces the water through the membrane against the natural osmotic gradient, thereby increasingly concentrating the dissolved solids in the water on one

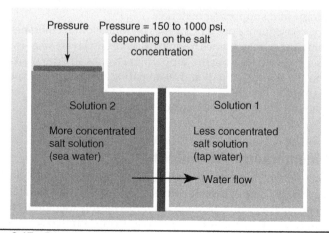

FIGURE 3-17   Reverse osmosis.

Salt	Concentration (%)	Approximate Osmotic Pressure (psig; bar)
Sodium chloride (NaCl)		
	0.5	55 (3.8)
	1.0	125 (8.6)
	3.5	410 (28.3)
Sodium sulfate (Na$_2$SO$_4$)		
	2	110 (7.6)
	5	304 (21.0)
	10	568 (39.2)
Calcium chloride (CaCl$_2$)		
	1	90 (6.2)
	3.5	308 (21.2)
Copper sulfate (CuSO$_4$)		
	2	57 (3.9)
	5	15 (7.9)
	10	231 (15.9)

TABLE 3-5   Approximate Osmotic Pressures for Selected Salt

side (i.e., the feed) of the membrane and increasing the volume of water with a lower concentration of dissolved solids on the opposite side (i.e., the filtrate or permeate). The required operating pressure varies depending on the osmotic pressure of the constituents of the feed water (i.e., osmotic potential), as well as on membrane properties and temperature, and can range from less than 100 psi for some NF applications to more than 1,000 psi for seawater desalting. Table 3-5 provides approximate osmotic pressures for sodium chloride, sodium sulfate, calcium chloride, and copper sulfate.

Both NF and RO are pressure-driven separation processes that utilize semipermeable membrane barriers. NF differs from RO only in terms of its lower removal efficiencies for dissolved substances, particularly for monovalent ions such as sodium. This results in unique applications of NF, such as the removal of hardness (calcium and magnesium ions) at lower pressures than would be possible using RO. Consequently, NF is often called membrane softening or loose RO.

Transmembrane Pressure

Transmembrane pressures for diffusive membranes are much higher than for low pressure modules. TMPs for low pressure modules range from 5 to 40 psi (0.3 to 2.86 bar). TMPs for diffusive membranes can vary from 100 to 1,200 psi. (6.9 to 82.7 bar), depending upon the osmotic pressure of the contaminants.

Net Driving Pressure

The net driving pressure is the pressure available to drive the feed water through the membrane minus permeate and osmotic back pressure:

$$\text{NDP} = P_F - \frac{1}{2}(P_F - P_C) - P_P - \Delta\pi$$

where P_F = feed pressure
P_C = concentrate pressure
P_P = permeate or filtrate pressure
$\Delta\pi$ = change in osmotic pressure

Turbidity Effects

Turbidity is not an issue for NF or RO membranes. The differences between NF and RO are irrelevant with respect to the removal of particulate matter, and as a result, these two membrane processes are functionally equivalent. Because semipermeable NF and RO membranes are not porous, they have the ability to screen microorganisms and particulate matter in the feed water; however, they are not necessarily absolute barriers.

These membranes are designed for their feed waters to be turbidity free. The Silt Density Index (SDI) test (see Glossary and Chapter 5) and other associated tests are used to determine and monitor the fouling potential of the feed water. The standard SDI limit has become stricter now that low pressure membranes are used as pretreatment for diffusive membranes.

Integrity Testing

In membrane applications for wastewater treatment and reuse, the objectives are either to remove microbial pathogens—i.e., protozoa, bacteria, and viruses—or to pretreatment for other treatment processes for wastewater reuse, i.e., NF or RO. Direct integrity testing is not performed on diffusive membranes. A similar test using vacuum can be and sometimes is performed, although current spiral wound systems are not designed to conduct automatic integrity tests (ITs). For NF and RO membranes removing dissolved species, other online parameters are used to monitor integrity. Conductivity is a low-cost accurate parameter to monitor. Today, online TOC monitors and sulfide monitors are used to monitor integrity.

One strategy for high quality filtrates is to monitor for significant decline in parameters.

In situations where diffusive membranes are pretreated with low pressure membranes, direct integrity tests are performed on the low pressure membrane which guarantee that the permeate to the NF or RO meets effluent requirements.

NF and RO membranes are specifically designed for the removal of total dissolved solids and not particulate matter, and thus the elimination of all small seal leaks that have only a minor impact on the salt rejection characteristics is not the primary focus of the manufacturing process. Consequently, NF and RO spiral wound elements are not intended to be sterilizing filters; some passage of microorganisms may occur, attributed to slight manufacturing imperfections. For this reason, diffusive membranes are given a lower log removal rating by regulators for viruses than are UF membranes. A typical log removal value is 2 for an RO membrane and can vary from 1 to 4 for UF membranes. This has special significance when using RO membranes for high level water reuse applications.

Membrane Fouling

The membrane fouling potential of diffusive feed water is estimated by the Silt Density Index, a test method often used to determine the effectiveness of pretreatment processes to reduce the fouling tendency of feed water for dense membranes. The test method is described by two ASTM standards, D4189-95 and (the most recent) D4189-07.

The SDI of any NF or RO feed water is calculated using the rate of plugging of a 0.45 μm membrane filter. The test is described in detail in Chapter 5.

Until the last few years, spiral wound membranes were not considered effective in treating water with an SDI greater than 5. Today, with low pressure membrane pretreatment, SDI numbers less than 3 are usually required. The Membrane Filtration Guidance Manual (MFGM) compares NF and RO flux for a surface water with an SDI of 2 to 4 and a clean groundwater. The estimated NF and RO fluxes are 8 to 14 gfd and 14 to 18 gfd, respectively. Membrane pretreatment for NF and RO allows the NF and RO to operate at higher fluxes and a higher critical flux.

Temperature Effects

NF and RO membranes operate at much higher TMPs than low pressure membranes. In wastewater reuse applications, typical TMPs for NF and RO membranes range from 150 to over 300 psi. For brine and seawater applications, TMPs can get as high as 1,000 psi. Diffusive membrane operation must increase TMP to maintain constant flux to compensate for higher fouling or lower temperatures. The *Membrane Filtration Guidance Manual* (USEPA, 2005) gives the following temperature correction factor (TCF):

$$TCF = \exp\left[U\left(\frac{1}{T+273} - \frac{1}{298}\right)\right] \qquad (3.12)$$

where TCF = temperature correction factor
T = water temperature (°C)
U = manufacturer-supplied membrane-specific constant. In most cases, the membrane manufacturer will supply the TCF

Knowing the TCF, the required TMP at temperature T is calculated by

$$TMP_T = TMP_{25}/TCF$$

where TMP_T = transmembrane pressure (adjusted to temperature T; psi)

TMP_{25} = transmembrane pressure at 25°C reference temperature (psi)

TCF = temperature correction factor.

Equation 3.12 can also be used to adjust for reduced viscosity at higher temperatures.

Membrane Materials

Diffusive membrane materials are discussed in Chapter 5.

Membrane Modules

A membrane module is the smallest discrete filtration unit in a membrane system; just as with low pressure modules, there are a number of different types of membrane materials, modules, and associated systems utilized by diffusive membranes. In general, NF and RO use spiral wound membranes. The term *spiral wound* refers to the module in which the membrane media are manufactured.

Module construction typically involves potting or sealing the membrane material into a corresponding assembly. Spiral wound modules are also manufactured for long-term use, although the design of the membrane filtration systems that utilize spiral wound modules requires that the modules be encased in a separate pressure vessel independent of the modules themselves.

In addition to hollow fiber and spiral wound modules, there are several other types of less common configurations that may be used in membrane filtrations systems. These configurations include HFF, cassette, tubular, and plate and frame modules. These configurations are sometimes used in membrane filtration systems, if coupled with bioprocesses such as membrane bioreactors.

Semipermeable HFF membranes were the original hollow fiber membranes and were developed for desalting (i.e., RO) applications. With the development of widely used porous hollow fiber MF and UF membranes for particulate filtrations with much larger fiber diameters, the semipermeable variety gradually became known as hollow fine fiber membranes. HFF membranes are bundled length wise and shaped into a U (called a U-tube), which is potted in a cylindrical pressure vessel. Feed water enters an HFF module via a perforated tube in the middle of the vessel and flows radially outward through the membrane bundle. The water that permeates the membrane is collected in the fiber lumen and exits the element at the open end of the U-tube.

The remaining water that does not permeate into the fiber lumen carries concentrated salts and suspended solids out of the pressure vessel through the concentrated port. Typically, hollow fine fibers are only about 40 μm in diameter (inside), allowing a very large number of fibers to be contained in a single pressure vessel and maximizing the available membrane surface area per unit volume in the pressure vessel. However, the high packing density also significantly increases the potential for fouling and reduces the flux to levels well below those possible using spiral wound membranes, more than offsetting the advantage in increased surface area. HFF membranes are still most commonly used in seawater desalting applications, particularly in the Middle East.

CHAPTER 4

Low Pressure Membrane Technology— Microfiltration and Ultrafiltration

Introduction

This chapter covers the low pressure membranes—microfiltration (MF) and ultrafiltration (UF)—used for water reuse applications.

Excellent brief histories of UF and MF development are presented by Pearce (2007) and the American Water Works Association (AWWA). Subcommittee on Periodical Publications (2008). See Table 4-1 for a chronology of significant events in membrane development.

Low pressure membrane filtration was first developed in the 1950s, when MF membranes were commercialized for sterilization applications. The first UF membranes were used for protein separation with molecular weight cutoffs (MWCOs) of 3 to KDa. These membranes were too tight for treating water and wastewater, even using tubular and hollow fiber formats. Permeabilities were too high for consideration in the municipal water and wastewater markets. UF membranes used today for water reuse are typically in the 0.01 to 0.024 μm range, nominal—i.e., MWCOs between 80 and 150 KDa. Pearce reports the first large-scale municipal UF plant installed in 1988. The low pressure membranes used in municipal drinking water were first installed in North America in the early 1990s. The applications were for removing turbidity and microorganisms from surface and groundwater. In 1993, a 3.6 mgd (13.6 million liters per day [MLD]) system was installed at Saratoga, California (AWWA, 2008).

1784	Discovery of osmosis by Jean-Antoine Nollet
1960	Development of first reverse osmosis (RO) membrane
1960s	Development of asymmetric membranes; commercialization of low pressure membranes
1970s	Commercialization of Thin Film Composite (TFC) RO membrane
1976	Water Factory 21—a 5 million gallon per day water reuse plant using conventional pretreatment to RO
1980s	Commercialization of nanofiltration (NF) membranes
1990s	MF/UF membrane systems for water treatment in the US take off
1993	First large-scale filtration plant (Saratoga, CA—3.6 mgd)

Adapted from Pearce (2007) and AWWA Manual 53 (2005).

TABLE 4-1   Significant Dates in Membrane Development

Membrane systems were seen as one process in an integrated system that combined membranes with pre and post technologies, i.e., pretreatment technologies such as sedimentation, post technologies such as granular activated carbon (GAC), and advanced oxidation such as UV-Ozone (AWWA Manual 53, 2005). Table 4-2 lists the integrated membrane treatment alternatives.

More stringent regulations in the water and wastewater markets, along with market competition, continue to be the drivers behind the increasing popularity of membranes. The publication of the final Long Term 2 Enhanced Surface Water Treatment Rule and Stage 2 Disinfectants and Disinfection Byproducts Rule have pushed more utilities to look at the membrane alternative for rule compliance. Stricter discharge regulations are forcing municipalities and industry to consider reuse, to avoid violating their national pollution discharge elimination (NPDES) permits. Conventional treatment alone cannot meet these regulations all of the time. Competition in the marketplace has pushed technological innovations and process improvements that have greatly enhanced membrane process efficiency. One example is chemically enhanced backwash and its derivatives, now in use by many low pressure membrane manufacturers. These techniques, when combined with high performance membranes, have demonstrated operating flux increases ranging from 30% up to 60%. If we compare typical membrane flux and operating transmembrane pressure (TMP) from just eight years ago to typical values achieved today in the marketplace, there is a tremendous increase across the board.

There is no question that a membrane system able to operate at high flux is more efficient than one operating at low flux, and that operating at lower TMPs has a significant effect on operating cost. The effect of flux on system costs includes capital—both membranes and associated system hardware— footprint, and operation and maintenance—membrane

Integrated Membrane Systems
Preliminary Membrane Treatment System
Membrane filtration → UV disinfection + hydrogen peroxide
Membrane filtration → ozone biological filtration
Membrane filtration → NF or RO
Intermediate Membrane Treatment System
Conventional pretreatment → membrane filtration → NF or RO
Lime softening → membrane filtration → NF or RO
Final Membrane Treatment System
Preoxidation → membrane filtration
Adsorption → membrane filtration
In-line coagulation → membrane filtration
Conventional pretreatment → membrane filtration
Lime softening → GAC→ membrane filtration
Lime softening → conventional pretreatment → membrane filtration
Ozone → contact flocculation (powdered activated carbon) → membrane filtration

Source: AWWA (2005).

TABLE 4-2   Integrated Membrane Treatment Alternatives

replacement, and chemicals to a lesser degree. It has been suggested that the single most important factor dictating plant costs is membrane flux (Chellam, Serra, and Wiesner, 1998). This still holds true to large degree today.

Flux is not the only area of improvement. Low pressure membranes are able to operate at extremely low TMPs. One manufacturer's MF membrane competes with new ceramics on life, with very few broken fibers.

Water Quality

MF and UF are the two processes most often associated with the term *membrane filtration*. The nominal pore sizes of MF and UF membranes used in water and wastewater are somewhere between 0.001 and 0.2 μm (AWWA, 2005). Both are designed to operate at ambient temperature using a sieve mechanism to remove particular matter, including turbidity (for drinking water), suspended solids (for wastewater), or both (for wastewater reuse), protozoa such as *Cryptosporidium* and *Giardia*, bacteria such as the coliform group (including *E. coli* and the organisms that cause Legionnaires' disease), viruses, and particulate inorganics such as iron and manganese.

MF and UF are characterized by their ability to remove suspended or colloidal particles via a sieving mechanism based on the size of the membrane pores relative to that of the particulate matter.

MF and UF Removal Efficiency

All membranes have a distribution of pore sizes, and this distribution will vary according to the membrane material and manufacturing process. When a pore size is stated, it can be presented in microns (μm) as either nominal (i.e., the average pore size) or absolute (i.e., the maximum pore size). MF membranes used in wastewater recycling and reuse are generally considered to have a pore size range of 0.1 to 0.2 μm (nominally 0.1 μm), although there are exceptions, as MF membranes with pores sizes of up to 10 μm are available. Although MWCO is widely accepted as the means to classify UF membranes, pore sizes generally range from 0.01 to 0.05 μm (nominally 0.01 μm) or less, decreasing to an extent at which the concept of a discernible pore becomes inappropriate—a point at which some discrete macromolecules can be retained by the membrane material. In terms of a pore size, the lower cutoff for a UF membrane is approximately 0.005 μm.

Although the concept of using the nominal or absolute pore size is sometimes used in reference to the filtration capabilities of membrane material, this concept is overly simplistic and does not fully characterize the removal efficiency of a membrane. For example, the mechanisms of filtering particles close in size to the pore size distribution of a membrane become more complex than sieving, as particles smaller than most pores may be removed through probabilistic interception through the depth of the filter media. In addition, for some membrane materials, particles may be rejected through electrostatic repulsion and adsorption to the membrane material. The filtration properties of the membrane may also depend on the formation of a cake layer during operation, as the deposition of particles can obscure the pores over the course of a filter run, thus increasing the removal efficiency.

Currently, no established standard method exists for characterizing and reporting pore sizes for the various membrane filtration processes. As a result, the meaning of this information varies among membrane manufacturers, limiting its value. The concept of pore size has no significance for NF and RO membranes, which are semipermeable and do not have pores. The concept of pore size also does not address the integrity of the manufactured membrane module assembly, which could potentially pass particles larger than the indicated pore size. Table 4-3 compares challenge testing of MF versus UF performance in removing *Giardia* cysts, *Cryptosporidium* oocysts, MS-2 bacteriophage, particle counts, and turbidity.

Overlap exists as to the designation of whether a membrane is an MF or UF membrane. In some cases a particular membrane can be

	MF	UF
Giardia cysts	4.5 to 7 log	5 to 7 log
Cryptosporidium oocysts	4.5 to 7 log	5 to 7 log
MS-2 bacteriophage virus	0.5 to 3.0 log	4.5 to 6 log
Particles >2 µm	<10/mL	<10/mL
Particles 2 to 5 µm	<10/mL	<10/mL
Particles 5 to 15 µm	<1/mL	<1/mL
Average turbidity	0.01 to 0.03 ntu	0.01 to 0.03 ntu

Source: Pall Coporation 2004.

TABLE 4-3 Typical Summary of Removals, MF vs. UF

considered either a UF or an MF membrane, based upon pore size. Table 4-4 (Table 3-1 in Chapter 3) illustrates the overlap.

Because some UF membranes have the ability to retain larger organic macromolecules, they have been historically characterized by MWCO rather than by a particular pore size. The concept of the MWCO (expressed in daltons, a unit of mass) is a measure of the removal characteristic of a membrane in terms of atomic weight (or mass) rather than size. Thus, UF membranes with a specified MWCO are presumed to act as a barrier to compounds or molecules with a molecular weight exceeding the MWCO. Because such organic macromolecules are morphologically difficult to define and are typically found in solution rather than as suspended solids, it may be convenient in conceptual terms to use MWCO rather than a particular pore size to define UF membranes when discussed in reference to these types of compounds. Typical MWCO levels for UF membranes range from 10,000 to 500,000 Da, with most membranes used for wastewater, recycling, and reuse treatment at approximately 100,000 Da. However, UF membranes remove particulate contaminants via a size exclusion mechanism and not on the basis of weight or mass; thus, UF membranes used for wastewater, recycling, and reuse are also characterized according to pore size with respect to microbial and particulate removal capabilities.

A critical distinction between MF and UF is that UF can consistently achieve greater than 3-log virus reduction in clean water. Most UF membrane systems meet the Safe Drinking Water Act's virus log removal requirement of 4 log. MF cannot remove viruses to any significant degree in clean water but with coagulant addition or treatment of a wastewater is capable of greater than 2-log virus removal. MF in most cases requires an additional virus disinfection step.

With a nominal pore size of 0.1 µm, the MF membrane represents an absolute barrier to protozoan cysts, oocysts, and bacteria. If the water is

Membrane Process	Membrane Driving Force	Separation Mechanism	Typical Pore Size	Typical Reuse Pore Size (μm)	Materials
Microfiltration	Pressure vacuum difference or vacuum in open vessels	Sieve	Macropores (> 50 nm)	0.08 to 0.2	Ceramic (various materials), polypropylene (PP), polysulfone (PS), polyvinylidene fluoride (PVDF), polyethersulfone (PES)
Ultrafiltration	Hydrostatic pressure difference or vacuum in open vessels	Sieve	Mesopores (2 to 50 nm)	0.005 to 0.2	Aromatic polyamide, ceramic (various materials), cellulose acetate, polypropylene, polysulfone (PS), polyvinylidene fluoride
Nanofiltration	Hydrostatic pressure difference	Sieve or solution/ diffusion + exclusion	Micropores (< 2 nm)	0.001 to 0.01	Cellulosics, aromatic polyamide, polysulfone, PVDF, thin-film composite
Reverse osmosis	Hydrostatic pressure difference	Solution/diffusion + exclusion	Dense (< 2 nm)	0.0001 to 0.001	Cellulosic, aromatic polyamide, thin-film composite

Adapted from Asano et al., (2007) p. 427.

TABLE 4-4 General Characteristics of Wastewater Reuse Membrane Processes

turbid, as is usually the case in wastewater reuse applications, MF can remove viruses up to 2.5 log. If the water is clean—i.e., distilled—and the membrane is not fouled, its ability to remove viruses is negligible.

In water reuse applications, MF is usually considered an intermediate between UF and multimedia granular filtration, with pore sizes ranging from 0.08 to 0.2 μm. For drinking water the range is 0.1 to 0.2 μm. Hollow fiber MF used for wastewater reuse falls in the range of 0.1 to 0.4 μm (0.4 μm is a company the membrane bioreactor [MBR] business) (see Chapter 6). It is an effective barrier for particles, bacteria, and protozoan cysts. It operates at pressures between 10 and 30 psi (0.7 and 2.1 bar).

The major distinction between MF and UF performance in wastewater reuse applications is that MF is not an effective barrier to viruses. UF has been shown in pilot challenges using water specified in the *Membrane Filtration Guidance Manual* (US Environmental Protection Agency [USEPA], 2005) to achieve 6-log virus reduction. Typical log removals reported in the industry vary between 1 and 4 log. These data were generated with new membranes. Little to no data are currently available to verify UF virus log removal after six months, one year, or two years of operation; this is a concern when using UF without pretreatment, coagulant, or chlorine to achieve desired virus removal. Viruses have very low tolerance to chlorine. At the typical temperature range (0°C to 20°C; 32°F to 68°F) and pH range (6 to 9) of drinking water, the *Ct* value (the product of disinfectant, concentration, and exposure time) to achieve 4-log virus reduction is from 3 to 12 mg/L min. MF generally cannot remove more than 0.5 log of viruses in clean water or 1 log in turbid waters. Rejection in dirtier water, including secondary effluent, has not yet been verified in a full-scale trial. The University of New Hampshire showed up to 3-log removal by MF when operating at high TMPs (Dwyer, 2003). The reader is directed to the membrane manufacturers for removal data for their membranes under specific conditions.

UF membranes operate at pressures between 4 and 50 psi (0.3 and 3.5 bar) and retain particulates, bacteria, protozoa, viruses, and organic molecules larger than their MWCO. When integral, UF membranes offer absolute barriers to cysts, bacteria, and most viruses. Operating pressures are usually lower than for NF and RO and higher than for MF. Some UF membranes are able to run at lower pressure than MF membranes. Fluxes in general are lower than for MF, and power consumption may be higher, depending upon mode of operation. UF membrane systems meet virus log removal requirements. Most do not tolerate oxidants well and offer a single barrier approach to protozoa and viruses with redundancy. Pearce (2007) reports that the full spectrum of pore sizes for UF is approximately 0.001 to 0.02 μm, with a typical removal capability on the range of 0.01 to 0.02 μm. The reader is directed to the membrane manufacturers to obtain pore size and pore size distribution data.

In the United States, the majority of systems are MF. In Europe, the majority of systems are UF. In treating secondary effluent, MF can achieve greater than 2-log virus removal with direct coagulation, and higher with coagulants. For most reuse applications, virus inactivation is not an issue.

M/F Filtration Configurations

Dead-End Filtration

The USEPA's *Membrane Filtration Guidance Manual* (2005) defines *dead-end filtration* as term commonly used to describe "the deposition mode hydraulic configuration of membrane filtration systems;" also synonymous with "direct filtration." This document further identifies *deposition mode* as "a hydraulic configuration of membrane filtration systems in which contaminants removed from the feed water accumulate at the membrane surface (and in microfiltration [MF]/ultrafiltration [UF] systems are subsequently removed via backwashing)."

In dead-end filtration, fluid is introduced in a direction normal to the membrane under an applied pressure. Particles and macromolecules that are too large to pass through the membrane pores accumulate at the membrane surface or in the depth of the filtration medium. The accumulation of contaminant results in a differential pressure increase or flux reduction and must be physically or chemically removed from the membrane surface on a regular basis. This filtration mode is represented in Fig. 4-1.

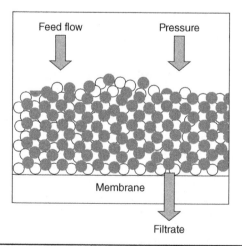

FIGURE 4-1   Dead end filtration.

Dead-end filtration applies to both disposable and regenerable cartridges, including specially designed MF and UF modules in the Pall
product line. A simplified diagram, shown in Fig. 4-2, illustrates the
similarities in configuration and operation of a dead-end disposable
filter device and an MF/UF hollow fiber water module. In both cases,
pressurized fluid enters the housing through an inlet port and passes
through the filter pores in a direction normal to the membrane surface. Contaminants removed from the feed water accumulate at the
membrane surface, as described in the USEPA's definition of a dead-
end filtration device. Both designs have an additional multipurpose
port that can be used as an air vent or for membrane flushing or
cleaning purposes.

Unlike with a cross-flow device, fluid flow through hollow fiber
dead-end filters is typically in an outside to inside flow direction at
very low velocities (<0.1 m/s). Particles and foulant materials are
retained on the outer membrane surface and throughout the cross

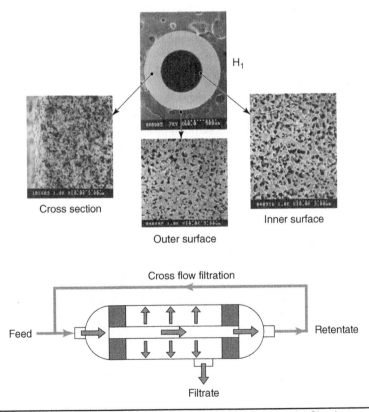

FIGURE 4-2 (a) Cross section of high crystalline PVDF hollow. (b) Simple
crossflow device. (*Pall Corporation, 2011.*)

section until they are removed by a chemical cleaning or regeneration process. Although there is a small tangential component of the fluid velocity over some part of the filter area, the low fluid velocity encountered in these dead-end flow devices is insufficient to maintain contaminants in suspension and prevent membrane plugging. With a few exceptions, most MF and UF processes utilize systems designed around hollow fiber modules. Hollow fiber membrane filtration systems are designed and constructed in one or more discrete water production units, also called racks, trains, or skids. A unit consists of a number of membrane modules that share feed and filtrate valving, and each unit can usually be isolated from the rest of the system for testing, cleaning, or repair. A typical hollow fiber system is composed of a number of identical units that combine to produce the total filtrate flow.

Most of the currently available hollow fiber membrane systems are proprietary, such that a single supplier will manufacture the entire filtration system, including the membranes, piping, appurtenances, control system, and other features. What this means is that each membrane manufacturer offers a module or family of module designs different from those of the other manufacturers. The reader is directed to membrane manufacturers' websites and to the individual manufacturers, or to operating water and wastewater plants to inspect and decide which module configuration and membrane material are best suited for a given application.

Pearce (2007) reports—and it is obvious to anyone experienced in membrane treatment—that one manufacturer developed the proprietary concept of a multiple horizontal design for UF membranes based on the concept used for RO, i.e., four 1.5 m (4.9 ft) elements installed into a 6 m (19.7 ft) pressure vessel. This design can offer more efficient utilization of space, lower operating flux with associated lower chemical use, and reported lower TMPs. The concept uses large diameter headers, which feed from both ends, with a dead end in the center of the vessel. The impetus to develop a standard MF module continues with no success. Pearce's analysis of the concept is that high fluxes used in UF systems cause a situation in which the middle elements are starved for flow and cause uneven fouling of the membrane. All other pressure modules are configured vertically to allow for air release; air expulsion is more difficult in a horizontal configuration. Lastly, longer clean-in-place (CIP) soak times are required if chemicals cannot be circulated. All of these limitations may or may not impact design. In large plants with many racks, the limitations may offer operational challenges. The author has no data on this topic.

Although each manufacturer's system is distinct, all of the hollow fiber membrane systems fall into one of two categories—pressure driven or vacuum driven—according to the driving force for operation. In a pressure-driven system, pressurized feed water is piped directly to the membrane unit, where it enters the module and is filtered through the membrane. Typical operating pressures range from 3 to 40 psi (0.2 to 2.8 bar). Most applications require designated feed pumps to generate the required operating pressure, although there are some water treatment plants that take advantage of favorable hydraulic conditions to operate an MF or UF system via gravity flow.

Pressure vs. Vacuum

Both MF and UF systems can be operated in either pressure or vacuum mode at ambient temperature. In pressure systems, water is forced through the membrane pores using a pressure pump.

The earlier system designs were considered to be the pressure type; that is, feed water is forced through the membrane pores by a pressure pump. A few years after the pressure type system was proven to be commercially successful, submerged membrane systems, also known as vacuum systems, were introduced.

While all hollow fiber systems employ pressure as a fundamental driving force, a vacuum-driven system is distinguished by its utilization of negative pressure and, consequently, its significantly different design and configuration. Unlike pressure-driven systems, in which each membrane module incorporates a pressure vessel, vacuum-driven systems utilize hollow fiber modules that are submerged or immersed in an open tank or basin. While the ends are fixed, the lengths of the hollow fibers are exposed to the feed water in the basin. The filtrate is drawn through the membrane using a vacuum pump. Submerged (vacuum) systems offer one distinct advantage over pressure systems: they are able to catch a much higher concentration of solids, e.g., suspended solids, turbidity, and, in MBRs, mixed liquor suspended solids. For that reason, submerged systems are the preferred configuration for MBRs and applications with high turbidity or high suspended solids. An emerging application for MF and UF membrane submerged systems is treating MF or UF backwash wastewaters and conventional sand granular media backwash. By design, vacuum-driven membrane filtration systems cannot be operated via gravity, nor in an inside-out mode. However, a favorable hydraulic gradient might enable the use of a gravity-based siphon to generate the suction required to drive the filtration process in a vacuum-driven system. In some cases with a substantial hydraulic gradient, the large amount of available head could be used to generate the power for suction pumps via on-site turbines—which is a consideration depending upon the siting of the system to the wastewater plant.

Pressure membranes offer the following advantages over vacuum systems:

- Greater range in operating pressure
- Higher design fluxes
- No flux reduction in cold-water applications

- Safer chemical cleaning operations
- Better security against accidental or deliberate contamination
- Easier monitoring and repair of leaks
- Smaller building size
- No permanent hoisting equipment required

The most obvious advantage is that the pressure systems can rely on a great reserve of pressure, should upset conditions occur in the system. Vacuum systems can only operate within the range of atmospheric pressure, so there is much less of a safety reserve, should something go wrong within the treatment process; pressure systems, by contrast, typically operate well below the pressure rating of the equipment. Vacuum systems may operate at average pressures approaching 70% (10 to 11 psig; 0.69 to 0.76 bar) of the maximum pressure differential available. Pressure systems typically operate at 40% to 50% (16 to 20 psig; 1.1 to 1.4 bar) of the maximum transmembrane pressure when tested on the same waters. This builds an intrinsic safety factor into a design, allowing the pressure system to better cope with process upset, changes in influent water quality, or any other changes in plant operation.

The reliance on atmospheric pressure further hampers vacuum systems if they are to be installed at elevations higher than 4,000 ft above sea level, where atmospheric pressure starts to be noticeably reduced. For example, at 4,000 ft above sea level, the atmospheric pressure is reduced from 14.7 to 12.7 psia (1.0 to 0.9 bar). Lower atmospheric pressure means less available differential pressure, translating into lower fluxes, bigger footprint, and higher capital and operating costs.

Design Flux

Pressure systems frequently operate at much higher fluxes than vacuum systems. In applications where the feed water is coagulated and settled, pressure systems typically operate at fluxes higher than 80 gfd (136 lmh). Two pilot skids achieved fluxes between 80 and 100 gfd (136 and 170 lmh) during the pilot phase at St. Helens, Oregon, and Barberton, Ohio. Vacuum systems typically peak at 40 to 45 gfd (2.8 to 3.1 lmh).

Flux Reduction in Cold Water

Another advantage of pressurized membrane systems is the capability to treat water in cold climate areas. During winter, the feed water temperatures drop, which increases viscosity and the driving force required to force water through a membrane pore. This significantly affects vacuum systems, since their maximum differential pressure available is 14.7 psia. It forces a vacuum system to be designed at much lower fluxes. For pressurized membrane systems, cold water is not a problem. By the very nature of operation, pressure systems operate in a much wider differential pressure range (0 to 40 psig [0–2.8 bar]). This allows a pressure system to provide consistent water flows, regardless of feed water temperature.

Because the feed water is contained in an open basin, the outside of the fibers cannot be pressurized above the static head in the tank. Therefore, a vacuum of approximately −3 to −12 psi (−0.2 to −0.8 bar) is induced at the inside of the fibers via pump suction. The water in the tank is drawn through the fiber walls, where it is filtered into the lumen.

Membrane Materials

MF and UF membranes are porous, with distinct pores, and remove dissolved organic and inorganic contaminants. MF and UF membranes may be constructed from a wide variety of materials, including cellulose acetate, polyvinylidene fluoride (PVDF), polyacrylonitrile (PAN), polypropylene, polysulfone (PS), polyethersulfone (PES), and other polymers. Each of these materials has different properties with respect to surface charge, degree of hydrophobicity, pH and oxidant tolerance, strength, and flexibility.

The two most widely used membrane materials today are PVDF and PES. The principal MF manufacturers use PDVF. UF manufactures also use PVDF manufactured by a different process and PES. Cellulose acetate (CA) and polysulfone (PS) are also used. One company offers both CA and PS. Other companies offer both PS and PES Pearce (2012). Independent studies confirm that material selection is critical for long-term mechanical and chemical stability of membranes. Zondervan and colleagues (2007) used a pressure pulse unit to simulate hydraulic cleaning of a PES membrane and concluded that the fouling status of the membrane was a major aging factor.

Arkhangelsky et al. (2007) studied the effect of PES and cellulose acetate membranes with sodium hypochlorite (a standard cleaning chemical in the water industry). The results indicated that the mechanical strength of the membranes deteriorated with hypochlorite cleaning. The ultimate tensile strength, ultimate elongation, and Young's modulus of the membranes decreased as the exposure (concentration times contact time) to hypochlorite was increased. Another evaluation by Pellegrin and colleagues (2011) found that exposure of PS membranes to sodium hypochlorite had a negative effect on mechanical properties.

The selection of a specific polymer will have a significant impact on mechanical and chemical stability of the membrane. In service, exposure to temperature extremes may cause the polymer materials used in low pressure MF and UF membranes to exhibit glassy (brittle) or rubbery (ductile) characteristics.

Table 4-5 summarizes the glass transition (T_G) and melting point (T_M) temperatures for common polymers used in the manufacture of MF and UF membranes.

It can be observed that for the most common polymeric materials used for UF and MF membranes in water applications, PVDF has glass transition temperatures well below 0°C (32°F), while the others have T_G values above 60°C (116°F). This implies that with the exception of polyethylene and PVDF, polymeric materials would be in the glassy region and exhibit a rigid and brittle nature. If PES, PS, PAN, and polyvinyl alcohol were used, the hydraulic cleaning regimes that are required to maintain membrane permeability would be limited. Alternatively, polyethylene and PVDF membranes would be in the rubbery region and would exhibit a higher tolerance for rigorous hydraulic cleaning (Liu, 2007). In an integrity study conducted by Childress et al. (2005), PVDF, PS, and PAN hollow fiber membranes were evaluated. The physical integrity of each membrane type was assessed over a 10-month period. The PVDF membrane demonstrated nearly total removal of coliforms, while there was significant passage of coliforms through the PS and PAN membranes. The report concluded that the probability of measuring coliforms exceeded 90% for the PS membrane, which was indicative of fiber breakage.

In recent years, the trend has been to use PVDF membranes for new water plant installations. According to a survey by the American Water Works Association Research Foundation (AWWARF) in 2003, PVDF, MF, and UF membranes comprised over one-half of the installed capacity in million gallons per day in water treatment plants. Other membrane materials (PES, PS, polypropylene, and cellulose acetate) comprised from 8% to 15% of the installed capacity (Adham et al., 2005). The dominance of PVDF membranes may be related to the high strength and chemical resistance of these products. Mechanical strength is a consideration, since a membrane with greater strength can withstand higher TMP levels, allowing for greater operational flexibility. The use of a membrane capable of higher pressures allows detection of smaller holes using pressure-based direct integrity testing. A membrane with bidirectional strength allows cleaning operations or integrity testing to be performed from either the feed or the filtrate side of the membrane.

Material properties influence the exclusion characteristics of a membrane as well. A membrane with a particular surface charge may achieve enhanced removal of particulate or microbial contaminants of the opposite surface charge due to electrostatic attraction. A membrane can be characterized as being hydrophilic (i.e., water attracting) or hydrophobic (i.e., water repelling). These terms describe the ease with which membranes can be wetted, as well as the propensity of the material to resist fouling to some degree. A hydrophilic membrane is less prone to fouling from constituents found in

Polymer	Chemical Structure	T_G (°C)	T_M (°C)
Cellulose acetate		67 to 68	150 to 230
Polyacrylonitrile (PAN)		80 to 110	320
Polyethersulfone (PES)		225	N/A
Polyethylene	$(CH_2)_n$	−90 to −30	137 to 145
Polysulfone (PS)		190 to 250	N/A
Polyvinyl alcohol		65 to 85	230 to 260
Polyvinylidene fluoride (PVDF)	$(CH_2CF_2)_n$	−50 to −35	160 to 185

TABLE 4-5 T_G and T_M Values for Common Polymeric Materials Used in MF and UF Membranes

wastewater, but exhibits lower mechanical strength and oxidant resistance and is limited in the flux maintenance cleaning strategies used to clean the membrane.

Chemical and Oxidant Compatibility

It is critical to consult with the membrane manufacturer prior to implementing any form of chemical pretreatment.

In a non-solvent-induced phase separation, phase separation occurs when a nonsolvent is introduced into the solution of polymer and solvent. In thermally induced phase separation, on the other hand, the polymer material solidifies to a porous product through varying the temperature of the material solvent during the process. Other production factors influence the pore development and polymer morphology.

In general, some MF and UF membranes are not compatible with disinfectants and other oxidants. High crystalline PVDF membranes are the most oxidant resistant in the industry, and can handle high concentrations of chlorine, potassium, permanganate, and hydrogen peroxide. However, some MF and UF membranes that are incompatible with stronger oxidants such as chlorine may have a greater tolerance for weaker disinfectants such as chloramines, which may allow for a measure of biofouling control without damaging the membranes. Certain types of both MF and UF membranes require operation within a certain pH range; this depends to some degree on the charge of the polymer relative to the charge associated with the membrane. A polymer with a charge opposite to that of the membrane is likely to cause rapid and potentially irreversible fouling. PVDF has the highest oxidant tolerance, with excellent foulant resistance and cleanability.

Hollow Fiber Modules

In 2005, there were approximately six primary suppliers of MF and UF membranes designed for use in municipal drinking water treatment applications that were currently active in the North American marketplace. In most cases, each membrane manufacturer also builds its own proprietary system around its membranes, using a custom module design, operational configuration, instrumentation and control, etc., such that membrane modules are not interchangeable between different manufacturers' systems. Most suppliers manufacture a single type of system (i.e., MF or MF/UF). These systems are operated in the direct (dead-end) filtration mode with minimal recirculation. During operation, the feed water flow is normal to the membrane surface and so suspended particulates and fouling species are retained on the membrane surface. Resistance to flow and the accumulation of solids on the shell side of the membrane surface result in an increase in the TMP, which is the effective pressure across the membrane.

Most hollow fiber modules used in wastewater reuse applications are manufactured to accommodate porous MF or UF membranes and designed to filter particulate matter. As the name suggests, these modules are composed of hollow fiber membranes, which are long and very narrow tubes that may be constructed of any of the various membrane materials described previously.

Hollow fiber MF membranes are organic polymeric tubes (fibers) usually less than a millimeter in diameter and are enclosed in a module. The fibers are sealed at the bottom end of the filter module in such a manner as to direct flow streams to the outside or shell side of the fiber. The fibers are sealed at the top end of the filter module to allow filtered water to exit from the inside (i.e., the lumen side) of the fiber. The water to be treated is pumped into the module and exits from the open ends of the lumens. A vertical configuration allows the use of gravity to separate air and water in the process. The number of hollow fibers housed in a module can be in the thousands. Module length usually ranges from 1 to 2 m. Feed water contacts the shell of the hollow fiber and product water collects on the fiber lumen. A pump upstream of the module pressurizes the shell side of the MF fiber. A hollow fiber membrane made from high crystalline PVDF provides oxidant resistance to essentially any oxidant and can therefore be used after oxidation by chlorine, potassium permanganate, or hydrogen peroxide without the need to neutralize the oxidant prior to the membrane. Cationic polyelectrolytes, if overdosed, tend to bind with the membrane.

Hollow fiber MF and UF membrane systems are operated in one of three configurations: cross-flow, dead-end, or submerged. In the positive pressure cross-flow configuration, a portion of the feed is recycled to the head of the process. Since the recycle stream is higher in solids, a bleed stream or waste stream is incorporated to reduce solids accumulation.

In the positive pressure dead-end mode, the system operates like a cartridge filter with no recycle or bleed streams.

In the submerged mode, the membranes are not encased in a module but are submerged in a tank, and a vacuum is applied to create the driving force. The reader is directed to manufacturers of submerged systems to evaluate the better scheme for controlling flow and solids removal.

MF and UF systems are operated in either a constant flux or constant pressure mode. Most MF and UF systems operate in a constant flux mode and use variable speed pumps to meet the increasing pressure requirements as the membranes foul. In a constant pressure mode of operation, membrane flux—the volumetric flow per unit area—declines over the filtration cycle. Flux ranges from 35 to 75 gfd (60 to 128 lmh) on settled wastewater. On some treated waters, fluxes in excess of 75 gfd (128 lmh), computed for the outside-in configuration, have been documented using a 0.1 μm PVDF

MF system. At these fluxes, operating TMPs range from 3 psi (0.2 bar) for clean membranes to greater than 43.5 psi (3 bar) for some fouled membranes. Typical values range from 7 to 30 psi (0.5 to 2.1 bar). Process recovery is typically 95% for settled wastewater and can be as high as 99% when an MF system is used to treat the primary MF reject.

MF and UF membranes are typically cleaned chemically every one to three months. PVDF, PES, and PS membranes can be cleaned with mineral acids, strong bases, and chelating agents such as citric acid, as well as oxidants such as sodium hypochlorite. The fibers may be bundled in one of several different arrangements. In one common configuration used by many manufacturers, the fibers are bundled together longitudinally, potted in a resin on both ends, and encased in a pressure vessel that is included as a part of the hollow fiber module. These modules are typically mounted vertically, although horizontal mounting may also be utilized. One alternate configuration is similar to spiral wound modules in that both are inserted into pressure vessels that are independent of the module itself. These modules (and the associated pressure vessels) are mounted horizontally. Another configuration, in which the bundled hollow fibers are mounted vertically and submerged in a basin, does not utilize a pressure vessel. A typical commercially available hollow fiber module may consist of several hundred to over 10,000 fibers. Although specific dimensions vary by manufacturer, approximate ranges for hollow fiber construction are as follows:

- Outside diameter: 0.5 to 2.0 mm
- Inside diameter: 0.3 to 1.0 mm
- Fiber wall thickness: 0.1 to 0.6 mm
- Fiber length: 1 to 2 m

A cross section of a symmetric hollow fiber is shown in Fig. 4-2. Hollow fiber membrane modules may operate in either an inside-out or an outside-in mode. In inside-out mode, the feed water enters the fiber lumen and is filtered radially through the fiber wall. The filtrate is then collected from outside of the fiber. In outside-in operation, the feed water passes from outside the fiber through the fiber wall to the inside, where the filtrate is collected in the lumen. Although the inside-out mode utilizes a well-defined feed flow path that is advantageous when operating under a cross-flow hydraulic configuration, the membrane is somewhat more subject to plugging as a result of the potential for the lumen to become clogged. The outside-in mode utilizes a less well-defined feed flow path but increases the available membrane surface area for filtration per fiber and avoids potential problems with clogging of the lumen.

Both inside-out and outside-in operating modes for hollow fiber modules are utilized, but high operating costs have driven the standard mode to outside-in. When a hollow fiber module is operated in inside-out mode, the pressurized feed water may enter the fiber lumen at either end of the module, while the filtrate exits through a filtrate port located at the center or end of the module. In outside-in mode, the feed water typically enters the module through an inlet port located in the center and is filtered into the fiber lumen, where the filtrate collects prior to exiting through a port at one end of the module. Most hollow fiber systems operate in dead-end or direct filtration mode and are periodically back-washed to remove the accumulated solids. Note that submerged hollow fiber membranes operate in outside-in mode but do not utilize pressure vessels.

Hollow Fiber (MF and UF) Systems

With a few exceptions, most MF and UF processes utilize systems designed around hollow fiber modules. Hollow fiber membrane filtration systems are designed and constructed in one or more discrete water production units, also called racks, trains, or skids. A unit consists of a number of membrane modules that share feed and filtrate valving, and each unit can usually be isolated from the rest of the system for testing, cleaning, or repair. A typical hollow fiber system is composed of a number of identical units that combine to produce the total filtrate flow.

Most of the currently available hollow fiber membrane systems are proprietary, such that a single supplier will manufacture the entire filtration system, including the membranes, piping, appurtenances, control system, and other features. The manufacturer also determines the hydraulic configuration and designs the associated operational subprocesses—such as backwashing, chemical cleaning, and integrity testing—that are specific to the particular system. As a result, there are significant differences in the proprietary hollow fiber membrane systems produced by the various manufacturers, and the membranes and other components are not interchangeable. Although each manufacturer's system is distinct, all of the hollow fiber membrane systems fall into one of two categories—pressure driven or vacuum driven—according to the driving force for operation.

Pretreatment

MF and UF membranes are microporous and do not remove dissolved contaminants without pretreatment, e.g., oxidation, powdered activation carbon, or coagulation—full conventional treatment or direct coagulation with post-treatment such an ion exchange, NF or RO, or advanced oxidation such as ozone, UV-ozone, or UV-peroxide.

A number of different chemicals may be added as pretreatment for MF or UF, depending on the treatment objectives for the system. For example, lime and soda ash may be added for softening applications; coagulants may be added to enhance removal of total organic carbon with the intent of minimizing the formation of disinfection by-products or increasing particulate removal; disinfectants may be applied for either primary disinfection or biofouling control; and various oxidants may be used to oxidize metals such as iron and manganese for subsequent filtration. It is important to ensure that the applied pretreatment chemicals are compatible with the particular membrane material used. As with conventional media filters, presettling may be used in conjunction with pretreatment processes such as coagulation and lime softening. While an MF or UF system may be able to operate efficiently with the in-line addition of lime or coagulants, direct coagulation or presettling in association with these pretreatment processes can enhance membrane flux and increase system productivity by reducing the solids loading, thus minimizing frequency of backwashing and chemical cleaning.

Either a separate coagulation/flocculation process or direct coagulation is used to reduce the dissolved concentration of wastewater. Powdered activated carbon is often used. Oxidation of dissolved iron and manganese is required to change the species from dissolved to particulate. Oxidant selection depends upon the iron or manganese species and whether or not they are sequestered.

The extent of organic removal is a function of pH, coagulant choice, mixing or detention time, temperature, and mixing velocity—which are the same for conventional treatment except that only a pin floc greater than the MF or UF pore size is required. The efficiency of powdered activated carbon is a function of the type of carbon, detention time, and dose.

Arsenic is removed by MF and UF when an iron-based coagulant is used. The species of arsenic may require oxidation before treatment.

If properly operated, secondary effluent produces a consistent effluent—feed water to the membrane. The frequency of upset of conventional activated sludge will impact the type and extent of pretreatment required. Typical effluent quality is 20 mg/L of suspended solids and 20 mg/L of biochemical oxygen demand. Filamentous bacteria and sludge carryover can occur, changing the feed water to the membrane.

Applications

MF and UF systems are used to treat surface waters and groundwaters under the influence to meet the requirements of the Safe Drinking Water Act. The filtrate turbidity is less than or equal to 0.1 ntu and the system filtrate is disinfected with regard to *Cryptosporidium*, *Giardia*, and coliform bacteria—that is, zero, but reported as less than

2 plate forming units (pfu)/100 mL. This performance is consistent at all times, regardless of influent turbidity or microorganism concentration. MF and UF systems are used to remove iron and manganese from surface waters, groundwaters, seawater, or secondary effluent to less than 0.05 mg/L. Together with an iron salt, MF systems remove arsenic to 10 ppb or less. The Silt Density Index is always less than 3 and usually less than 2. The systems provide pretreatment to NF and RO systems for removing nutrients such as nitrates. MF and UF systems can remove phosphates to less than 0.1 mg/L with the addition of a trivalent salt. These systems can be used together with coagulants to reduce total organic carbon and dissolve organics.

Figures 4-3 to 4-6 demonstrate typical MF and UF applications.

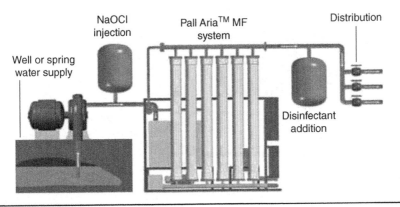

FIGURE 4-3   Iron removal.

Soluble Iron and Manganese are removed by oxidization. MF provides "the barrier" and removes the precipitated iron/manganese to < 0.05 mg/L.
PVDF membranes can handle a variety of oxidants.

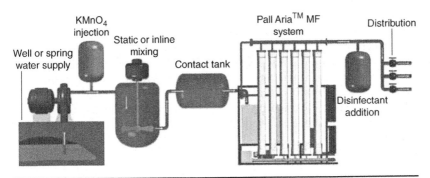

FIGURE 4-4   Manganese removal.

Soluble Arsenic is removed using ferric hydroxide to complex the arsenic. MF provides "the barrier" and removes the arsenic floc particles.

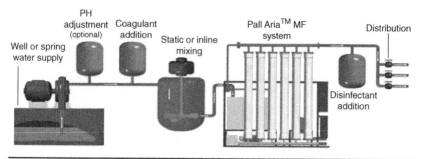

FIGURE 4-5   Arsenic removal from ground water.

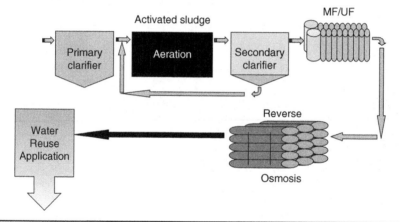

FIGURE 4-6   MF/UF treatment of secondary effluent.

Membrane Systems

With a few exceptions, most MF and UF processes utilize systems designed around hollow fiber modules. Hollow fiber membrane filtration systems are designed and constructed in one or more discrete water production units, also called racks, trains, or skids. A unit consists of a number of membrane modules that share feed and filtrate valving, and each unit can usually be isolated from the rest of the system for testing, cleaning, or repair. A typical hollow fiber system is composed of a number of identical units that combine to produce the total filtrate flow.

Operation

MF and UF systems are generally operated in either constant flux or constant pressure modes, in direct (dead-end) filtration with minimal

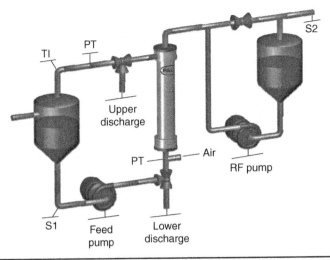

TI

PT

S2

Upper
discharge

PT — Air

RF pump

S1   Feed
pump

Lower
discharge

FIGURE 4-7   Product filtration mode—forward flow. (*Adapted from Escabar, 2010.*)

recirculation. During operation (see Figure 4-7), the feed water flow
is normal to the membrane surface and so suspended particulates
and fouling species are retained on the membrane surface. Resistance
to flow and the accumulation of solids on the shell side of the membrane surface result in an increase in the TMP, which is the effective
pressure across the membrane. Under constant pressure operation,
the membrane flux will decline over the course of a filtration cycle as
the membranes foul. Most MF and UF systems, however, are operated at constant flux using variable speed pumps to meet increasing
pressure requirements caused by membrane fouling, which restricts
the passage of water. Operating TMPs may span from approximately
3 psi (0.2 bar) to as high as 43.5 psi (3 bar) in some cases, although
more typical values range from 7.3 psi (0.5 bar) to 29 psi (2 bar).

There are three different general operating configurations for MF
and UF systems. As the membranes become fouled over the course of
operation, a combination of both backwashing and chemical cleaning
is used to remove foulants to the maximum practical extent, thus ideally restoring operating parameters such as flux (for constant pressure operation) and TMP (for constant flux operation) to baseline
levels. For most MF and UF systems, backwashing is typically conducted every 30 to 60 min for a duration of 1 to 5 min and consists of
water, air, or a combination of both. Backwashing may also be initiated based on a TMP set point or a predetermined volume of filtrate
produced. Chemical cleaning is conducted at a point at which the
ability of routine backwashing to restore the desired system flux
and/or TMP is significantly diminished. Important considerations
for chemical cleaning include the type and dose of chemical(s) used,
frequency, duration, temperature, rinse volume, recovery and reuse

of cleaning agents, and neutralization and disposal of residuals. A variety of chemicals may be used for cleaning, including detergents, acids, bases, oxidizing agents, sequestering agents, and enzymes. Chlorine is also commonly used in the chemical cleaning process for both disinfection and oxidation (for compatible membrane materials) at a wide range of doses (2 to 2,000 mg/L).

Figure 4-8 shows typical TMP, flux, temperature, feed turbidity, and filtrate turbidity. Integrity testing is also a critical feature of MF and UF systems, enabling the detection and isolation of breaches that would otherwise compromise the effectiveness of the membrane as a barrier for pathogens and particulate matter. Almost all commercially available MF and UF systems designed for municipal drinking water treatment are equipped with the ability to conduct a pressure decay test using air, which directly challenges the membrane barrier by pressurizing the system modules and monitoring the rate of decay over a short period of time (about 5 min).

Decay rates exceeding a given benchmark threshold value (specific to each site-specific system) indicate an integrity breach, with the magnitude of the response directly proportional to the magnitude of the breach. Less sensitive indirect methods are used to monitor some aspects of filtrate water quality, such as turbidity or particle counts, that are generally indicative of membrane integrity on an ongoing basis during operation between periodic applications of a more sensitive direct test (e.g., the pressure decay test), as the current direct integrity test methods do not permit simultaneous production of filtered water during testing.

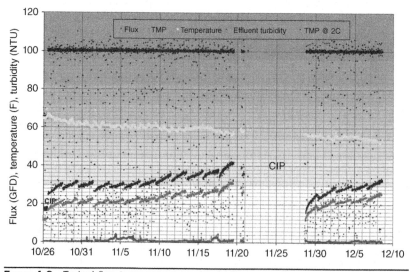

FIGURE 4-8   Typical flux, temperature, and turbidity versus time curve.

For a membrane system to effectively produce water, fouling species—which include particulates, microorganisms, chemical precipitates, and other particulates—must be effectively removed from the hollow fiber surface. Techniques to accomplish this involve physical steps—e.g., reverse filtration—or chemical steps which necessitate that the module be taken off-line and out of service.

Reverse Filtration (Backwash)

Fouling species are removed from the PVDF membrane surface by the reversal of the direction of flow across the membrane, or reverse filtration (RF). See Fig. 4-9. The efficiency of the process is affected by the velocity, volume, and duration of the RF. RF is generated by either water alone or water with oxidants, acids, or compressed air (air scrubbing). In both air and liquid backwash, water will dislodge the fouling material. The shear force at the membrane surface is greater during air scrubbing, as it increases the velocity of water on the membrane surface. The duration of the RF in air scrubbing is usually short. Dislodged material can be flushed from the module with feed water. Oxidants are sometimes added to the RF water to enhance the efficiency of the liquid backwash. In the industry, this is called chemically enhanced backwash. Strong oxidants, such as sodium

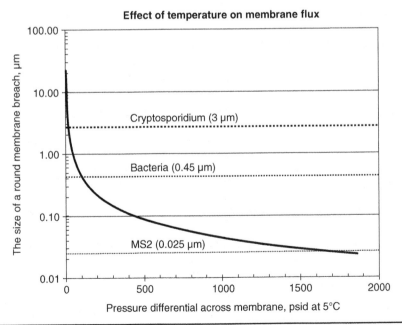

FIGURE 4-9 Relationship of pressure differential across membrane and size of membrane breach. (*Li, 2010.*)

hypochlorite, can disrupt the structure of the fouling material and facilitate its removal. Alternatively, the recirculation of small air bubbles on the membrane surface can scour and disrupt the fouling material mechanically.

Chemical Cleaning

RF and air scrubbing provide a short-term strategy for removing fouling materials. Although an effective RF or air scrubbing strategy will retard the rate of membrane fouling, eventually the membrane must be chemically cleaned to restore its TMP. The typical duration between chemical cleanings is 4 to 6 weeks. Cleaning protocols for MF membranes vary according to the membrane tolerance to cleaning chemicals. In general, a low pH solution (pH of 2 to 3) removes cationic species and a high pH solution (pH of 11 to 12) removes organic material. The action of the caustic cleaning solutions can be enhanced by the addition of nonionic surfactants (provided the surfactants can be rinsed to meet regulatory requirements), which help to disperse the organic particles without binding to the membranes, and chelating agents, which are particularly useful for the disruption of cationic polymer bridging in biologically fouled membranes.

Processes such as backwashing, chemical cleaning, and integrity testing are specific to a particular system. As a result, there are significant differences in the proprietary hollow fiber membrane systems produced by the various manufacturers, and the membranes and other components are not interchangeable.

In general, most NF and RO membranes and some MF and UF membranes are not compatible with disinfectants and other oxidants. High crystalline PVDF membranes are the most oxidant resistant in the industry, and can handle high concentrations of chlorine, potassium, permanganate, and hydrogen peroxide. However, some MF and UF membranes that are incompatible with stronger oxidants such as chlorine may have a greater tolerance for weaker disinfectants such as chloramines, which may allow for a measure of biofouling control without damaging the membranes. Certain types of both MF and UF and NF and RO membranes require operation within a certain pH range. Chemicals used to raise pH include lime, sodium hydroxide, and sodium bicarbonate. The mineral acids, hydrochloric acid and sulfuric acids, are used to lower pH. Depending upon water quality, inorganic coagulants affect pH.

MF and UF membrane systems usually operate within a relatively narrow range of TMPs, which may limit the ability to increase the TMP in order to maintain constant flux as the water temperature decreases. Because the membrane modules can be damaged if the TMP exceeds an upper limit, as specified by the manufacturer, it may not be possible to operate the system at a TMP that is sufficient to meet the required treated water production during colder months if demand remains high. As a result, additional treatment capacity (i.e., increased membrane area or number of membrane modules) is incorporated into the design of the system, such that the water treatment production requirements can be satisfied throughout the year.

The MF or UF membrane manufacturer may have an alternate preferred method of determining the design membrane area for a particular application based on temperature, membrane material, or other site- or system-specific factors. It is recommended that the utility collaborate with the state, the membrane manufacturer, and its consulting engineer (if applicable) to select the most appropriate method for determining the required area. Note that the addition of membrane area to compensate for low-temperature flow will also help the system to meet higher flow demands during warm weather without operating at an exceedingly high membrane flux. Temperature compensation for MCF systems, if necessary, can be determined using the methodology for MF and UF systems.

System Recovery

Unlike in RO and NF, the recovery of the MF process is not limited by the formation of scale or the precipitation of salts. Recovery is defined as the ratio of the volume of product water (permeate) produced to raw water treated:

$$R = \frac{Q_P}{Q_F} \times 100\%$$

where R = recovery (%)
  Q_P = permeate flow (gpm or m/h)
  Q_F = feed flow (gpm or m/h).

More simply,

$$\text{Recovery (\%)} = \frac{\text{product flow}}{\text{feed flow}} \times 100\%$$

Recovery in the MF process is a function of the permeate flow rate, length of the filtration cycle, and volume of water used in backwash. For surface waters, the filtration cycle is usually 20 to 30 min, with a process recovery of 95% to 97%. For secondary effluent, recovery ranges from 80% to 95%, depending upon the quality of the secondary effluent and the selection of pretreatment.

Integrity Testing of Low Pressure Membranes

A unique feature of low pressure membrane processes is the ability to perform integrity testing. Membrane integrity, defined as the wholeness or freedom from defects of a membrane, is an important issue in drinking water purification. The USEPA published the *Membrane Filtration Guidance Manual* in 2005, a significant portion of which discusses membrane integrity testing. Although the manual is primarily intended for drinking water membrane processes, the same principles are applicable to wastewater reuse applications.

There are several key issues in membrane integrity:

- What are the target contaminants to be removed by the membrane processes?

- What are the tests to verify the membrane integrity?

- How much of the removal of the target contaminants by membranes can be demonstrated and verified by those tests?

- How often should the tests for the verification of membrane integrity be performed?

- How reliable are the results of the tests for the verification of membrane integrity?

- How can a membrane breach be located quickly in a system containing many membranes?

As low pressure membranes are primarily used for the removal of particulate substances and pathogenic microbes from the water, water quality parameters such as turbidity, total suspended solids, particle counts, and counts of pathogenic microbes are the obvious choices for target contaminants removed by membranes. Therefore, the goal of membrane integrity testing should be to verify the removal of those contaminants.

There are generally two categories of membrane integrity testing methods: (1) tests based on the fluid movement across the membranes and (2) tests based on monitoring the filtrate quality of membrane processes. The first category includes testing methods such as pressure decay, forward flow, vacuum decay, water replacement flow, and others. The second category includes online turbidity monitoring, online particle counts and their derivatives, and microbial testing. In the USEPA's guidance manual, integrity testing methods are classified as direct integrity testing and indirect integrity testing (USEPA, 2005). Direct integrity testing includes all the methods based on fluid movement across membranes plus the marker method (i.e., injecting a marker substance upstream of the membrane and measuring its removal across the membranes), while indirect integrity testing includes methods based on water quality, such as online turbidity monitoring and particle counts.

Two distinct differences between direct and indirect integrity testing are the sensitivity and testing frequency of the methods. Generally speaking, direct integrity testing has higher sensitivity than indirect testing. On the other hand, indirect testing usually can provide continuous membrane integrity monitoring, while direct testing is typically performed on a periodic basis (e.g., daily or even weekly). Therefore, direct and indirect integrity testing methods are mutually supplementary. Both direct and indirect integrity testing are required when a low pressure membrane system is for the purpose of compliance with the Long Term 2 Enhanced Surface Water Treatment Rule (USEPA, 2006).

There are specific performance criteria for direct integrity testing: resolution (the smallest integrity breach that is detectable by the testing), sensitivity (the highest log reduction value that can be achieved by the testing), and testing frequency. In addition, control limits for a membrane system must be established for both direct and indirect integrity testing.

In membrane applications for wastewater treatment and reuse, the objectives are either to remove microbial pathogens—i.e., protozoa, bacteria, and viruses—or to pretreat for other treatment processes for wastewater reuse, i.e., RO. Although the objectives and performance criteria for membrane integrity testing could be different from those for drinking water applications, the basic principles outlined in the USEPA's guidance manual are still applicable.

A relevant issue for wastewater treatment is how to define the resolution of the membrane integrity testing. Most low pressure membrane systems use the pressure decay method as the standard for direct integrity testing. Essentially, the pressure decay method involves pressurizing one side of a membrane at a predetermined pressure and measuring the pressure decay over time as an indicator of membrane integrity. For systems applying the pressure decay method, the resolution of the testing is determined based on the relationship between testing pressure and the size of the membrane breach, as illustrated in Fig. 4-9.

As indicated in Fig. 4-9, the required pressure differentials across the membrane to meet the criteria for the removal of *Cryptosporidium*, bacteria, and viruses are approximately 14.5, 97, and 1,750 psid (1.0, 6.7, and 120.6 bar), respectively, at 5°C. Obviously, it is not feasible to use the pressure decay method for integrity testing if the goal is to completely remove viruses or even bacteria, as the pressure is too high to be practical. Therefore, developing a membrane integrity testing method that is capable of meeting the resolution criterion for bacteria and virus removal would be very beneficial. Another issue is the sensitivity of the test. As all methods that are able to monitor membrane integrity continuously are generally of low sensitivity, and methods of relatively high sensitivity are unable to be performed on a continuous basis, the current membrane plants have to shut down periodically to perform integrity testing that is sufficiently sensitive. This operation mode is the result of balancing the risk to water quality due to membrane integrity breach and plant productivity. To resolve this issue, a method that is both sensitive and continuous is highly desirable. Innovations in methods of monitoring membrane integrity are greatly in need.

If intact, a membrane offers essentially 100% removal efficiency for targeted microbial pathogens. If the membrane is not integral—and depending upon the number of fibers compromised—microbial contaminants will pass through the membrane. Therefore, it is important to know if a compromise to the membrane has occurred.

Residuals Characteristics and Management

The primary residuals associated with MF and UF systems are the wastewaters generated from flux maintenance—i.e., backwashing and air scrubbing—and chemical cleaning operations, i.e., chemically enhanced backwash and clean in place (CIP) operations. Encased MF and UF systems operating in a cross-flow mode also generate concentrated waste streams which are similar to backwash residuals. With the exception of those systems operating in cross-flow mode (or a similar variant with a bleed stream), MF and UF systems typically produce residuals on an intermittent basis as a result of regular backwashing; however, for larger systems, the combination of multiple units backwashing at staggered intervals may generate a waste stream on a more continuous basis. Backwash residual characteristics depend on the quality of the membrane system feed water as well as on the recovery of the membrane system. At typical recoveries of 85% to 95%, the solids in the feed are concentrated by a factor of 7 to 20 in the backwash residuals.

Backwash flows generally represent about 95% to 99% of the total residuals volume from MF and UF systems; the remainder is generated from the cleaning processes used to control fouling. Waste cleaning solutions reflect the chemicals used in the cleaning process combined with the extracted foulants. The volume of CEB waste is typically very low; for example, a daily CEB generally represents less than 0.2% to 0.4% of the plant flow. Residuals from a CIP conducted monthly (i.e., a typical industry benchmark interval) will normally be less than 0.05% of plant flow. These volumes may be reduced if the cleaning solutions are recycled or reused. The characteristics of typical backwash and chemical cleaning wastes are summarized in Table 4-4.

Disposal of MF and UF water treatment residuals is generally governed by the same regulations that control disposal of conventional treatment residuals, as summarized in Table 4-6. Additional state and/or local regulations that protect sensitive ecosystems, such as cold-water fisheries or pristine watersheds, may also be applicable.

The selection of residuals disposal methods depends on several site-specific factors (AWWA, 2005): climate; availability of land for facilities; size of the MF or UF installation; feasibility; and federal, state, and local requirements. According to a recent survey of 65 MF/UF installation sponsored by the United States Bureau of Reclamation (Mickley, 2003), the most common methods for backwash residuals

Disposal Method	Regulation	Representative Discharge Limitation
Surface water discharge	NDES permitting under the Clean Water Act	pH of 6 to 9 TSS < 30 mg/L plus raw water TSS Chlorine < 0.2 mg/L BODS < 30 mg/L Minimum 90% removal of TSS and cysts
Groundwater discharge (only if aquifer is closely linked to surface water)		
Groundwater discharge	Underground injection control Permitting under the Safe Drinking Water Act	Varies with water quality in the receiving aquifer and nature of geology State rules will likely apply
Percolation ponds and leaching fields		
Any other methods that may affect groundwaters		
Sewer or public owned treatment works discharge	Industrial pretreatment program under the Clean Water Act	pH of 6 to 9 TSS < 400 to 500 mg/L Chlorine < 10 mg/L BODS < 400 to 500 mg/L Nothing that will harm infrastructure or operations or result in NPDES violations
Landfill disposal	Resource Conservation and Recovery Act	Pass paint filter test (typically > 20% solids) and toxicity test (TCLP) Residuals that fail are classed as hazardous and must be properly disposed of
Land application of solids	Solid Waste Disposal Act	Limitations on cumulative loadings of metals

Source: AWWA (2008) p. 96.

TABLE 4-6   Disposal Methods and Regulations for MF and UF Water Treatment Residuals

disposal were surface water discharge (38%), sewer discharge (25%), land application, such as percolation ponds or irrigation (22%), and recycling (14%). Treatment was provided at a small number of locations to meet specific discharge requirements. Only one small installation (0.12 mgd; 0.45 MLD) in Oklahoma used an evaporation pond.

The most common methods of CIP residuals disposal were sewer discharge (37%), land disposal (25%), and surface water discharge (11%). Blending, neutralization, and other treatment were frequently provided in conjunction with these disposal methods. Additional detailed information regarding the management of membrane residuals is provided in the 2003 AWWA Membrane Residuals Management Subcommittee report *Residuals Management for Low-Pressure Membranes*.

Disposal of MF and UF secondary effluent residuals is more limited, since the flux maintenance wastewaters will contain, in addition to the contaminants from water treatment, high concentrations of microorganisms and organics. Disposal to a surface water or percolation pond is not an option. Sewer discharge and backwash recovery with subsequent disposal of the recovered concentrate—the MF backwash is concentrated another 7 to 10 times—to an anaerobic process should be considered. Depending upon the backwash characteristics and volume of reject, conventional thickening, gravity belt thickening, and dewatering processes should be considered.

CHAPTER 5

Diffusive Membrane Technology— Nanofiltration and Reverse Osmosis

Introduction

The removal of dissolved constituents, dissolved inorganic solids, total dissolved solids (TDS), dissolved organics, and dissolved organic carbon (DOC) from membrane treated settled secondary effluent is accomplished using the pressure-driven processes of nanofiltration (NF) and reverse osmosis (RO). Nanofiltration and reverse osmosis are classified as diffusive membranes and will be referred to as such in this book. This chapter addresses them in more detail.

An excellent and in-depth academic treatment of RO is provided by Dr. Michael E. Williams (2003). RO was developed in the 1950s to remove dissolved salts from seawater. Attainable fluxes at that time were about 0.1 gfd (0.17 lmh). The maximum life guarantee for an RO membrane was 3 y. Both hollow fiber and spiral wound modules were in development. Today, fluxes exceeding 18 gfd (30.61 lmh) are achievable, and membranes are guaranteed for 5 y. Both NF and RO systems use spiral wound modules for most water reuse applications.

Terminology and Definitions

Figure 3-4, membrane processes and characteristics, is also Figure 5-1. It is a general schematic of the various flows into and out of a membrane process. A diffusive membrane is either a nanofiltration membrane or a reverse osmosis membrane.

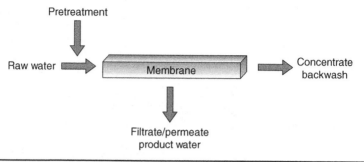

Figure 5-1   Membrane processes and characteristics.

Feed: The input solution—for wastewater reuse, most likely micro-filtration (MF) or ultrafiltration (UF) treated secondary effluent—to the NF or RO membrane system.

Feed Water Temperature: A critical design parameter. It has significant effects on feed pump pressure requirements, hydraulic flux balance between stages, permeate quality, and solubility of sparingly soluble salts. As a rough rule of thumb, every 10°F decrease in feed temperature increases the feed pump pressure requirement by 15%. The hydraulic flux balance between stages (in other words, the amount of permeate produced by each stage) is impacted by temperature. When water temperature increases, the elements located in the front end of the system produce more permeate, which results in reduced permeate flow by the elements located at the rear of the system. A better hydraulic flux balance between stages occurs at colder temperatures. At warmer temperatures, salt passage increases due to the increased mobility of the ions through the membrane. Warmer temperatures decrease the solubility of calcium carbonate. Colder temperatures decrease the solubility of calcium sulfate, barium sulfate, strontium sulfate, and silica. Minimum and maximum temperatures must be known for proper design. NF and RO systems are rated at 25°C (77°F). Lower temperatures will result in permeate flow that is lower than in the design.

Permeate: The portion of feed which passes through the membrane.

Concentrate: The portion of feed which does not permeate the membrane but retains ions, dissolved organics, dissolved inorganics, and any colloidal particles which are rejected by the membrane.

System Recovery: The ratio of permeate flow to feed flow of the NF or RO system.

Element Recovery: The ratio of permeate flow to feed flow of an individual element within a system.

Osmotic Pressure: The pressure phenomenon resulting from the difference of salt concentrations across an RO membrane. Increasing TDS levels result in increased osmotic pressure. The RO feed pump has to generate sufficient pressure to overcome this osmotic pressure before permeate is produced. A rough rule of thumb is that 1,000 ppm of TDS equals 11 psi of osmotic pressure. A brackish water at 550 ppm TDS produces 5 psi osmotic pressure. A seawater at 35,000 ppm TDS produces 385 psi osmotic pressure.

Temperature Correction Factor: See Chapter 3 for general predictive equations. Websites such as www.rotools.com and others are available to predict flux change.

Transmembrane Pressure (TMP): See Chapter 3. TMPs for dense membranes are a function of the osmotic pressure and vary from 75 to 300 psi (5.2 to 20.7 bar; estimated by this author); TMPs for NF and RO membranes in general range from 100 to 1,200 psi (6.9 to 82.7 bar). Unlike low pressure membranes, which may operate in a narrow TMP range, NF and RO systems are designed to allow for increased pressure drops. TMP is increased to maintain constant flux if the temperature decreases or the membrane fouls. Also, unlike with low pressure membranes, cleaning fouled NF and RO membranes reduces membrane rejection. Too many cleanings can cause the need for membrane replacement ahead of schedule, impacting the cost of operation.

Cross-Flow Filtration: According to the *Membrane Filtration Guidance Manual* (US Environmental Protection Agency [USEPA], 2005), cross-flow filtration (also referred to as *tangential flow filtration*) is "the application of water at high velocity and tangential to the surface of a membrane to maintain contaminants in suspension. The suspension mode hydraulic configuration is typically associated with spiral-wound nanofiltration (NF) and reverse osmosis (RO) systems."

An applied pressure forces a portion of the fluid through the membrane to the filtrate side. Particles and macromolecules that are too large to pass through the membrane pores are retained on the upstream side of the membrane as retentate. In a properly designed cross-flow device, contaminant does not build up on the membrane surface, due to the sweeping action of the tangential flow. This concept is shown in Figure 5-2.

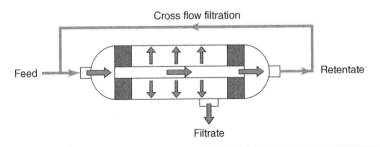

FIGURE 5-2 Simple cross-flow device.

A simplified diagram of a cross-flow device with hollow fiber membranes is depicted in Figure 5-2. In operation, fluid flows through each tubular opening (in this example, in an inside to outside mode) at high velocities that are sufficient to prevent contaminant buildup on the membrane surface.

The applied pressure forces filtrate through the pores of the membrane perpendicular to the direction of the feed flow. The concentrated retentate is continuously recycled back to the feed stream, ultimately increasing the feed concentration. Turbulent flow on the membrane surface is required to keep the solids in suspension and to prevent fiber plugging. Cross-flow velocities (averaged over the feed cross-sectional area) in the range of 1 to 7 m/s (3.3 to 23 ft/s) are essential for sustained operation. In addition to maintaining a turbulent cross-flow, a large bore design is typically needed to handle high solids levels without plugging. Cross-flow filtration can concentrate solids and semisolids very effectively because the filter configuration and operating conditions are designed to keep the solids suspended rather than in the filter matrix.

Silt Density Index (SDI): An empirical test developed for membrane systems to measure the rate of fouling of a 0.45 μm filter pad by the suspended and colloidal particles in a feed water. This test involves the time required to filter a specified volume of feed at a constant 30 psi at time zero and then after 5 min, 10 min, and 15 min of continuous filtration. Membrane manufacturers will warranty their membranes at specified SDI values. A typical RO element warranty, for example, might list a maximum SDI of 4.0 at 15 min for the feed water. An SDI of 2 to 3 is desirable today for desalting membranes. RO membranes used for wastewater recycling may warranty to a different SDI.

If the SDI test is limited to only 5 or 10 min readings due to plugging of the filter pad, the user can expect a high level of fouling for the RO system. Deep wells typically have SDIs of 3 or less and turbidities less than one with little or no pretreatment. Surface sources typically require pretreatment for removal of colloidal and suspended solids to achieve acceptable SDI and turbidity values.

The SDI is an empirical, dimensionless measure of particulate matter in water and is generally useful as a rough gauge of the suitability of a source water for efficient treatment using NF or RO processes. It is extensively used in the industry to determine and monitor the effectiveness of pretreatment to an NF or RO system. The SDI value of the feed water has a role in determining the design and operating flux of the NF or RO system. The ASTM Standards D4189-95 and D4189-07 detail the procedure for determining SDI.

Figure 5-3 shows a schematic of an SDI apparatus.

SDI measurements are taken by filtering a water sample through a 0.45 μm flat sheet filter with a 47 mm diameter at a pressure of 30 psi. The time required to collect two separate 500 mL volumes of filtrate is

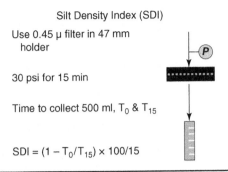

FIGURE 5-3   Simple schematic of an SDI apparatus.

measured, and the resulting data become the inputs to a formula used to calculate SDI. Water samples that contain greater quantities of particulate matter require longer to filter and thus have higher SDI values. Until recently, a general rule of thumb was that spiral wound NF and RO modules were not effective for treating water with an SDI of 5 or greater, as this quality of water contains too much particulate matter for the nonporous, semipermeable membranes, which would foul at an unacceptably high rate. Thus, some form of pretreatment to remove particulate matter was required for SDI values exceeding 5 in NF and RO applications. Today, in many cases, the required SDI is less than or equal to 3.

NF and RO membrane module manufacturers can usually provide a rough estimate of the range of anticipated operating fluxes based on the type of source water, which is roughly associated with a corresponding range of SDI values.

SDI is computed by the following equation:

$$\text{SDI} = 100 \times \frac{1 - (T_i / T_f)}{T_t}$$

where T_i = time in minutes to collect the initial 500 mL
  T_f = time in minutes needed to collect 500 mL after being online for 15 min
  T_t = total time of the test, 15 min

SDI is only one measure of water quality; there are a number of site- and system-specific water quality and operational factors that combine to dictate the flux for a given system. Thus, the ranges cited in Table 5-1 should only be used as a rough guideline. Caution should also be exercised when interpreting SDI results, as measurements can vary from test to test and analyst to analyst, as well as with both temperature and the specific type of membrane used. Escobar-Ferrand and colleagues (2009) showed that the membrane specified in the 07 ASTM standard can adsorb organics in the wastewater and provide an SDI

SDI	Estimated NF/RO Flux (gfd)	Source Water
2–4	8–14	Surface Water
<2	14–18	Groundwater

TABLE 5-1   Estimated NF/RO Membrane Fluxes as a Function of SDI

value higher than actual. When using SDI to select or monitor NF or RO pretreatment of secondary effluent, the effect of SDI membrane selection and the ASTM standard used should be considered.

SDI results are given for comparable conditions. SDI determination is a batch process; it is not conducted continuously online and is not typically utilized as gauge of water quality or system performance during daily operation in the way turbidity or conductivity monitoring are often employed. SDI should be used together with online turbidity measurements.

The *Membrane Filtration Guidance Manual* (USEPA, 2005) gives a general rule that dense membranes will foul when the SDI of the feed water is 5 or greater. Membrane manufacturers consider this when determining the warranty and usually provide a rough estimate of the range of anticipated operating fluxes based on the type and SDI of the source water. Now that membrane pretreatment is slowly replacing conventional filtration system (CFS) and cartridge filters as pretreatment for secondary effluent, SDIs less than or equal to 3—and many times less than 2—are more often required.

The *Membrane Filtration Guidance Manual* (USEPA, 2005) provides information that suggests that a correlation exists between SDI and achievable flux for NF and RO systems; this is shown in Table 5-1.

Stiff Davis Saturation Index (SDSI):   SDSI, in a similar fashion to LSI, is a method of reporting the scaling or corrosion potential of high TDS seawater based on the level of saturation of calcium carbonate. The primary difference between SDSI for high TDS seawater and LSI for low TDS brackish water is the effect that increasing ionic strength has on increasing solubility. The solubility of sparingly soluble salts increase with higher TDS and ionic strength, based on the theory that a denser ion population interferes in the formation and/or precipitation of the sparingly soluble salt.

Feed Water Quality

Feed water to an NF or RO system, in addition to meeting SDI requirements, must consider the following:

- **Iron.** Dissolved iron (ferrous iron, Fe^{2+}) should be less than 0.1 mg/L. MF systems can reduce iron to < 0.05 mg/L when properly oxidized.

- **Silica.** Silica can be either colloidal or reactive. The solubility of silica is 120 mg/L at pH 7 and 25°C (77°F). Silica will precipitate on the NF or RO membrane if its concentration exceeds its solubility. It must be either removed, reduced in concentration so that the process does not concentrate it, or chelated.

- **Barium.** Barium is highly insoluble and will combine with sulfates and other anions and precipitate on the NF or RO membrane.

- **Strontium.** Although more soluble than barium, strontium is highly insoluble, with a high inorganic fouling tendency.

NF and RO Flux

Membrane flux is the quantity of permeate produced per square foot of membrane, expressed in gfd. The following example calculates the flux of an RO system:

A system has a permeate flow of 100 gpm, 24 elements of 8 in., a throughput per element of 4.17 gpm, and an element membrane area of 350 ft². The first step is to convert gpm to gpd:

$$\frac{4.17 \text{ gal}}{\text{min.}} \times \frac{1440 \text{ min.}}{\text{day}} = 6,000 \text{ gpd}$$

Flux = flow/area = 6,000 gpd/350 ft² = 17 gfd (28.9 lmh)

In membrane applications, *critical flux* is that sustainable flux below which fouling does not occur and above which fouling occurs. The critical flux is dependent upon cross-flow velocity and design flux.

Membrane Materials

Common membrane materials used today are polyamide thin-film composites, cellulose acetate, and cellulose triacetate. Thin-film composite membranes show superior strength and durability and higher rejection rates than cellulose acetate and cellulose triacetate membranes. They also are more resistant to microbial attack, high pH, and high TDS. Cellulose acetate and cellulose triacetate membranes tolerate chlorine better. Sulfonated polysulfone membranes are chlorine tolerant and can withstand higher pH. These membranes are used on soft water and high pH water and are often used when high nitrates are a concern.

Modules

Spiral wound and hollow fiber are the two major module configurations used for NF and RO applications. Hollow fine fiber membranes (discussed in Chapter 3), because of the very tight packing, require cleaner water quality than spiral wound membranes—very low SDIs. With SDI's less than 3 and many times less than 1 or 2 thanks to MF or UF pretreatment of secondary effluent, hollow fine fiber membranes are a consideration in future membrane water reuse applications.

Tubular and plate and frame are used in industrial applications. Plate and frame (disc tube) NF or RO is considered the best available technology for landfill leachate treatment. Today, spiral wound membranes are most used in wastewater reuse applications.

The *Membrane Filtration Guidance Manual* (USEPA, 2005) provides an excellent general description of NF and RO modules. Spiral wound modules were developed as an efficient configuration for the use of semipermeable membranes to remove dissolved solids, and thus are most often associated with NF/RO processes. Figure 5-4 illustrates a typical spiral wound pressure vessel for NF/RO applications. The basic unit of a spiral wound module is a sandwich arrangement of flat membrane sheets, called a *leaf*, wound around a perforated central tube. One leaf consists of two membrane sheets placed back to back and separated by a fabric spacer called a permeate carrier. The layers of the leaf are glued along three edges, while the unglued edge is sealed around the perforated central tube. A single spiral wound module 8 in. (203 mm) in diameter may contain up to approximately 20 leaves, each separated by a layer of plastic mesh called a spacer that serves as the feed water channel. Feed water enters the spacer channels at the end of the spiral wound element in a path parallel to the central tube. As the feed water flows across the membrane surface through the spacers, a portion permeates through either of the two surrounding membrane layers and into the permeate carrier, leaving behind any dissolved and particulate contaminants that are rejected by the semipermeable membrane. The feed channel spacer is designed to induce turbulence and reduce concentration polarization. The filtered water in the permeate carrier travels spirally inward around the

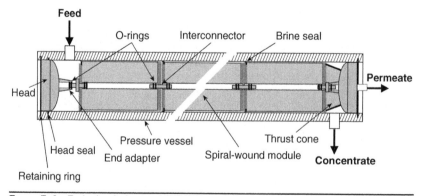

FIGURE 5-4 Typical spiral-wound (NF/RO) module pressure vessel.

element toward the perforated central collector tube. The water in the feed spacer that does not permeate through the membrane layer continues to flow across the membrane surface, becoming increasingly concentrated in rejected contaminants. This concentrate stream exits the element parallel to the central tube through the opposite end from which the feed water entered.

Recovery is a function of the feed concentrate path length. To operate at acceptable recoveries, spiral wound systems usually connect three to six elements in a pressure tube. The number can be as high as six or seven, and systems can be manufactured to contain more. In this configuration, the concentrate stream of the first element is the feed for the second, and so on. The concentrate stream from the last element exits the tube as waste and is the reject or concentrate. The permeate stream from each element is collected in a tube and exits as the permeate. A single pressure tube with four to six elements connected in series can be operated up to 50% recovery.

A diagram of a spiral wound element is shown in Fig. 5-5.

Virtually, all NF and RO membrane processes for supplied wastewater treatment in the United States utilize systems designed for spiral wound membrane modules.

A group of pressure vessels operating in parallel collectively represent a single stage of treatment in an NF or RO spiral wound system. The total system recovery is increased by incorporating multiple stages of treatment in series, such that the combined concentrate (or reject) from the first stage becomes the feed for the second stage. In some cases in which higher recovery is an objective, a third stage may also be used. This configuration is sometimes referred to as *concentrate staging*. Because some fraction of the feed to the first stage has been collected as filtrate (or permeate), the feed flow to the second stage will be reduced by that fraction. As a result, the number of total pressure vessels (and hence the number of modules) in the second

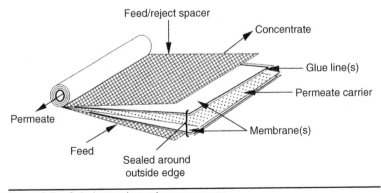

FIGURE 5-5 Spiral-wound membrane process.

stage is also typically reduced by approximately that same fraction. Similar flow, module, and pressure vessel reductions are propagated through all successive stages.

Although the potential system recovery is a function of the feed water quality, as a rough approximation a two-stage design may allow recoveries up to 75%, while the addition of a third stage can potentially achieve recoveries up to 90%. Although concentrate staging is most often used in drinking water applications, another arrangement called *permeate staging* may also be employed. In this configuration, the filtrate (or permeate) from a stage (rather than the concentrate) becomes the feed water for the subsequent stage. While this arrangement is more commonly employed in ultrapure water applications (typically in industry), it may also be used for drinking water treatment when the source water salinity is very high, such as with seawater desalination. In these cases, the product water must pass through multiple stages to remove a sufficient amount of salinity to make the water potable.

The combination of two or more stages in series is called an array, which is identified by the ratio of pressure vessels in the sequential stages. An array may be defined by the ratio of either the actual number or relative number of pressure vessels in each stage. For example, a 32:16:8 array expressed in the actual number of pressure vessels may be alternatively called a 4:2:1 array in relative terms. Two-stage arrays, such as 2:1 and 3:2 (relative), are most common in drinking water treatment, although the specific array required for a particular application is dictated in part by the feed water quality and targeted overall system recovery. Figure 5-6 illustrates the configuration of a typical 2:1 (relative) array, showing both plan and end perspective views.

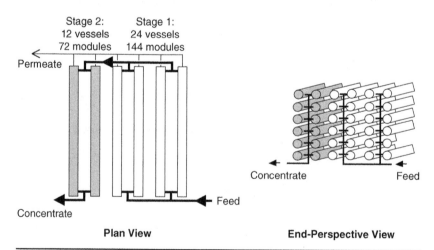

Plan View   **End-Perspective View**

FIGURE 5-6   Typical 2:1 (relative) array of pressure vessels.

As with hollow fiber systems, spiral wound membrane systems are designed and constructed in discrete units that share common valving and can be isolated as a group for testing, cleaning, or repair. For spiral wound systems, these uniform units are typically called *trains*, or alternatively *racks* or *skids*. NF and RO treatment processes consist of one or more trains that are typically sized to accommodate a feed flow of about 5 mgd (13.2 mld) per train. A schematic of a typical NF or RO system is shown in Fig. 5-7.

Unlike hollow fiber systems, spiral wound membrane filtration systems are not manufactured as proprietary equipment. With the exception of the membrane modules, spiral wound systems are generally custom-designed by an engineer or an original equipment manufacturer to suit a particular application. Although the membrane modules are proprietary, standard-sized spiral wound NF and RO modules share the same basic construction, and thus membranes from one manufacturer are typically interchangeable with those from others.

Pretreatment

Pretreatment is applied to the feed water prior to its entering the NF or RO membrane system. Pretreatment is used to prevent physical, chemical, or microbial damage to the membranes. Different types of pretreatment can be used in conjunction with any given membrane filtration system, as determined by site-specific conditions and treatment objectives. For diffusive membranes, pretreatment includes prefiltration: removal of particulate and suspended solids to an SDI less than 4, minimally, and optimally less than 2 or 3. MF and UF pretreatment provides feed water at all times of less than 0.1 ntu and SDI below 3. If the secondary effluent is high in organics, coagulants or powdered activated carbon can be used in conjunction with the MF or UF to reduce the organic loading in the NF or RO feed water. Destruction of any oxidants that may affect the fragile NF or RO membrane—usually by adding sodium bisulfate—occurs in pretreatment. A granular activated carbon bed can be used to remove both organics and chlorine. Chemical conditioning, using an antiscalant to prevent scaling or precipitation of inorganics such as barium, strontium, or silica on the membrane, is added as chemical pretreatment.

Table 5-2 compares secondary effluent to quaternary effluent. Figure 5-8 illustrates the pretreatment step in the membrane process.

Prefiltration

NF and RO utilize nonporous semipermeable membranes that cannot be backwashed and are almost exclusively designed in a spiral

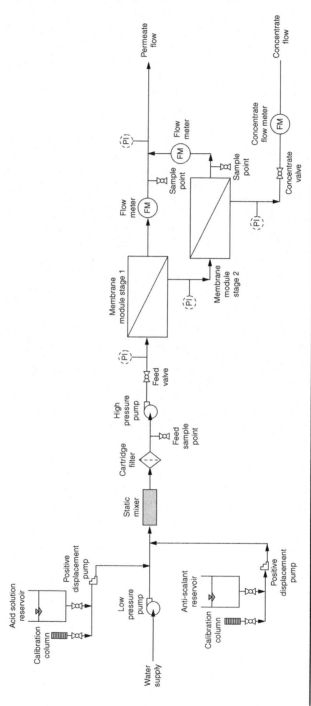

Figure 5-7 Typical NE/RO process diagram.

160

Parameter Projected Average mg/L (except where noted)	Secondary Effluent	MF/UF Filtrate/Permeate
BOD5	20	<5
COD	37 (max 42)	< 10
TOC	12	< 4
TSS	< 10	< 0.1
TKN	4	< 4
NH3-N	0.1 (max 1.0)	< 0.1
P (total)	< 3 (max 10)	< 0.10
Ortho-P	< 2	< 0.10
Turbidity	< 10	< 0.1

TABLE 5-2   Comparison of Secondary US Quaternary Effluent Quality

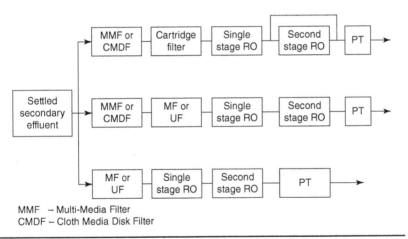

MMF – Multi-Media Filter
CMDF – Cloth Media Disk Filter

FIGURE 5-8   Conventional and membrane pretreatment for RP processes.

wound configuration for wastewater recycling and reuse; each requires more prefiltration. These systems require much finer prefiltration to minimize exposure of the membranes to particulate matter of any size. Spiral wound modules are highly susceptible to particulate fouling, which can reduce system productivity, create operational problems, reduce membrane life, and in some cases damage or destroy the membranes. If the feed water has a turbidity less than approximately 1 ntu—today less than 0.1 ntu—or an SDI less than approximately 5—today 3 or less—cartridge filters with ratings ranging from about 5 to 20 μm are commonly used in the NF and RO prefiltration scheme as insurance.

Prefiltration of secondary effluent for NF and RO treatment consists of a 100 to 500 μm backwashable strainer after the final clarifier or CFS and prior to the MF. If UF is used, dissolved air flotation (DAF) or additional pretreatment to the UF may be required. If CFS alone is used, it would be followed by 5 μm cartridge filtration prior to the NF or RO. Prefiltration can be applied either to the membrane filtration system as a whole or to each membrane unit separately. The particular pore size associated with the prefiltration process (where applicable) varies depending on the type of membrane filtration system and the feed water quality.

Using membrane prefiltration for another is becoming common. This type of treatment scheme is commonly known as an integrated membrane system. Typically, this involves the use of MF or UF as pretreatment for NF and RO in applications that require the removal of particulate matter and microorganisms as well as some dissolved contaminants, such as iron, manganese, barium, strontium, and/or DOC. One of the most significant advantages of an integrated membrane system treatment scheme is that the MF or UF filtrate is of consistently high quality with respect to particulate matter and often allows the NF or RO system to maintain stable operation by reducing the rate of membrane fouling. MF and UF systems can constantly provide water with SDIs less than 2. Of course, this assumes no fiber breakage.

Chemical Conditioning

Chemical conditioning may be used for a number of pretreatment purposes, including pH adjustment, disinfection, biofouling control, scale inhibition, reduction of total and dissolved organic carbon, and oxidation of dissolved species. Some type of chemical conditioning is almost always used with NF and RO systems, most often the addition of an acid (to reduce the pH) or a proprietary scale inhibitor—antiscalant—recommended by the membrane manufacturer to prevent the precipitation of sparingly soluble salts such as calcium carbonate ($CaCO_3$), barium sulfate ($BaSO_4$), strontium sulfate ($SrSO_4$), and silica species (e.g., SiO_2).

Software programs that simulate NF and RO scaling potential based on feed water quality are available from the various membrane manufacturers. In some cases, such as for those NF and RO membranes manufactured from cellulose acetate, the feed water pH must also be maintained within an acceptable operating range to minimize the hydrolysis (chemical deterioration) of the membrane. Chlorine or other disinfectants are added before MF or UF pretreatment to disinfect or to control biofouling. Because most NF and RO membrane materials are readily damaged by oxidants, it is important that any disinfectants be added upstream and neutralized with a reducing agent prior to contact with such membranes.

Coagulation and lime softening are pretreatment for low pressure membrane treatment. For secondary effluent treatment, NF softening may be preferable to lime soda softening. An MF or UF system may be able to operate more efficiently treating effluent softened with lime soda—which would be an additional step after final clarification and would allow the MF or UF membrane to operate at a higher flux, decreasing the capital cost of the membrane pretreatment.

Certain types of NF and RO membranes may require operation within a certain pH range. Coagulants and lime are incompatible with many NF and RO membranes but are typically compatible with most types of MF and UF membranes. Polymers are incompatible with NF and RO membranes and generally with MF and UF membranes as well, although this depends to some degree on the charge of the polymer relative to the charge associated with the membrane. A polymer with a charge opposite to that of the membrane is likely to cause rapid and potentially irreversible fouling.

The primary goal of chemical conditioning after treatment is the stabilization of the NF or RO filtrate with respect to pH, buffering capacity, and dissolved gases. Most chemical conditioning is associated with NF and RO systems, because the removal of dissolved constituents that is achieved by these processes has a more significant impact on water chemistry than the filtering of suspended solids alone. For example, because NF and RO pretreatment often includes acid addition to lower the pH and, consequently, increase the solubility of potential inorganic foulants, a portion of the carbonate and bicarbonate alkalinity in the water is converted to aqueous carbon dioxide, which is not rejected by the membranes. The resulting filtrate can thus be corrosive, given the combination of a low pH, elevated carbon dioxide levels, and minimal buffering capacity of the filtrate. Other dissolved gases, such as hydrogen sulfide, will also readily pass through the semipermeable membranes, further augmenting the corrosivity of the filtrate and potentially causing turbidity and problems with taste and odor. These issues are discussed more in Chapter 7.

Most NF and RO reuse or recycling applications involve wastewater reuse, brackish wastewater reuse, and seawater desalination. Table 5-3 compares constituent removal ranges expected by NF and RO. Note that rejection rates depend upon membrane properties and pressure.

Constituent	Rejection Rate for NF	Rejection Rate for RO
Total dissolved solids	40% to 60%	90% to 98%
Total organic carbon	90% to 98%	90% to 98%
Color	90% to 96%	90% to 96%
Hardness	80% to 85%	90% to 98%
Sodium chloride	10% to 50%	90% to 99%
Sodium sulfate	80% to 95%	90% to 99%
Calcium chloride	10% to 50%	90% to 99%
Magnesium sulfate	80% to 95%	95% to 99%
Nitrate	10% to 30%	84% to 96%
Fluoride	10% to 50%	90% to 98%
Arsenic (+5)	< 46%	85% to 95%
Bacteria	3 to 6 log	4% to 7%
Protozoa	76 log	>7%
Viruses	3 to 5 log	4 to7%

TABLE 5-3   Range of NF Rejection Rates for Typical Constituents

CHAPTER 6

Membrane-Coupled Bioprocesses

Introduction

Membrane-coupled bioprocesses (MCBs) marry membranes (usually microfiltration [MF] or ultrafiltration [UF]) with an aerobic or anaerobic bioprocess, e.g., dispersed growth aerobic process (conventional sludge), high rate anaerobic digestion process—upflow anaerobic sludge blanket (UASB) process, Mobilized Film Technology (MFT) process, expanded granular sludge bed (EGSB) process, or—just recently—direct treatment of primary effluent. Membrane bioreactors are a subset of MCBs.

The membrane function is to separate the aerobic or anaerobic biomass, mixed liquor suspended solids (MLSS), or the suspended solids in primary effluent, from the treated wastewater. The membrane is coupled to the bioprocess in a number of process schemes.

Conventional Activated Sludge–Low Pressure Membrane Process

In the conventional activated sludge process, settled secondary effluent or tertiary effluent is further treated by either a pressurized or a vacuum MF or UF membrane. The secondary treatment can be any activated sludge process—e.g., completely mixed, step aeration, extended aeration, oxygen activated sludge process, or any fixed film secondary treatment process—with or without nitrogen removal or phosphorus removal (see Fig. 6-1).

Phosphorus removal can be biological or chemical. The MF or UF membrane process will remove precipitated phosphorus. Various pre- and posttreatment technologies are often coupled to MF or UF. These MCB systems are effective for almost any flow rate, from small packaged plants to very large wastewater treatment plants treating millions of gallons of municipal wastewater each day. Equalization may or may not be used in the process and may or may not be aerated.

165

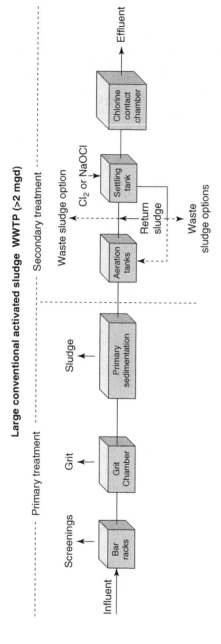

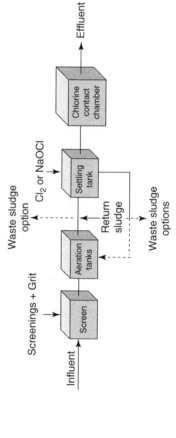

Large conventional activated sludge WWTP (>2 mgd)

Small conventional activated sludge WWTP (>2 mgd)

Figure 6-1 Typical conventional activated sludge.

Grit chambers, aerated grit chambers, or centrifugal grit removal may be used. Fine screens may be used, depending upon the existing design or engineer's preference for grit removal.

The first membrane-coupled bioprocess of this configuration was posttreatment of secondary or tertiary treatment wastewater using MF or UF hollow fiber technology in either a submerged or a pressure mode.

The conventional activated sludge–low pressure membrane (CAS–LPM) process has the following characteristics:

- All secondary treatment processes are included—e.g., lagoons, fixed film bioprocesses, and all activated sludge processes.

- Concentration of mixed liquor suspended solids (MLSS) is that of conventional secondary process and varies from 1,500 to 5,000 mg/L (oxygen activated sludge), with 2,000 to 4,000 mg/L typical.

- Conventional pretreatment, i.e., grit removal and primary clarification, or screens (< 2 mm) can be used.

- The sludge recycle rate Q_R is in the range of conventional systems.

No additional tanks above those required for the nutrient removal biological processes are usually required, if pressurized membranes are used.

An emerging application of coupling secondary or tertiary municipal wastewater treatment with MF membranes is to provide an alternative design consideration for municipal wastewater plants faced with hydraulic expansion requirements. No UF membranes are currently used, but they are an option. Solids carried over from shortened detention time in the final clarifiers, caused by increased flow, are absorbed by the MF membranes. The MF permeate is a consistent 0.1 µm quality regardless of the solids loading to the membrane; it can then be discharged to the surface water, blended with the secondary effluent for discharge to a stream to meet National Pollutant Discharge Elimination System (NPDES) requirements, reused for Title 22 requirements, or combined with nanofiltration (NF) or reverse osmosis (RO) membranes for industrial process water requirements such boiler feed water. Figure 6-2 illustrates this alternative. The low pressure membrane is followed by RO. Additional treatment, e.g., ion exchange, or electrodeionization (EDI) is used to meet industrial water standards for silica and metals. Design considerations for the viability of this option include the manufacturer's membrane flux, associated energy requirements, flux maintenance requirements, and frequency and expected quality of settled secondary effluent versus the economics of expansions of the plant. The commercial viability of this alternative is driven by land and space availability as well as timing, since a membrane

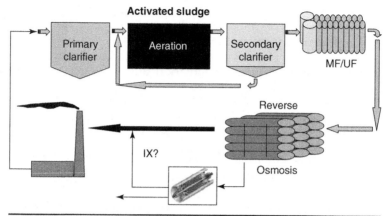

FIGURE 6-2 Conventional activated sludge–low pressure membrane process with RO followed by ion exchange.

solution can in most cases be implemented in a shorter time than wastewater plant expansion.

Nitrogen nutrient removal in this scheme is usually performed biologically in the secondary effluent biological process or combined with nitrate removal in a fixed film multimedia filter.

Phosphorus removal, if required, can also be accomplished either biologically or chemically with the addition of an alum or ferric chloride coagulant and using the low pressure membrane to remove phosphorus to achieve values below 0.1 mg/L.

The MF or UF effluent is disinfected—i.e., free of organisms—if membrane integrity is maintained, and requires little to no further disinfection prior to discharge to surface water. Reuse of the MF effluent (without blending) is an option, as the water quality exceeds California Title 22 requirements.

Sequencing Batch Reactor–Low Pressure Membrane Process (SBR–LPM)—Aqua-Aerobic Systems' AquaMB Process

A logical extension of the CAS-LPM process is Aqua-Aerobic Systems's AquaMB™ Process. This proprietary process couples Aqua-Aerobic Systems' time managed biological processes (such as the AquaSBR sequencing batch reactor), cloth media disk filtration (CMDF), and finally MF or UF membrane filtration. Each sequencing batch reactor (SBR) includes a settle and decant phase to enable the CMDF to receive a flow of secondary treated effluent. The AquaSBR incorporates multiple reactors that alternate between the fill and draw sequence. This sequencing allows one reactor to be filling while the other is decanting to the downstream CMDF and MF or UF.

The SBR–LPM process has the following characteristics:

- Typical MLSS design concentrations are 4,500 to 5,500 mg/L, which is higher than conventional sludge but lower than the high MLSS typical of a membrane bioreactor.

- Sludge recycle is not required in an SBR.

- Biomass is separated using decantation and CMDF prior to hollow fiber MF or UF.

- Membrane fluxes are higher than with a membrane bioreactor (MBR): 50 to 75 gfd (85 to 128 lmh) versus 10 to 15 gfd (17 to 25.5 lmh).

- Footprint is larger than for conventional SBR but smaller than for continuous flow systems; SBRs usually function in a single tank—in this case two tanks—but separate sedimentation is not required.

- Effluent quality is the same as or better than that of an MBR because the MF or UF membrane receives water that meets Title 22 standards.

- Nutrient removal is not a membrane function but a biological function. Total nitrogen less than or equal to 3 to 5 mg/L is typical.

- Aeration requirements are less than for MBR and similar to those for a conventional SBR, since membranes do not require continuous aeration.

- Power consumption is higher than for conventional treatment that does not include tertiary treatment, but lower for MBR.

Figures 6-3 and 6-4 illustrate the AquaMB Process (a registered trademark of Aqua-Aerobic Systems).

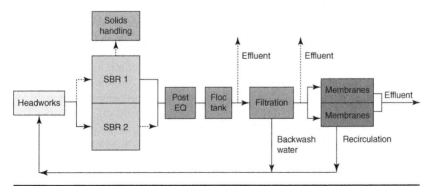

Figure 6-3 The AquaMB Process simplified process flow diagram.

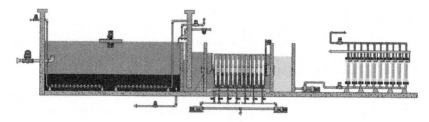

FIGURE 6-4   The AquaMB Process. (*Courtesy Aqua-Aerobic.*)

High Rate Anaerobic Coupled Bioprocesses

MCBPs also include high rate anaerobic bioprocesses such as UASB, MFT, and EGSB (a variant of UASB combined with an LPM).

Anaerobic bioprocesses differ from aerobic processes in two respects: the absence of oxygen and the generation of methane as a by-product. Anaerobic waste treatment is done through anaerobic digesters, anaerobic filters, and innovative proprietary processes described later.

The reader is directed to the anaerobic literature. A good introduction is Metcalf and Eddy/AECOM (2007).

UASB is an anaerobic version of the solids contact clarifier—wastewater flows through an anaerobic sludge blanket. Flocculants are used and an acclimation period is required. EGSB is relatively new technology, a hybrid between fluidized bed and UASB reactors. Detention times for UASB and EGSB are less than 24 h versus the 15- to 30-day retention times in conventional anaerobic digesters. Combining a high rate anaerobic process with membranes allows recovery of biogas and production of reuse water from high strength organic wastewaters. Depending upon the wastewater strength and possible markets for the water and the biogas, the anaerobic process is tied to either conventional activated sludge followed by a low pressure membrane, or a robust membrane process previously described PFD followed by RO and/or advanced oxidation, if required.

The Mobilized Film Technology (MFT) process (by Ecolab) is a proprietary high rate, plug flow, anaerobic treatment system designed to treat wastewaters containing high concentrations of biodegradable soluble organic materials—up to 350,000 mg/L COD—within a very small footprint. High concentrations of active microorganisms—10 to 50 times higher than CAS MLSS—attached to and maintained within a bed of small diameter biological support media, are effectively used to degrade high strength wastewaters. The diameter of the media provides the maximum surface area for microbial growth within a small reactor volume.

MFT maintains a consistently low effluent concentration—low enough to meet most municipalities' discharge limits without the need for additional polishing—and has been successfully coupled to an MF membrane.

Membrane Bioreactor Process

The first MCB was the membrane bioreactor MBR. The MBR process couples conventional activated sludge with MF or UF membranes to separate MLSS, and eliminates the final clarifier. Membrane pore sizes range from 0.04 to 0.45 µm. Membrane platforms are flat sheet, tubular, or hollow fiber. Flat sheet and hollow fiber formats dominate. Membranes may be submerged—over 90%, or pressure. External configurations, i.e., not submersed in the MLSS, pressure driven hollow fiber alternatives are now available. Membrane polymers are PVDF, PE, or PES.

The membrane is either submerged in the aeration tank's MLSS or configured in a separate membrane tank to reduce the fouling activity of the MLSS. The MBR can be used together with CAS in a hybrid configuration.

Both the CAS-LPM configuration and the AquaMB Process use conventional process knowledge to reduce FOG and other fouling chemicals and surfactants prior to the membranes to prevent plugging, coating, and irreversible fouling of the membrane. MBRs utilize a 2-mm screen. Conventional grit and grease removal processes are used as pretreatment to 2-mm screen.

The following are characteristics of MBRs.

- The first MBRs were designed to operate at very high MLSS concentrations of 12,000 to 20,000 mg/L. Today, 9,000 to 15,000 mg/L (9,000 mg/L average) is typical with a trend toward lower MLSS concentrations versus 2,000 to 5,000 mg/L for CAS and 6,000 to 8,000 mg/L for the AquaMB Process.

- Higher MLSS result in lower hydraulic retention times in the bioreactor, allowing for smaller aeration basins. The current trend to reduce MLSS will lengthen hydraulic detention time and increase footprint accordingly.

- Complete nitrification and denitrification is provided with the use of pre-anoxic zones or potentially post-anoxic zones to affect a Bardenpho process, if stringent nitrogen limitations dictate its use (Wallis-Lage et al., 2008).

- Unique pretreatment requirements include a very fine—less than 1 mm—screen, with 2 mm screen, typical for solids and grit removal, and FOG reduction to prevent plugging and fouling of the membrane. Large MBR systems use conventional pretreatment.

- Hollow fiber membrane manufacturers require a 0.5- to 2-mm screen opening, whereas plate membrane manufacturers require only 2- to 3-mm openings. This range in sizes may seem small, but the impacts on capital and operation and maintenance costs can be significant (Hunter and Cummings, 2008).

- Solids recycle rates are very high, up to 5 times the flow.

- Aeration requirements and power cost are the highest of all MCBPs. Pearce (2011) reports 0.3 kWh/m^3 and higher.

- Experience with immersing membranes in the aeration tank proved to be not as effective as a separate tank.

- Two main process configurations are available. Submerged MBRs include membrane modules that are integral to the biological reactor and contain membranes that are either immersed directly in the bioreactor or in a separate tank. External or sidestream MBRs are configured so that the membrane modules are a separate unit process, external to the reactor and necessitating an intermediate pumping step.

- MBRs provide the same water quality as other MCBPs.

- Final clarification step is substituted with the membrane filtration process, thus permitting complete disinfection of treated effluent and allowing for the large clarifier tanks, characteristic of CAS treatment, to be replaced with compact membrane modules (see Fig. 6-5).

- Smallest footprint—3 times the treatment capacity of a conventional plant with the same area.

MBR effluent quality is shown in Table 6-1.

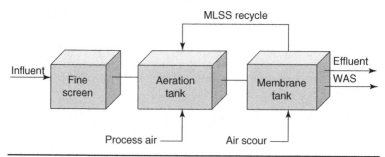

Figure 6-5 Conventional MBR process flow diagram.

	Extended Aeration/ Filtration	Membrane Bioreactor
BOD mg/l	10	<2
TSS mg/l	10	<2
TN, mg/l	10	10
Turbidity, NTU	2	<0.2
Coliform, mpn	2.2	<2.2
Pathogens/Viruses	Little reduction	2–3 log
Metals/Organics	Similar	Similar

Source: A comparison of effluent water quality highlights the benefit and capability of the membrane in removing particulate material. It is important to note that with respect to nutrients, it is the bioreactor design/configuration that dictates effluent water quality and not the membrane. One advantage not captured in this table is the improved efficiency of disinfection that is realized with mbr effluent due to reduced particles and particle shielding.

TABLE 6-1 Comparison of Effluent Quality

Without adequate pretreatment to reduce fats, oils, and greases (FOG) and to remove grit, trash, hair, and other fibrous materials, a number of adverse effects can occur, such as clogging and fouling of the MBR membrane modules. Poor pretreatment can also increase the risk of sludge accumulation and eventual damage to the membrane, leading to a reduction of the hydraulic capacity of the plant and degradation of the effluent quality (Côté et al., 2006).

The treatment processes and design configurations fundamental to MBR systems allow for higher levels of nutrient reduction and enable MBR plants to produce consistent high-grade effluent in a space-efficient footprint. Indeed, depending on the solids retention time selected, an MBR offers the potential to remove more nanopollutants than conventional high rate activated sludge basins and may allow municipalities a stronger position to meet future effluent standards relating to endocrine disrupting compounds (Wallis-Lage et al., 2008).

Beyond these important advantages, more benefits of MBR technology include its operational simplicity, modular expandability, higher aesthetic value, and capacity for producing effluent that either is suitable for a range of reuse applications or can feed directly into an RO treatment process. Taken together, these advantages have been key drivers for the rapid expansion of MBR technology across the United States and overseas. In addition to the design flexibility and performance advantages of MBRs, the rapid growth in the use of this technology at higher flows has also resulted from recent reductions in membrane costs and steady increases in commodity costs such as steel and concrete (Zhao and Hoang, 2011). Further catalysts for MBR technology's significant growth over the past decade include more stringent water quality regulations and an increase in water reuse applications (Hirani et al., 2010).

While traditionally MBR technology was limited to smaller scale or satellite treatment options of 5 mgd or less, MBR plants are now being built for considerably higher capacities, in some cases greater than 40 mgd. The majority of municipal MBR installations are still designed to treat less than 5 mgd of flow, but a steady growth in large MBR installations has occurred in recent years, with a 300% increase in installed cumulative capacity occurring from 2004 to 2008 (Hirani et al., 2010).

Global growth rates of MBR installations between 9.5% and 12% are routinely quoted in reports produced by market analysis (Judd, 2011), and the global market for MBRs is forecast to reach US$888 million by the year 2017, driven by stringent effluent discharge norms, rising water scarcity, and enhanced emphasis on water reuse and recycling for freshwater conservation (Global Industry Analysts, 2011).

Despite its many benefits and the surge in its use, MBR technology also has its disadvantages. These include higher capital costs and greater energy requirements, leading to higher total life costs. In addition, the perceived risk related to fouling and the replacement costs of the membrane remain important limiting factors to its broad application (MBR-Network).

However, research and development efforts in the industry are focused on overcoming these limitations by improving the scouring efficiency of MBR technology and creating new MBR systems that require less energy and offer greater performance in even smaller footprints.

Municipal Wastewater Primary Effluent Coupled Low Pressure Membrane

The logical progression from the CAS-LPM process is the IMANS System. This emerging MCBP couples conventionally settled primary effluent with MF membrane technology. The process was developed by Carollo engineers in California (USA) and is shown in Fig. 6-6.

The process uses MF followed by RO or NF technology to treat primary effluent in combination with high rate anaerobic treatment. The membrane technologies concentrate biochemical oxygen demand while producing high quality effluent suitable for reuse. The MF membrane treats primary effluent to 0.1 ntu feed to the RO. The MF reject is mixed with primary sludge and fed to the high rate anaerobic process to generate biogas.

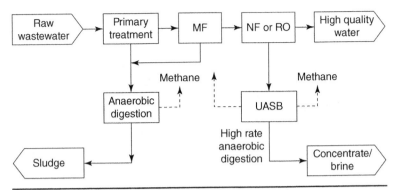

Advantages of the process over CAS and other MCBPs are lower power requirements (as CAS is eliminated), greater biogas generation, and lower biosolids disposal.

This technology is in its infancy and offers a way forward for the future.

CHAPTER 7

Design of Membrane Systems for Water Recycling and Reuse

Introduction

The American Water Works Association Subcommittee (2008) offers an excellent summary of design concepts for microfiltration (MF) and ultrafiltration (UF) facilities which translate well to membrane reuse facilities. Water reuse case studies are included at the end of the chapter. Selected sections are written as a membrane specification.

Membrane Application Flow Schemes

Figures 7-1 to 7-7 provide a number of process flow schemes to meet a variety of finished water treatment goals. Membranes are not usually selected to meet National Pollution Discharge Elimination System (NPDES) discharge requirements—the water quality is too good. Membranes are employed, whether MF or UF alone or together with nanofiltration (NF) or reverse osmosis (RO), to meet specific goals.

System Design Considerations

In the process of planning and implementing a membrane filtration system, there are several important issues that can have a particularly significant impact on system design and operation, thus warranting special consideration. These issues include membrane flux, temperature compensation, water quality, design recovery, integrity testing and broken fiber determination, pretreatment considerations, clean in place, chemically enhanced backwash and neutralization considerations, and system reliability. Each of these subjects is discussed briefly in the following subsections.

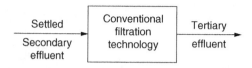

FIGURE 7-1   Tertiary treatment—conventional treatment.

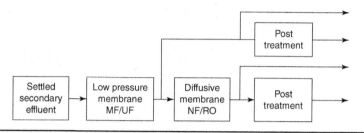

FIGURE 7-2   Quaternary membrane treatment—with and without post treatment.

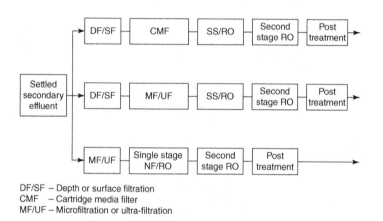

DF/SF – Depth or surface filtration
CMF   – Cartridge media filter
MF/UF – Microfiltration or ultra-filtration

FIGURE 7-3   Typical diffusive membrane process flow diagrams.

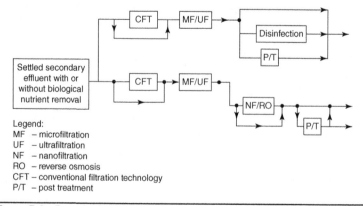

Legend:
MF   – microfiltration
UF   – ultrafiltration
NF   – nanofiltration
RO   – reverse osmosis
CFT – conventional filtration technology
P/T – post treatment

FIGURE 7-4   Process flow schemes.

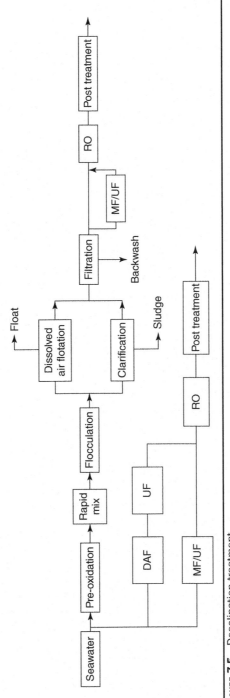

Figure 7-5 Desalination treatment.

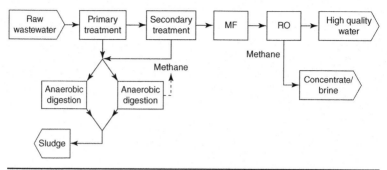

Figure 7-6   Current state-of-the-art for water reclamation. (*Judy et al., 2000.*)

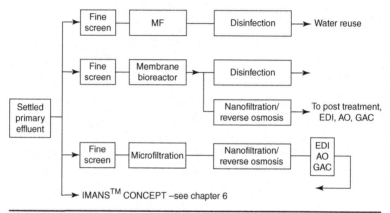

Figure 7-7   Current and future wastewater reuse schemes.

Membrane Flux

Flux—the flow per unit of membrane area, as defined in earlier chapters—is one of the most fundamental considerations in the design of a membrane filtration system, since this parameter dictates the membrane area necessary to achieve the desired system capacity and thus the number of membrane modules required. Because the membrane modules represent a substantial component of the capital cost of a membrane filtration system, considerable attention is given to maximizing the membrane flux without inducing excessive reversible fouling, thereby minimizing the number of modules required.

Typically, the maximum flux associated with a particular membrane filtration system is determined during the pilot testing phase, and specified by the State based on the pilot results, specified by an engineer according to his/her experience, a manufacturer's demonstrated experience on a particular water, or a combination of these inputs. Independent of maximum flux, pilot testing is also commonly used to determine a reasonable operating range that balances flux with backwash and chemical cleaning frequencies. Because higher fluxes accelerate fouling, backwashing and chemical cleaning must usually be conducted more frequently at higher fluxes. The upper bound of the range of acceptable operating fluxes (provided this bound does not exceed the state mandated maximum) is sometimes called the *critical flux*, or the point at which a small increase in flux results in a significant decrease in the run time between chemical cleanings. A membrane filtration system should operate below this critical flux to avoid excessive downtime for cleaning and the consequent wear on the membranes over time from increased chemical exposure.

The flux through a membrane is influenced by a number of factors, including pore size, porosity, membrane type for MF—i.e., hollow fiber, spiral wound, etc.—membrane material, and water quality. Factors such as estimated membrane life, fouling potential, frequency of fiber breaks, frequency and effectiveness of chemical cleaning, chemical use, and energy requirements to maintain a given flux should also be considered. These are covered in earlier chapters.

The instantaneous flux rate is the filtrate flow of a membrane train measured at any time the train is in production divided by the active membrane surface area (feed side). Instantaneous flux is also calculated as the net flux (gfd) divided by the percent online service factor. For example, a system with a net flux of 25 gfd (42.5 lmh) and a service factor of 96.7% has a calculated instantaneous flux of $25/0.967 = 25.85$ gfd (43.9 lmh).

Net flux rate is the daily average net filtrate produced in gallons per day—in this case *not* including finished water used for backwash, chemically enhanced backwash (CEB), etc.—divided by the active membrane surface area (feed side) in square feet. The units of net flux are also gfd. The net flux rate is equal to the instantaneous flux rate multiplied by the percent online service factor. For example, for a system with a 25.85 gfd instantaneous flux rate and a service factor of 94.70%, the calculated net flux is $25.85 \times 97.7\% = 24.5$ gfd (46.7 lmh).

Temperature Compensation

Like other water quality parameters—such as turbidity and, for NF and RO systems, total dissolved solids (TDS)—the temperature of the feed water also affects the flux of a membrane filtration system. It impacts MF and UF design as well as NF and RO design. Secondary effluent is usually consistent in temperature and will fall within a narrow range. Northern climates have the greatest impact. At lower temperatures water becomes increasingly viscous; thus, lower temperatures reduce the flux across the membrane at constant transmembrane pressure (TMP) or alternatively require an increase in pressure to maintain constant flux. The means of compensating for this phenomenon vary with the type of membrane filtration system used. General viscosity-based means of compensating for temperature fluctuations for both MF and UF systems as well as NF and RO

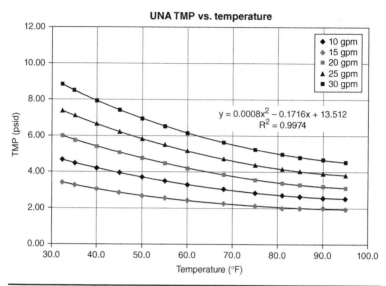

UNA TMP vs. temperature

$$y = 0.0008x^2 - 0.1716x + 13.512$$
$$R^2 = 0.9974$$

- ◆ 10 gpm
- ◆ 15 gpm
- ■ 20 gpm
- ▲ 25 gpm
- ■ 30 gpm

TMP (psid)

Temperature (°F)

Figure 7-8 Typical manufacturer curve of temperature versus TMP.

systems are described later, although membrane manufacturers may have a preferred product-specific approach.

Temperature correction factors are described in earlier chapters. Membrane manufacturers develop their own temperature correction factors specifically for each membrane or class of membranes (see Fig. 7-8). These curves should be made available and consulted during the design process.

MF and UF membrane systems usually operate within a relatively narrow range of TMPs, which may limit increasing the TMP in order to maintain constant flux as the water temperature decreases. This is especially true of vacuum systems. Because the membrane modules can be damaged if the TMP exceeds an upper limit, as specified by the manufacturer, it may not be possible to operate the system at a TMP that is sufficient to meet the required treated water production during colder months if demand remains high. In vacuum systems, the TMP cannot exceed the design value. As a result, additional treatment capacity (i.e., increased membrane area or number of membrane modules) is incorporated into the design of the system so that the water treatment production requirements can be satisfied throughout the year.

Source Water Quality and Variability

Source water quality, characteristics (see Chapter 2), and variability tie in directly with finished water goals. Because water quality can have a significant impact on membrane flux, feed water quality is also a primary design consideration for membrane filtration systems. Poorer water quality will require lower fluxes, which in turn increase the necessary membrane area and required number of modules, augmenting both the cost and the size of the system. However, pretreatment can often improve feed water quality at a lower cost than additional membrane area. Conversely, better water quality will allow higher fluxes, reducing the required membrane area, the size of the system, and the capital cost. Typically, membrane flux is determined through pilot testing. In the absence of pilot test data, it is important to have some understanding of how critical water quality parameters such as SDI, turbidity, organic carbon, and dissolved solids affect the flux. The influence of each of these parameters on flux is briefly described in this section. Temperature also has a significant impact on membrane flux.

Silt Density Index

SDI is an empirical, dimensionless measure of particulate matter in water and is generally useful as a rough gauge of the suitability of source water for efficient treatment using NF or RO processes. The ASTM Standard D 4189-95 details the procedure for determining SDI. In general, SDI measurements are taken by filtering a water sample through a 0.45-μm flat sheet filter with a 47-mm diameter at a pressure of 30 psi. The time required to collect two separate 500-mL volumes of filtrate is measured, and the resulting data become the inputs to a formula used to calculate SDI. Water samples that contain greater quantities of particulate matter require longer to filter and thus have higher SDI values. As a general rule of thumb, spiral wound NF and RO modules are not effective for treating water with a SDI of 5 or greater, as water of this quality contains too much particulate matter for the nonporous, semipermeable membranes, which would foul at an unacceptably high rate. Thus, some form of pretreatment to remove particulate matter is generally required for SDI values exceeding 5 in NF and RO applications. NF and RO membrane module manufacturers can usually provide a rough estimate of the range of anticipated operating fluxes based on the type of source water, which is roughly associated with a corresponding range of SDI values.

However, SDI is only one measure of water quality, and there are a number of site- and system-specific water quality and operational factors that combine to dictate the flux for a given system. Caution should also be exercised when interpreting SDI results, as measurements can vary from test to test and from analyst to analyst, as well as with both temperature and the specific type of membrane used. Consequently, it is important that the results be given for comparable conditions when evaluating SDI data. Note that because SDI determination is a batch process, it is not conducted continuously online and thus is not typically utilized as gauge of water quality or system performance during daily operation in the way turbidity or conductivity monitoring are often employed.

Turbidity

Turbidity is a measure of the scatter of incident light caused by particulate matter in water. Because turbidity is widely used as a performance gauge for conventional media filters, among the various types of membrane filtration systems it is most often used as an assessment tool for MF and UF, since these systems are specifically designed to remove particulate matter. Higher turbidity measurements are indicative of greater quantities of suspended solids and thus the potential to cause more rapid membrane fouling. Therefore, water with higher turbidity is usually filtered at lower fluxes to minimize fouling and the consequent frequency of backwash and chemical cleaning. In some cases when turbidity levels are extremely elevated, it may be more economical to provide pretreatment for an MF or UF system to reduce the solids loading to the membranes.

Organic Carbon

Another water quality constituent that influences membrane flux is the organic carbon content (see Chapter 2), which is typically expressed in terms of either TOC or dissolved organic carbon (DOC). Organic carbon in the feed water can contribute to membrane fouling, either by adsorption of the dissolved fraction onto the membrane material or by obstruction by the particulate fraction. Thus, lower fluxes may be necessary if membrane filtration is applied to treat water with significant organic carbon content. The tendency for a membrane to be affected by TOC is partially influenced by the nature of the organic matter in the water. TOC can be characterized as either hydrophilic or hydrophobic in composition, and studies suggest that the hydrophobic fraction contributes more significantly to membrane fouling. The character of the organic carbon content can be roughly quantified by measuring the specific ultraviolet (light) absorbance (SUVA) of the water.

Because TOC is more commonly measured than DOC, SUVA is sometimes estimating using values for TOC in place of those for DOC. Higher SUVA values tend to indicate a greater fraction of hydrophobic organic material, thus suggesting a greater potential for membrane fouling. Generally, SUVA values exceeding 4 L/mg-m are considered somewhat more difficult to treat. However, organic carbon (as well as turbidity) can often be removed effectively via coagulation and presettling, particularly if it is more hydrophobic in character, thus minimizing the potential for membrane fouling and facilitating operation at higher fluxes. Coagulation can also be conducted in-line (i.e., without presettling) with MF and UF systems. Pretreatment using the injection of powdered activated carbon may also reduce DOC in the membrane feed; however, because spiral wound membrane modules cannot be backwashed, powdered activated carbon should not be used in conjunction with NF or RO systems unless provisions are made to remove the particles upstream.

Dissolved Solids

TDS and the particular species of dissolved solids present in the membrane feed are both critical considerations for NF and RO systems (see Chapter 2)—species such as silica, calcium, barium, and strontium, which can precipitate as sparingly soluble salts, can cause scaling and a consequent rapid decline in flux under certain conditions. Scaling is typically controlled using pretreatment chemicals such as an acid to lower the pH and/or a proprietary scale inhibitor. However, the total quantity of dissolved solids of all species also influences system operation, as the net driving pressure required to achieve a target flux is related to the osmotic pressure of the system, which is directly proportional to the TDS. Thus, as the TDS increases, so does the required feed pressure. TDS in secondary effluent is rarely a consideration and is usually less than 500 mg/L.

TDS is generally not a significant consideration for MF, UF, and Membrane Cartridge Filtration (MCF) systems, since these processes do not remove dissolved solids. In some cases, however, the use of upstream oxidants may cause the precipitation of iron or manganese salts (either unintentionally or by design as a pretreatment process), which could accelerate membrane fouling.

Bench-Scale Testing

The raw water analysis in Table 7-1, combined with source water history and a bench scale jar test membrane evaluation, can provide enough preliminary design information to scope the project and select the membrane and nonmembrane options. A description of how to perform jar testing is provided in Jar Testing Procedures. Please note that these are only a starting point for the development of the project testing procedures. Bench scale testing is not intended to replace piloting but provides detailed piloting targets if time and budget allow.

A 10- to 20-mL minute capacity bench scale hollow fiber module used in conjunction with jar testing provides the best design information in the least time for the least cost. The data gathered do not address membrane performance over time; this is accomplished with a full-scale module pilot operating for at least two cleanings. Obviously, the best results are achieved if the pilot can be operated over the entire change of seasons. Even though secondary effluent is one of the most consistent feed water sources for MF and UF membranes, evaluation for at least a month in a pilot situation is warranted.

Type of Membrane	Element Construction	Mem. Area per Element	Typical	Flux Max	Recovery per State
MF/UF	Hollow Fiber	Up to 1,000 sf	30–50 gfd	120 gfd	80–98%
NF/RO	Spiral Wrap	Up to 500 sf	10–15 gfd	25 gfd	40–80%

TABLE 7-1   Design Criteria for Membrane Process

Design Recovery

Membrane system recovery is the total net volume of membrane filtrate (filtrate volume minus the volume of filtrate or plant finished water used for backwash, chemical soak, CEB, cleaning in place [CIP], and all other treated membrane filtrate uses) divided by the total volume of feed water entering the membrane trains from the prefilter system over a 24-h period, expressed as a percentage. It is also equal to 1 minus the waste flow divided/feed water volume then multiplied by 100 to convert to a percentage. For example, a system with a combined 0.5 mgd (1.9 mld) waste volume from the membranes and 12.5 mgd (47.3 mld) feed water to the membrane system has a system recovery of $(12.5 - 0.5)/12.5 \times 100 = 96.0\%$ or $[1 - (0.5/12.5)] \times 100 = 96.0\%$ or $(47.3 - 1.9)/47.3 \times 100 = 96.0\%$.

The prefilter (strainer or screen) system recovery is the total volume of prefiltered feed water delivered to the membrane trains divided by the total volume of prefiltered influent entering the prefilter system over a 24-h period. It is expressed as a percentage.

The overall net membrane filtration system recovery is the net total volume of filtrate (MF/NF) or MF/RO–delivered to the plant's finished water storage tanks (filtrate volume minus the volume of filtrate or plant finished water used for backwash [BW], CIP, CEB, and other uses) divided by the total volume of feed water entering the prefilter system over a 24-h period. It is also expressed as a percentage.

The online service factor is the average percentage of the time per day that filtrate is produced, considering downtime for backwashing, CEB, tank deconcentrations and drains, membrane integrity testing, and any other activity that requires a stoppage of filtrate production. For example, if a train is off-line for backwashing, CEB, and integrity testing for a total of 75 min during an average day, the online service factor for that train is $(1,440 - 75)/1,440 \times 100 = 94.8\%$.

Integrity Testing

Integrity testing of membrane reuse systems will only become more stringent as higher quality water reuse applications occur and as designers seek to lengthen NF and RO life and monitor broken fibers.

Integrity testing (IT) is a critical MF and UF system requirement, designed to protect the public from the introduction of *Cryptosporidium* and *Giardia* into the water supply. The membrane filtration system should include direct and indirect integrity monitoring systems. The following is an adaptation of many public bid specifications for MF integrity testing requirements.

A typical membrane IT specification for direct integrity test could read as: The direct integrity test (DIT) system should include an automated pressure hold test to provide a direct means of monitoring the integrity of the membrane modules and membrane system. It should operate automatically and be controlled by the membrane filtration system programmable logic controller (PLC). It can be of either the diffusive airflow or the pressure decay type, designed in accordance with ASTM 6908-03, *Standard Practice for Integrity Testing of Water Filtration Membrane Systems.*

The IT specification would be worded to include the 3 μm hole requirement system should have the following characteristics:

- Resolution: The direct integrity test must be responsive to an integrity breach on the order of 3 μm or less. Using calculation procedures outlined in the US Environmental Protection Agency's *Membrane Filtration Guidance Manual* (2005; MFGM) (UPSEA, 2005) (see Appendix N) the minimum applied test pressure must be calculated using the following conservative values unless the membrane manufacturer can provide and substantiate supporting information, as part of its proposal, the use of less conservative

  1. Pore shape correction factor = 1.0.

  2. Liquid membrane contact angle = 0.

  3. Surface tension at air liquid interface calculated at the minimum anticipated water temperature.

- Sensitivity: The direct integrity test must be able to verify a *Cryptosporidium* log removal value (LRV) equal to or greater than 5 logs, using calculations defined in the USEPA's MFGM.

- Frequency

  1. For the purposes of calculating the impact of off-line time on membrane system sizing, assume that each membrane train shall be direct integrity tested once per 24 h.

  2. The DIT shall have an operator-adjustable frequency of testing, with a range of 4 to 168 h.

- Control limits

  1. Automatically alarm a failure and automatically shut down the specific membrane train or the entire system, depending on the extent and severity of the failure and on the option selected by the owner.

  2. It is the owner's intent to operate the facility based on an LRV basis. Only scheduled quarterly (3 months) maintenance fiber repair events are planned for the facility. These events will be planned to keep individual trains' or cells' LRV above a value of 4.3. Therefore, the owner will consider an alert when an individual membrane unit's LRV falls below 4.3.

The membrane manufacturer is usually responsible for the detailed calculations and equations to calculate the membrane train LRV and the critical number of broken fibers from the DIT pressure decay test results that cause the LRV to fall below the State's requirement. The pressure decay test results shall be used to calculate the membrane train LRV as follows:

a. Equations and assumptions shall be selected by the MFSS in conformance with the *Membrane Filtration Guidance Manual* (USEPA, 2005).

b. Temperature, filtrate flow rate, and TMP shall be averaged over a filtration cycle for each filtration cycle in the previous membrane integrity test (MIT) cycle (between the last and current MIT).

c. The LRV shall be calculated for each filtration cycle. The MFSS's LRV and the MIT cycle.

Continuous Indirect Integrity Monitoring System

1. The membrane system should incorporate a continuous indirect integrity monitoring system.

2. Each membrane train shall be supplied with a laser turbidimeter for automatic monitoring of filtered water quality. The accepted laser turbidimeters are usually specified by the engineer. The membrane system PLC shall record turbidimeter, readings from each membrane train once per minute and shall average the data automatically to generate an average reading for each train every 15 min.

3. Control limits for laser turbidimeters

   a. An upper control limit (UCL) of 0.15 ntu is usually specifies. For any membrane train with two consecutive 15-min average readings over the UCL, automatically shut down the respective train, generate a system alarm, and initiate a direct integrity test on that train.

   b. The control system shall be compatible with monitoring at least two operator-defined intermediate control limit (ICL) values, such that two consecutive 15-min average turbidity readings above the ICL will generate an automatic alarm failure and optional shutdown of the specific membrane train, as selected by the operator.

4. Control limits for particle counters

   a. A UCL shall be incorporated into the control system as recommended by the membrane manufacturer. For any two consecutive 15-min average readings over the UCL, automatically generate a system alarm and initiate a direct integrity test on specific trains, shutdown of specific trains, or shutdown of the entire system, as selected by operator.

Determination of Minimum Number of Equivalent Broken Fibers

- The membrane supplier shall determine the minimum or critical number of equivalent broken fibers required to drop the LRV of an integral membrane system from the highest verifiable value (test sensitivity) to values of 4.3 and 4.0 log based on the following parameters:
  - Equivalent broken fibers shall have the following characteristics:
    (1) A fiber is fully cut through its diameter at the point along the fiber's length that results in the largest response in membrane unit LRV (typically at the interface between fiber and potting material).
    (2) The fully cut fiber described in subsection 1 is located in the membrane module that results in the largest response in the membrane unit LRV.
    (3) For the purpose of these specifications, fiber location within a module does not affect the LRV response, except as already specified.
  - The minimum number of equivalent broken fibers shall be determined based on the operating conditions that result in the largest response in membrane unit LRV per equivalent broken fiber:
    (1) Temperature
    (2) TMP (typically the highest operating TMP)
    (3) Operating water level (or back pressure)
    (4) Instantaneous flux (typically the lowest operating fluid flux)
    (5) And other design and operating variables, as determined by the MFSS, that result in the largest changes in LRV per equivalent broken fiber.

Typical RO element warranties list a maximum of 1.0 ntu for the feed water. MF and UF pretreatment, when integral, will meet this requirement.

Coagulant selection considerations are detailed in Chapter 2. The only foolproof way to select the best coagulant for an application is to use the information provided by a complete water analysis with jar testing using a bench scale membrane system.

The quality of the treated water is the same 24 h per day at less than 0.1 ntu, most times in the 0.02 to 0.05 ntu range. This is irrespective of the influent turbidity. If influent turbidity spikes, the permeate remains constant. Recovery—the percentage of treated water with respect to feed water—is especially high for hollow fiber MF on secondary effluent, ranging from 80% to 95% depending upon the membrane. If a second membrane system is used, recoveries as high as 98% can be achieved.

Pretreatment

Pretreatment is typically applied to the feed water prior to its entering the membrane system. Pretreatment is used to prevent physical damage to the membranes. Different types of pretreatment can be used in conjunction with any given membrane filtration system, as determined by site-specific conditions and treatment objectives. Pretreatment includes prefiltration, oxidation, and chemical conditioning.

Prefiltration

Prefiltration, including screening or coarse filtration, is a common means of pretreatment for membrane filtration systems that is designed to remove large particles and debris. Prefiltration can be applied either to the membrane filtration system as a whole or to each membrane unit separately. The particular pore size associated with the prefiltration process (where applicable) varies depending on the type of membrane filtration system and the feed water quality. For example, although hollow fiber MF and UF systems are designed specifically to remove suspended solids, large particulate matter can damage or plug the membrane fibers. For these types of systems, the pore size or micron rating of the selected prefiltration process may range from as small as 100 μm to as large as 3,000 μm or more, for sedimentation or dissolved air flotation, depending on the influent water quality and manufacturer specifications. Generally, hollow fiber MF and UF systems that are operated in an inside-out mode are more susceptible to fiber plugging and thus may require finer prefiltration.

Because NF and RO utilize nonporous semipermeable membranes—which cannot be backwashed, are almost exclusively designed in a spiral wound configuration for wastewater recycling and reuse, and require more prefiltration—these systems require much finer prefiltration to minimize exposure of the membranes to particulate matter of any size. Spiral wound modules are highly susceptible to particulate fouling, which can reduce system productivity, create operational problems, reduce membrane life, and in some cases damage or destroy the membranes. If the feed water has a turbidity less than approximately 1 ntu or a Silt Density Index (SDI) less than approximately 5 (today 3 or less), cartridge filters with ratings ranging from about 5 to 20 μm are commonly used for NF and RO prefiltration. However, if the feed water turbidity or SDI exceeds these values, a more rigorous method of particulate removal, such as conventional treatment (including media filtration) or MF or UF membranes, is recommended as pretreatment for NF and RO.

Using one type of membrane filtration prefiltration for another is more common, e.g., MF or UF as pretreatment for NF or RO. This type of treatment scheme is commonly known as an integrated membrane system. Typically, this involves the use of MF or UF as pretreatment for NF or RO in applications that require the removal of particulate matter and microorganisms as well as some dissolved contaminants, such as hardness, iron and manganese, or disinfection by-product (DBP) precursors. One of the most significant advantages of an integrated membrane system treatment scheme is that the MF or UF filtrate is of consistently high quality with respect to particulate matter and often allows the NF or RO system to maintain stable operation by reducing the rate of membrane fouling. MF and UF systems can constantly provide water with SDIs less than 2. If course, this assures no fiber breakage.

Clean in Place, Chemically Enhanced Backwash, and Neutralization Considerations

Separate tanks are provided for acid cleaning solution, caustic cleaning solution, and neutralization. The membrane supplier should provide controls for the chemical system through the membrane filtration system PLC(s). Membrane filtration system PLC(s) shall record the time between successive CEBs and between successive CIPs, and shall send an alarm when the next chemical cleaning is needed based on the operator entered time interval or an increase in TMP. Chemicals for cleaning and neutralization are provided by the consulting engineer in the specification.

Filtrate or plant finished water shall be provided to fill the CEB or CIP tanks. A design consideration is to ensure that the water quality of the filtrate meets the water quality requirements of the CEB and CIP makeup solutions.

Redundancy shall be provided so that a CEB and a CIP can be performed simultaneously on different membrane trains.

The CIP system shall be designed to remove all solids and dissolved compounds that have accumulated in the membrane modules (including on the membrane surface and within the membrane pores) so that the clean water permeability of the membrane is consistent.

The cleaning system shall be designed for automatic operation once a cleaning cycle has been initiated by the operator. The system shall include provisions for a heated CIP solution per manufacturer guidelines.

Equipment shall be provided to inject and mix chemicals to neutralize and/or dechlorinate spent CIP and CED solutions prior to discharge from the neutralization tank. Automatic pH and chlorine residual monitoring shall be provided so the plant operators can determine if discharges from the neutralization tank to the owner's residuals basins comply with NPDES permit discharge requirements. Instrumentation shall be provided to verify when a unit has been flushed sufficiently to go back into production. Controls should automatically shut down the membrane train if pH or chlorine residual levels are outside the set points.

CEB and CIP pumps shall be designed and supplied by MFSS for their intended use, with materials of construction compatible with the liquid being pumped. Redundancy for recirculating pumps of any kind should be provided.

Chemical Bulk Storage Tanks

Chemical bulk storage tank shall be provided for all chemicals required for the membrane filtration system. Bulk storage tanks shall provide sufficient storage volume for sufficient reserve to receive a full tanker truck delivery—usually about 4,000 gal plus 20%. System size determines the number of tanks.

Bulk storage tanks shall be constructed of materials suitable for the specific chemical to be stored.

Chemical Conditioning

Chemical conditioning may be used for a number of pretreatment purposes, including pH adjustment, disinfection, biofouling control, scale inhibition, coagulation, and oxidation. Some type of chemical conditioning is almost always used with NF and RO systems, most often the addition of an acid (to reduce the pH) or a proprietary scale inhibitor recommended by the membrane manufacturer to prevent the precipitation of sparingly soluble salts such as calcium carbonate ($CaCO_3$), barium sulfate ($BaSO_4$), strontium sulfate ($SrSO_4$), or silica species (e.g., SiO_2).

Software programs that simulate NF and RO scaling potential based on feed water quality are available from the various membrane manufacturers. In some cases, such as for those NF and RO membranes manufactured from cellulose acetate, the feed water pH must also be adjusted to maintain the pH within an acceptable operating range to minimize the hydrolysis (i.e., chemical deterioration) of the membrane. The addition of chlorine or other disinfectants may also be used as pretreatment for primary disinfection or to control biofouling. However, because some NF and RO membrane materials are readily damaged by oxidants, it is important that any disinfectants be added upstream and neutralized with a reducing agent prior to contact with such membranes.

A number of different chemicals may be added as pretreatment for MF or UF, depending on the treatment objectives for the system. For example, lime and soda ash may be added for softening applications; coagulants may be added to enhance removal of total organic carbon (TOC) with the intent of minimizing the formation of DBPs (a future reuse consideration) or increasing particulate removal; disinfectants may be applied for either primary disinfection or biofouling control; and various oxidants may be used to oxidize metals such as iron and manganese for subsequent filtration. It is important to ensure that the applied pretreatment chemicals are compatible with the particular membrane material used. As with conventional media filters, presettling may be used in conjunction with pretreatment processes such as coagulation and lime softening.

While an MF or UF system may be able to operate efficiently with the in-line addition of lime or coagulants, direct coagulation, conventional sedimentation presettling in association with these pretreatment processes can enhance membrane flux and increase system productivity by reducing the solids loading, thus minimizing the frequency of backwashing and chemical cleaning. The decision on whether to incorporate direct coagulation or conventional sedimentation depends upon the membrane's capabilities to handle solids and the economic trade-offs between increased flux maintenance and reduced recovery versus the capital and operating costs of a sedimentation basin.

Direct Coagulation vs. Sedimentation

Coagulation reduces membrane fouling by mainly reducing NOM and possibly small inorganic colloids in feed water. At the same time, coagulation generates a large amount of solids from metal hydroxides. As coagulation dose increases, the amount of solids generated increases. Therefore, at the core, the issue is at what point the impact of increased solids loading would offset the benefits of reduced TOC to the NF or RO membranes.

Sedimentation, if included in the process train, certainly increases capital costs of the project. Unless its benefit can be justified, adding sedimentation would not be accepted. The construction costs of clarifiers are one design input. One approach is to use two or more smaller clarifiers rather than one large clarifier. Costs to acquire the land to construct the clarifier are another consideration.

Given the fact that the capital cost of adding clarifier to the treatment train in a midsize to large plant (above 10 mgd; 37.85 mld) can carry a price tag in the millions of dollars, the cost savings in a small membrane system resulting from increased flux due to sedimentation must be carefully evaluated to justify the addition. It has been suggested that flux improvement of 20% to 25% is necessary to justify the inclusion of sedimentation (Howe and Clark, 2002).

With any form of chemical pretreatment, it is very important to understand whether any chemical under consideration for use is compatible with the membrane material. In addition to irreversible fouling and/or physical damage to the membranes, the use of an incompatible chemical may void the manufacturer's warranty. Some chemicals such as oxidants can be quenched upstream, while others such as coagulants and lime cannot be counteracted prior to membrane exposure. In general, most NF and RO membranes, as well as some MF and UF membranes, are not compatible with disinfectants and other oxidants. High crystalline polyvinylidene fluoride membranes are the most oxidant resistant in the industry, and can handle high concentrations of chlorine, potassium, permanganate, and hydrogen peroxide. However, some MF and UF membranes that are incompatible with stronger oxidants such as chlorine may have a greater tolerance for weaker disinfectants such as chloramines, which may allow for a measure of biofouling control without damaging the membranes. Certain types of both MF and UF membranes and NF and RO membranes require operation within a certain pH range. Coagulants and lime are incompatible with many NF and RO membranes but are typically compatible with most types of MF and UF membranes. Polymers are incompatible with NF and RO membranes, and generally not compatible with MF and UF membranes either, although this depends to some degree on the charge of the polymer relative to the charge associated with the membrane. A polymer with a charge opposite to that of the membrane is likely to cause rapid and potentially irreversible fouling. Chemical compatibility with various types of membrane materials is briefly discussed in Chapter 5; however, it is critical to consult with the membrane manufacturer prior to implementing any form of chemical pretreatment.

Posttreatment

Posttreatment requirements for membrane systems depend upon the ultimate end use of the water. For example, discharge back to surface water from MF or UF treatment may not require any posttreatment. The use of NF or RO for removal of nitrates, TOC, DBPs, or arsenic may or may not require chemical conditioning. Chemical conditioning is most often required when the treated water objective is drinking water.

The primary goal of chemical conditioning is the stabilization of NF or RO filtrate with respect to pH, buffering capacity, and dissolved gases. Most chemical conditioning is associated with NF and RO systems, because the removal of dissolved constituents that is achieved by these processes has a more significant impact on water chemistry than the filtering of suspended solids alone. For example, because NF and RO pretreatment often includes acid addition to lower the pH and, consequently, increase the solubility of potential inorganic foulants, a portion of the carbonate and bicarbonate alkalinity in the water is converted to aqueous carbon dioxide, which is not rejected by the membranes. The resulting filtrate can thus be corrosive, given the combination of low pH, elevated carbon dioxide levels, and minimal buffering capacity of the filtrate. Other dissolved gases, such as hydrogen sulfide, will also readily pass through the semipermeable membranes, further augmenting the corrosivity of the filtrate and potentially causing turbidity and taste and odor problems.

Degasification is commonly achieved via packed tower aeration (i.e., air stripping). Air stripping also increases dissolved oxygen levels, which may be very low in the case of an anaerobic ground water source. The pH of the water may subsequently be readjusted to typical finished water levels (i.e., approximately 6.5 to 8.5) by adding a base such as lime or caustic. Alkalinity (e.g., in the form of sodium bicarbonate) may also be added, if necessary, to increase the buffering capacity of the water. Alternatively, if the pH is raised prior to degasification (thus converting the dissolved carbon dioxide to bicarbonate), much of the alkalinity may be recovered. However, this posttreatment strategy also converts any dissolved hydrogen sulfide gas into dissociated sulfide, which may readily react with other dissolved species to produce sparingly soluble sulfide compounds that may precipitate.

While the use of membrane filtration does not specifically necessitate disinfection posttreatment as a result of process considerations, postdisinfection is generally required by regulation for primary and/or secondary disinfection for the reused recycled water. However, in some states (e.g., Connecticut) the use of membrane filtration may reduce primary disinfection requirements, thus helping to control DBP formation. Because membrane filtration is often the last major process in the treatment scheme, it is common to apply a disinfectant to the filtrate prior to its entry into a clearwell and/or the distribution system. This application is particularly important if disinfectants were either neutralized or not added at all prior to the membrane filtration process to avoid damaging oxidant-intolerant membranes.

For NF and RO membrane processes, if the disinfectant is applied prior to filtrate pH adjustment, postdisinfection may have the additional benefit of oxidizing sulfide to sulfate, thus reducing the potential for both sulfide precipitation and taste and odor concerns. Corrosion inhibitors may also be added prior to distribution, particularly for NF and RO systems that produce more corrosive water.

System Reliability

System reliability is an important consideration in the design of a membrane filtration system. Design capacities are typically maximum values and may not necessarily account for the possibility of one or more units being taken out of service for repair or routine maintenance. Even standard operational unit processes such as backwashing, chemical cleaning, and integrity testing that are normally accounted for in accurately sizing the facility and determining system capacity may be problematic if it becomes necessary to conduct these processes more frequently than was planned. For example, most membrane filtration systems are designed with sufficient storage and equalization capacity to meet average demand even when a unit is taken out of service for chemical cleaning. However, this storage would generally not allow the system to operate at capacity for extended periods (e.g., longer than about 24 h) with a unit of out service for repair. Reliability issues such as this may be particularly pronounced for smaller systems with fewer membrane units, since one out-of-service unit may significantly impact overall system capacity. Note that any state sizing and redundancy requirements should be considered in the design of the facility.

In order to ensure system reliability, it is common for membrane filtration systems to incorporate some measure of redundancy. There are two strategies that utilities commonly use to provide this redundancy: oversizing the membrane units at the design flux and providing a redundant membrane unit. Oversizing the membrane units at the design flux generally allows for operation at average flow as a minimum with one unit out of service and may allow for operation at or near system capacity in some cases. For example, a wastewater reuse water treatment plant with a rated capacity of 7 mgd (26.5 mld) may be designed with four membrane units rated at 2 mgd (7.6 mld) each, rather than 1.75 mgd (6.6 mld). Thus, with one unit out of service and the remaining three operating at capacity, the system can consistently produce as much as 6 mgd (22.7 mld) at the design flux. Although the design in this example would not allow the system to operate at maximum capacity, it is likely that the average flow could be met or exceeded.

A system could also be designed with greater excess capacity to allow for operation at the maximum output with one unit out of service (in this example, using four units at 2.33 mgd each [8.8 mld]), although this approach may be cost prohibitive, particularly for smaller systems, since the percentage that each unit must be oversized increases as the total number of units is reduced. Note that because this method of providing system redundancy is based on the assumption that the rated capacities are all specified with respect to the same design flux, the rate of fouling would not be increased by operating three of the four oversized units at capacity. In addition, continuing this example of a treatment plant with 7 mgd (26.5 mld) of permitted capacity consisting of four 2.33 mgd (8.8 mld) membrane filtration units, with all four units in operation the system could operate at reduced flux and still produce 7 mgd (26.5 mld) in accordance with its permit. This operation at reduced flux would decrease the rate of fouling and lower operating costs to partially offset the increase in capital cost associated with the extra unit capacity.

The second common method of adding system redundancy is by providing an additional membrane unit. Thus, using the same example of a water treatment plant permitted for a maximum rated capacity of 7 mgd (26.5 mld), the membrane filtration system would consist of five 1.75 mgd (6.6 mld) units rather than four. This design would always allow for operation at capacity with one unit either out of service or in standby mode. In this case, the standby unit can be rotated such that any time a unit is taken off-line for chemical cleaning, the standby unit is activated. The newly cleaned unit then reverts to standby mode until the next unit is due for chemical cleaning. This method of adding redundancy is more cost effective for larger facilities with a greater number of membrane units, such that the addition of an extra unit does not represent a large percentage increase in capital cost. This approach is recommended when the treatment facility utilizing membrane filtration is the primary source of drinking water for the public water system, such that operating the membrane system at capacity may be critical to the ability of the utility to satisfy customer demand. Alternatively, under this scenario all the units may be operated at lower flux to extend the interval between chemical cleanings.

Independent of the particular strategy used for providing some redundant capacity for the membrane filtration system, utilities should also provide some redundancy for any major ancillary mechanical equipment that may be utilized, such as pumps, compressors, and blowers. Utilities must comply with any state-mandated redundancy requirements for either membrane capacity or ancillary equipment.

Membrane filtration systems may also have some inherent redundancy if they are normally operated significantly below the maximum flux permitted by the state. In this case, if one membrane unit is taken out of the service, the utility can increase the temperature- and pressure-normalized flux through the remaining units to partially or fully compensate for the loss of production attributable to the out-of-service unit. However, if a unit is out of service for a prolonged period, the increase in flux may accelerate fouling in the remaining in-service units to an unacceptable rate.

Timely membrane replacement is another consideration for maintaining system performance and reliability. Membrane replacement is usually conducted on an as-needed basis, typically in cases in which either the membranes have been damaged or the flux has declined to an unacceptable level as a result of irreversible fouling. The useful life of membranes is commonly cited in the range of 5 to 10 y (a period generally consistent with manufacturer warranties). The use of membrane technology—particularly MF and UF—has't been in continuous operation for more than 5 to 10 y. Consequently, now there are field data available to document the typical useful life of membrane filtration modules. It is recommended that utilities keep a small number of surplus membrane modules on-site in the event that emergency replacement becomes necessary.

Residuals Treatment and Disposal

As with many water treatment processes, membrane filtration systems may generate several different types of residuals that must be treated and/or disposed of, including concentrate and both backwash and chemical cleaning residuals. The types of residuals that are generated vary with both the type of membrane filtration system and the hydraulic configuration in which the system is operated. For example, NF and RO systems produce a continuous concentrate stream and periodic chemical cleaning waste, but because these spiral wound membrane systems are not backwashed, no backwash residuals are generated. MF and UF systems are regularly backwashed and undergo periodic chemical cleanings, thus producing residuals from both these operational unit processes. However, only MF and UF systems that are operated in a cross-flow hydraulic configuration and waste the unfiltered flow (rather than recirculate or recycle to the plant influent)—the so-called feed and bleed mode—will generate a concentrate stream. Because cartridge filters operate in a deposition mode hydraulic configuration and are designed to be disposable, there are typically no residuals streams associated with MCF systems. Nevertheless, the spent filter cartridges (as well as all membrane filtration residuals) must be properly disposed of in accordance with any applicable state and local regulations, particularly if the cartridges have been used to filter any potentially hazardous materials.

The various potential residuals streams from membrane filtration systems—backwash residuals, chemical cleaning residuals, and concentrate—are discussed in the following subsections. Note that these discussions are not meant to represent a comprehensive review of residuals treatment and disposal, but rather a general overview of some of the primary considerations that should be taken into account when planning and designing a membrane filtration system.

Backwash Residuals

Among the various types of membrane filtration, only MF and UF systems employ backwashing, thus generating backwash residuals. Although the frequency varies on a site- and system-specific basis, backwashing is typically conducted every 15 to 60 min. Under normal operating conditions the backwash frequency should remain relatively consistent, allowing for the quantity of residuals generated to be estimated fairly accurately.

As a general rule of thumb, the residual stream produced from backwashing MF and UF membranes has a concentration of suspended solids that is approximately 10 to 20 times greater than that of the feed water. Although MF and UF systems remove approximately the same types of feed water constituents as conventional media filters, the volume and characteristics of the residuals may be significantly different. In many current applications of MF and UF for municipal water treatment, filter aids such as coagulants and polymers are not necessary. In these cases, the amount of solids removed in the backwash process may be significantly less than that for a comparable conventional filtration plant. In addition, disposal of these coagulant- and polymer-free MF and UF backwash residuals may be less problematic. However, in some applications of MF and UF, coagulants may be added in-line (i.e., without presettling) to help facilitate the removal of TOC, which is not normally removed to a significant degree by MF or UF. In such applications, the characteristics of the MF or UF backwash residuals will generally be more similar to those for conventional media filtration. Disposal options for MF and UF backwash residuals are not similar to those for conventional water treatment plants, which typically include

- Discharge to a suitable surface water body
- Discharge to the sanitary sewer
- Treatment with supernatant recycle and solids disposal

The discharge of backwash residuals used on secondary effluent to surface water bodies is not a consideration due the nature of the backwash. Discharge to a sanitary sewer is also likely to be subject to state and/or local regulations, and to require a permit. Moreover, the potential to utilize one of these options may be complicated if the residuals include chemical wastes. In addition to the use of coagulants added to the feed, some backwash procedures utilize chlorine or other chemicals. Small amounts of chlorine may be quenched, and acids or bases can be neutralized prior to discharge, but larger amounts or other types of chemicals added to the backwashing processes may require additional treatment or preclude discharge to a surface water body or sanitary sewer.

On-site treatment options for MF and UF backwash residuals are also similar to those that might be used with conventional media filtration and include clarification, sedimentation lagoons, gravity thickening, centrifuging, belt filter presses, and a combination of these processes. A second stage of MF or UF may also be utilized to further concentrate residuals and increase process recovery. If a sedimentation process is used to treat MF or UF backwash residuals, the addition of a coagulant may be necessary to improve the settling characteristics of the solids if coagulant is not already applied in the MF or UF pretreatment process. With on-site treatment, the supernatant is generally recycled to the treatment plant influent, while the concentrated solids are transported off site for landfilling or other means of disposal. As with discharging, the addition of chlorine or other chemicals to the backwash process may complicate residuals treatment.

It is important to note that backwash residuals concentrate any pathogenic organisms that are present in the feed water, as well as other suspended solids. The potential treatment of this stream should be taken into consideration if these residuals prohibit discharge into a surface water receiving body.

Chemical Cleaning Residuals

MF, UF, NF, and RO membranes undergo periodic chemical cleaning, and thus the systems generate spent chemical waste as a by-product of these processes. As with the backwashing process for MF and UF, the frequency of chemical cleaning varies on both site- and system-specific basis. Although chemical cleaning is conducted much less frequently than backwashing, the frequency is also more difficult to predict (see Chapter 4 for general estimates). Generally, MF and UF systems are cleaned no more frequently than once every month, for efficient operation and the minimization of system downtime, although it is not uncommon for these systems to operate for much longer without requiring chemical cleaning. The use of daily CEB lengthens the CIP interval but adds a more frequent disposal issue. The cleaning frequency for NF and RO may vary from 3 months to 1 y or longer, depending on the feed water quality and the effectiveness of feed water pretreatment for minimizing fouling. However, because chemical cleaning is a relatively infrequent batch process, estimating the quantity of residuals generated is not as critical for day-to-day operation, as is the case with backwashing.

Chemical cleaning residuals are generally treated on-site and discharged to a sanitary sewer, subject to state and/or local regulations. Oxidants such as chlorine used in the chemical cleaning process can be quenched prior to discharge, and acids and bases can be neutralized. The use of other chemicals, such as surfactants or proprietary cleaning agents, may complicate the process of obtaining regulatory approval for discharge and could require additional treatment.

Note that the rinse water applied to the membranes after the cleaning process may also represent a chemical waste and thus could require treatment prior to discharge. Although the rinse water increases the volume of the chemical cleaning residuals, this increase can be balanced somewhat by the recovery and reuse of a significant portion of the cleaning solutions. In some cases, as much as 90% of the applied cleaning solutions can be reused, reducing residuals treatment and disposal costs as well as chemical usage.

Concentrate

The term *concentrate* is usually associated with the continuous waste stream of concentrated dissolved solids produced by NF and RO processes. This waste stream is typically 4 to 10 times more concentrated than the feed with respect to suspended and dissolved constituents, and represents about 15% to 25% of the total feed flow, although it can exceed 50% in some cases. As a result, concentrate disposal is a significant logistical and regulatory concern for utilities and is often a critical factor in the planning and design of an NF or RO facility.

There are a number of methods for concentrate treatment and disposal, including

- Sanitary sewer discharge
- Land application/irrigation

Because there are complicating factors associated with each of these options, no single option is ideal or most appropriate for every application. In many cases, sanitary sewer discharge is the least expensive option. Discharge to the sanitary sewer may cause total dissolved solids (TDS) discharge considerations, since the wastewater treatment process does not typically affect dissolved solids concentrations to a significant degree, and the treatment plant effluent may ultimately be discharged to a surface water receiving body. High salinity or major ion toxicity may also preclude land application of the concentrate if levels exceed the threshold tolerances of the irrigated crops. Bioaccumulation of metals has also been cited as a potential concern for land application of NF or RO concentrate. Deep well injection was at one time considered an effective and commonly used technique for concentrate disposal, although this method risks the escape of brackish water into less saline or freshwater aquifers and may have unknown long-term environmental effects. It is no longer an option. The use of evaporation ponds is generally limited to areas with low precipitation and high evaporation rates, as well as an abundance of inexpensive and available land.

Another option for dealing with NF and RO concentrate is zero liquid discharge, which involves sufficiently concentrating the residual stream through the use of such technologies as crystallizers and evaporators to allow remaining solids to be landfilled. While zero liquid discharge is typically an expensive option, it does offer a number of advantages, including the ability to avoid the discharge permitting process and the ability to be utilized at any location independent of factors such as the proximity to a suitable surface water body or available land for evaporation ponds. In addition, zero liquid discharge maximizes facility recovery and has minimal environmental impact. Although still fairly uncommon, this option has become increasingly feasible relative to other methods as environmental and discharge regulations have become more stringent.

The considerations noted in this subsection represent only a cursory discussion of the issues associated with the various options for concentrate treatment and disposal. Note that all of these options are subject to applicable federal, state, and local regulations.

Some MF and UF systems that are operated in a suspension mode hydraulic configuration and waste the unfiltered flow generate a continuous concentrate stream. This stream differs from that associated with

NF and RO systems in two significant ways: MF and UF systems concentrate suspended rather than dissolved solids, and the concentrate stream represents only a small fraction of total feed flow. (Note that NF and RO membranes represent a barrier to particulate matter, and thus these systems will also concentrate suspended solids; however, because suspended solids rapidly foul the semi-permeable membranes, which cannot be backwashed, most particulate matter is typically removed with prefilters.) An MF or UF concentrate stream has characteristics somewhat similar to those of backwash residuals, and therefore can be considered comparable for the purposes of treatment and disposal. Alternatively, these two residuals streams may be blended together.

Guidelines for Applying Polymers in Membrane Treatment

Guidelines

- Polymers are often used to enhance coagulation and sedimentation. Many times the polymer will carry over to a membrane process. The following guidelines are author's recommendations on the use and impact of polymers on membranes.

- In-line and direct filtration: avoid the addition of *any* polymer.

- Membrane filtration following sedimentation:

    - Avoid nonionic and anionic polymers if possible.

    - Avoid cationic polymers with molecular weight > 500 KDa if possible.

    - The dose of cationic polymer should not exceed the following

        (1) 1 mg/L (preferably ≤ 0.5 ppm) or

        (2) The mass ratio of cationic polymer to the primary coagulant (e.g., polyaluminum chloride) ≤ 0.03.

Notes

1. This is a general guideline. Always check with the membrane manufacturer.

2. Typical polymers would be poly (dimethyl) diallyl ammonium chloride (polyDADMAC), poly acrylamide (PAM), and poly acrylic acid (PAA). For the polymers used for other applications in a water treatment plant, such as sludge dewatering, the characteristics of the polymer need to be compared to those that have been tested in order to assess their impacts.

3. For enhanced coagulation, the benefit of adding polymers to improve DOC and UV_{254} reduction is insignificant when the dose of primary coagulant approaches the optimal dose.

Case Study 7.1: Singapore Public Utilities Board NEWater Project, Republic of Singapore[*]

The Republic of Singapore has a population of about five million people. Although rainfall averages 98 in. (250 cm) per year, Singapore has limited natural water resources because of its small size of approximately 270 mi² (700 km²). Reclaimed water (referred to by the local utility as NEWater; see Fig. CS 1-1) is an important element of Singapore's water supply portfolio.

Currently, there are five NEWater treatment plants in operation, all of which include nearly identical treatment processes. Feed water to the treatment plants is activated sludge secondary effluent. The advanced water treatment processes include microscreening (0.3-mm screens), MF (0.2-μm nominal pore size) or UF, RO, and UV disinfection. Chlorine is added before and after MF to control membrane biofouling. The reclaimed water is either supplied directly to industry for nonpotable uses or discharged to surface water reservoirs, where the water is blended with captured rainwater and imported raw water. The blended water is subsequently treated in a conventional water treatment plant of coagulation, flocculation, sand filters, ozonation, and disinfection prior to distribution as potable water.

The NEWater factories all produce high quality product water with turbidity less than 0.5 ntu, TDS less than 50 mg/L, and total organic carbon less than 0.5 mg/L. The water meets all USEPA Environmental Protection Agency and World Health Organization drinking water standards and guidelines. Additional constituents monitored include many organic compounds, pesticides, herbicides, endocrine-disrupting

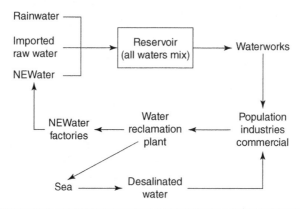

Figure CS 1-1   Schematic of the Singapore NEWater system. (*From Ong and Seah, 2003.*)

---

[*]*Source:* National Academy of Sciences (2012).

compounds, pharmaceuticals, and unregulated compounds. None of these constituents have been found in the treated water at health-significant levels.

The NEWater facilities at Bedok and Kranji went into service in 2003 and have since been expanded to their current capacities of 18 mgd and 17 mgd (68,000 and 64,000 m³/d), respectively. A third NEWater factory, the Seletar Water Reclamation Plant, was placed in service in 2004 and has a capacity of 5 mgd (19,000 m³/d). The fourth NEWater factory (Ulu Pandan) has a capacity of 32 mgd (121,000 m³/d) and went into operation in 2007. A fifth facility, the Changi NEWater Factory, was commissioned in two stages: the first 15-mgd (57,000 m³/d) phase was commissioned in 2009, with an additional 35-mgd (130,000 m³/d) phase commissioned in 2010. Once completed, these five plants will have a combined capacity of 122 mgd (459,000m³/d).

Most of the reclaimed water from the NEWater factories is supplied directly to industries. These industries include wafer fabrication, electronics, and power generation for process use, as well as commercial and institutional complexes for air-conditioning cooling purposes. Less than 10 mgd (38,000 m³/d) of NEWater currently is used for potable reuse via discharge to raw water reservoirs, accounting for slightly more than 2% of the total raw water supply in the reservoirs. However, the contribution of NEWater to the potable water supply is expected to increase in the coming decades.

The capital costs for all of the NEWater factories averaged about $6.03/kgal per year capacity (or $1.59/m³ per year). Annual operation and maintenance costs for the water are about $0.98/kgal ($0.26/m³) produced. The Public Utilities Board charges industries and others $2.68/kgal ($0.71/m³) for NEWater on a full cost recovery approach. This includes the capital cost, production cost, and transmission and distribution cost (A. Conroy, Singapore Public Utilities Board, personal communication, 2010).

Case Study 7.2: Water Reuse for Drought-Proof Industrial Water Supply in San Diego

A 200,000-gpd advanced MF system treats secondary wastewater effluent for reuse as process water for Toppan Electronics, Inc., in San Diego, California. The treated water replaces potable water purchased from the San Diego water system. Secondary wastewater effluent is always available as a source of water in water-short Southern California.

The new MF equipment is a Pall Aria small flow system, which provides consistently high removals of solids, pathogenic cysts, and bacteria from the secondary effluent to provide the right type of feed water to Toppan's existing RO system. Testing at other water reuse facilities has shown that MF significantly improves RO system operation by increasing allowable fluxes by 20%, reducing operating costs by 40%, and increasing the interval between chemical cleanings for the RO system by 4 times compared to conventional lime settling and filtration of secondary effluent.

The compact Aria system allowed the 200,000-gpd MF unit to be installed in a small space outside a building and under a landing. The system has been successfully operating since May 2000 and is providing high quality feed water for the RO system.

The low installed cost means that the cost of reused water is similar to the cost of purchased potable water. Jerry Barnes, president of Toppan, was very pleased with the result and said, "Water reuse

FIGURE **CS 2-1**   Aria System installation at Toppan electronics.

FIGURE CS 2-2   Pall Aria Microfiltration system for easy installation and simple operation.

using the Pall Aria system helps the environment by conserving precious supplies of potable water, while giving Toppan an economical water supply to keep operating even during droughts.

Clark Dawson, a local consultant instrumental in getting the installation running, added, "This system delivers some of the cleanest water on the planet, made from some of the dirtiest water—sewage."

The Pall Aria system is a compact, skid-mounted MF unit available in three capacities: 60, 175, and 350 gpm. The compact design allows the units to fit through a standard 3-ft door to simplify installation at existing facilities.

Case Study 7.3: Cleaner, Purer Water—Membrane Separation Provides Recovery of High Value Products and Transforms Wastewater into a Renewable Water Resource

Tom Wingfield and James Schaefer

Availability of water is variable based on geography, as freshwater sources and populations are unevenly distributed. The idea that water is available in unlimited supply has led to inefficient usage for years. The need for cleaner water is growing substantially beyond drinking water, since supplies of fresh water are finite and demand is increasing. Communities and commercial interests vie for this resource. The large, central treatment facilities of yesteryear are reaching their design capacities and have nowhere to expand.

Demand for Pure Water

Regulations governing water treatment tighten yearly. At the same time, new generations of more sensitive water monitoring and measurement devices reveal that previously undetected contaminants exist in drinking water supply systems. Conventional gravity-dependent technologies such as clarifiers and loose media filters do not reliably remove many of the target pathogens that can affect public health. The performance of the older technologies is also heavily dependent upon the appropriate application of water treatment chemicals.

Industries dependent upon water spend millions of dollars per year for chemicals to treat water. It is easily one of the highest costs for many treatment systems. In addition, the public's awareness of the potential health effects of some water treatment chemicals is growing.

Because of the escalating costs for freshwater and stringent discharge standards, manufacturers are looking for cost-effective methods to treat wastewater and reuse it when possible. Water reuse represents a practical and reliable means of extending water supplies in areas with water shortages. Advanced treatment systems including MF and RO economically produce very high quality water for aquifer recharge and industrial uses, such as boiler feed water or ultrapure water for electronics manufacturing.

New modular membrane treatment units are reducing costs to make water reuse practical for irrigation of parks and high value plantings and crops, and to protect watersheds that supply drinking water from treated wastewater discharges. A line of easy-to-operate, modular MF systems are being successfully applied to a wide range of reuse applications, as summarized in Table CS 3-1. Although pleated, backwashable MF is a growing competitor to hollow fiber systems, the following discussion includes only examples with the hollow fiber technology platform.

Owner/Location	Capacity	Feed Water	Reuse Application	System Description
Sonoma County Water Agency, California	3 mgd	Secondary effluent	Irrigation	4 MF racks, 58 modules each
Fountain Hills Sanitary District, Arizona	2 mgd	Tertiary effluent	Aquifer recharge	4 MF racks, 48 modules each
New York State Department of Corrections, Bedford Hills, New York	1 mgd	Secondary effluent	Discharge to watershed	3 MF racks, 40 module each
Toppan Electronics, San Diego, California	0.3 mgd	Tertiary effluent	Water processing, pre-RO	1 MF rack, 20 modules
City of Chandler, Arizona	1.7 mgd	Screened wastewater	Aquifer recharge, pre-RO	4 MF racks, 45 modules each

TABLE CS 3-1   Selected Water Reuse Installations

The Bedford Hills Correctional Facility and Toppan Electronics use small flow MF units to meet tough water quality standards at low cost. The large installations for Fountain Hills Sanitary District and Sonoma County Water Agency represent new applications for water reuse. They are designed to meet tough standards, provide flexibility, and meet increasing water demands.

Recycling Water for Wineries in Sonoma County

Sonoma County, in the heart of California's wine country, is reusing water to assure adequate supplies for drinking water, commercial, and agricultural uses. The 3-mgd water reuse facility treats secondary effluent from a partially aerated lagoon treatment plant near the county airport in Santa Rosa. The reused water irrigates fields near the airport and some of the many nearby vineyards. In the future, reused water may also replenish water in local geysers, which have been less active recently because of low water levels.

MF was chosen because it provides high quality filtered water with low suspended solids, which is required for vineyard irrigation. A detailed evaluation of capital and operating and maintenance costs over a 20-year period showed that MF, with its much lower operating and maintenance costs, was the most cost-effective alternative.

Aquifer Storage and Recovery in Arizona

The Fountain Hills Sanitary District 2-mgd MF installation is part of a system for conjunctive use of groundwater. During the fall, winter, and spring, the MF treats an average of 1.7 mgd of tertiary wastewater effluent, which is used to recharge the district's three aquifer storage and recovery (ASR) wells. During the summer, tertiary effluent is used directly for golf course irrigation, and water is pumped from the ASR wells to meet peak water demands.

This ASR system is one of only three such installations in the United States. This MF system was chosen because of its excellent turbidity removal, its ability to accept high chlorine residuals, and its ability to accept UF modules if the water must be disinfected without using chlorine. Feed turbidity ranges between 0.5 and 1.5 ntu, while the filtered turbidity is consistently 0.03 to 0.04 ntu. The extremely low turbidity improves ASR well longevity by minimizing the amount of solids pumped into the aquifer by the recharged water. A four-train 2-mgd system was designed such that expansion to 3 mgd involved only the addition of a fifth module rack and modules.

Watershed and Marine Protection in New York

During the past five to 10 years, New York City has evaluated a number of different treatment technologies to reduce contamination of its drinking water reservoirs by municipal and industrial discharges in the region. Submicron MF of secondary effluent was determined to be the best available technology to provide the level of water quality necessary for watershed protection. The first of many of these plants is at the Bedford Hills Correctional Facility in upstate New York.

All along the East Coast from Maine to Chesapeake Bay, effluent from wastewater treatment—especially nutrients such as nitrogen and phosphorus—damages marine fishing and shellfish industries. This "ounce of prevention" treatment method is dependent upon the controlled separation of contaminants inherent in a membrane treatment system. Not only does this measure serve to protect the public health, ecosystems, and habitat; it is also preserving jobs and a way of life. On the heavily populated eastern seaboard, this high quality effluent is also becoming a valuable resource for industrial makeup water and irrigation.

Macroelectronics Industry Conserving Water Supplies in California

Secondary wastewater effluent is a readily available source of water in water-short Southern California. In San Diego, treated water replaces potable water purchased from the San Diego water system for Toppan Electronics, Inc. An advanced MF system treats 0.3 mgd of secondary wastewater effluent for reuse as process water. The new MF equipment is a small flow system, providing consistently high removal of

Figure CS 3-1   2-mgd Pall Microfiltration system at Fountain Hills Sanitary district.

solids, pathogenic cysts, and bacteria from the secondary effluent and producing excellent feed water to Toppan's existing RO system.

A growing body of data at this and other water reuse facilities demonstrates that MF significantly improves RO system operation. Allowable fluxes increase by 20%, operating costs are reduced by 40%, and the time interval between chemical cleanings for the RO system is extended by 4 times compared to the conventional means of lime settling and filtration of secondary effluent.

The compact footprint of the MF system allowed the 0.3-mgd unit to be installed in a small space outside a building and under a landing. The system's simplified design allowed the units to fit through a standard 3-ft door at an existing facility. A low installed cost translates to a cost for reused water close to the cost of purchased potable water. The system has been successfully operating since May 2000. Reuse using the system helps the environment by conserving precious supplies of potable water while giving Toppan an economical water supply to keep operating even during droughts.

Water Reuse and Economic Development in Chandler, Arizona

The city of Chandler, Arizona, has managed to attract several large semiconductor manufacturers to the Phoenix area by providing the infrastructure necessary for their growth. A large part of that is the Chandler Industrial Process Water Treatment Facility (IPWTF), where approximately 80% of the high quality reclaimed water is used to recharge the local aquifer. Another portion of that water is used for landscape irrigation and for makeup water to cooling towers.

The manufacturer achieves a net water use of 1 mgd; this can be contrasted with a residential development occupying the same land area, which would use approximately 2 mgd. An added benefit is that this conservation reduces the net load on city services.

The IPWTF employs membrane filtration to achieve its goals. After initial pH adjustment of the raw wastewater from the manufacturing facility, the water passes through an MF unit prior to a filter cartridge en route to a spiral wound RO unit.

Prior to the introduction of MF into the process train, operating materials—such as replacement cartridge filters and treatment chemicals—ran $400,000 per year. A pre-RO MF step resulted in a 40% improvement in operating effectiveness due to drastically reduced frequencies of change-outs, extended intervals between RO cleanings, and longer life.

Recycle and Reuse Water: Membrane Filtration as a Practical Solution

Water is a precious resource in limited supply, whose use is subject to competition between industrial and municipal consumers. As a result, industry is looking for cost-effective alternatives to recycle and reuse water. Membrane filtration is a viable option to transform wastewater into a reusable resource. Membrane systems are revolutionizing the way water is treated by providing industrial and municipal users with a greater degree of control over their water quality while lowering operating costs. Membranes are gaining acceptance for a wide range of applications, including drinking water, process water, and reclaimed water from wastewater. Membrane separation provides recovery of high value products and transforms wastewater into a renewable water resource. Excellent performance, reliability, and competitive economics make membrane systems a good solution for integrated water management applications. Water reuse is no longer a practice relegated to arid and semiarid regions of the country. Today, advanced treatment and reuse of wastewater can be found wherever competition for water use is high (Pollution Engineering, 2002).

Case Study 7.4: Water Reuse via MF/RO—Integrated Microfiltration/Reverse Osmosis System Recycles Secondary Effluent Wastewater to Combat Water Scarcity

Challenge

A multinational oil refinery in Brisbane, Australia, required a substantial increase in water supply as a result of a major expansion. This increased demand was unable to be met by the existing potable water supply infrastructure.

Fresh water was becoming scarce in this Australian region, and the local municipality was struggling to meet the potable water supply demands of the refinery that was to be used for industrial purposes.

The refinery was about to undergo a substantial increase in production, which created an even higher demand for usable, clean water.

The municipality had plenty of secondary wastewater available, but the quality of the water was not sufficient to meet the refinery's needs.

Solution

By integrating advanced reverse osmosis (RO) technology with high-performance microfiltration (MF), Brisbane Water Authority was able to enter into a long-term contract to sell high quality recycled wastewater effluent from the local treatment plant to the refinery.

A fully automated dual membrane system capable of producing 3.7 million gal (14 million L) of high quality water per day produces water suitable for the refinery's requirement for boiler feed, cooling tower makeup, and other process uses.

Treated Secondary Effluent Wastewater

From the municipal wastewater treatment plant (WWTP), water is passed through two 200-μm automatic backwashing screens before entering the Pall Microfilter (MF) system. The MF system is comprised of six MF trains (five duty, one standby) utilizing the Pall Microza 0.1 μm polyvinylidene fluoride membrane to treat the feed water. The MF filtrate is then treated by a three-stage GE RO system comprising six trains (five duty, one standby). The product water from the RO system is chlorinated and stored in two 6-ML ponds before transfer on demand to the refinery site storage tank which is 2.5 mi. (4 km) away. All reject water from each membrane system is recycled to the WWTP.

The plant operates with minimal workers, as it is fully automated with a sophisticated PC-based SCADA system providing comprehensive operator interface to the plant.

Location:	Brisbane, Queensland, Australia
Commissioned:	October 2000
Application:	Municipal wastewater effluent reuse
Feed water type:	Secondary effluent followed by biological nutrient removal (BNR)
Feed water quality:	Cond 1,800 to 3,000 µS, Turb 1.6 ntu, HPC > 1,000
Product quality:	Cond 80 to 120 µS, Turb < 0.1 ntu, HPC < 100
Capacity:	14 mld of product water
Technologies:	Microfiltration, reverse osmosis, chlorination

The MF system utilizes the Pall Microza 0.1-µm polyvinylidene fluoride membrane in a six-train configuration including one standby. Each train contains 68 modules. Membrane flux is maximized through the use of a 5% to 10% recirculation flow, and the MF system achieves a 95% overall recovery. Automatic regular reverse filtration of the membranes, combined with air-scrubbing, ensures that clean in place (CIP) operations are only required on a monthly basis. A dedicated CIP system is provided to enable automatic cleaning of each MF. Filtrate from the MF plant is supplied to the RO via an intermediate storage tank.

MF filtrate is treated through a three-stage RO system. Each of the six RO trains (five duty, one standby) is configured in a three-stage 18:8:5 arrangement. Continuous disinfection of the plant for biofouling control is achieved through chloramination with online ORP monitoring of the RO feed. A dedicated CIP system enables automatic cleaning of each RO.

Reject streams from the MF and RO systems are returned to the head of the nearby WWTP. Permeate from the RO is chlorinated before storage in two 6-ML covered rubber lined ponds. Product water is transferred on demand to the nearby oil refinery by a dedicated 4-km pipeline.

The water reclamation plant is fully automated, with comprehensive online monitoring of the RO and MF systems as well as the product water quality.

Results

Despite operating in a water-scarce region of the world, the refinery now has a water supply of sufficient quality and quantity to meet its operating needs in a cost-efficient manner.

Parameter	Feed	Product
Flow (mld)	22.5	14
MF Recovery		95%
RO Recovery		83%
Conductivity (μS/cm)	1,800 to 3,000	80 to 120
Turbidity (ntu)	1.6	< 0.01
pH	6.8 to 8.0	
Free Chlorine (ppm)		0.3 to 0.5
Total Nitrogen (mg/L)		< 1.0
TOC (mg/L)	12	< 1.0
HPC (cfu/mL)	> 1,000	< 100

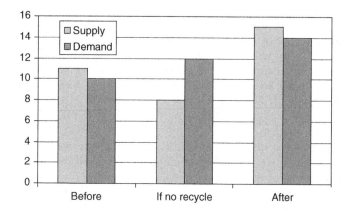

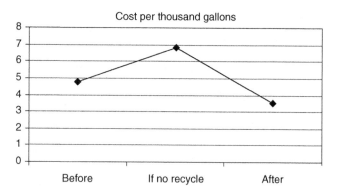

Case Study 7.5: Water Reclamation for Groundwater Recharge*

Water quality regulations for groundwater recharge vary from state to state. The best available method is considered to be the use of microfiltration (MF), reverse osmosis (RO), and advanced oxidation processes. Permeate from RO membrane processes has reduced salts and reduced microconstituents such as pharmaceuticals, personal care by-products, and endocrine disrupting chemicals.

One of the signature projects in this area, and a prototype for future facilities, is the Orange County Groundwater Replenishment (GWR) System in Southern California. Tetra Tech was a member of the design team for the GWR System, which was developed jointly by the Orange County Water District (OCWD) and the Orange County Sanitation District.

The GWR System provides state-of-the-art treatment for groundwater recharge to replenish the aquifer and to protect the aquifer from saltwater intrusion. The GWR System helps Orange County ensure that it is able to meet growing water demands and reduce its reliance on water imported from other sources.

Using an advanced purification process, the GWR System turns wastewater effluent that was being discharged into the ocean into water that is of higher quality than required by all state and federal drinking water standards. The purification process consists of three steps: MF to remove bacteria, small suspended particles, protozoa, and viruses; RO to separate out minerals and pollutants; and ultraviolet light and hydrogen peroxide disinfection for final sterilization.

The GWR System has the capacity to produce 70 mgd of near-distilled quality water. Approximately 35 mgd of the water is injected into an expanded seawater barrier to prevent ocean water from contaminating the groundwater supply. The remaining 35 mgd is pumped to OCWD's spreading basins in Anaheim, California, where it mixes with Santa Ana River water and other imported water and percolates into the groundwater basin. The water is then available for harvesting as water supply. OCWD estimates that the GWR System produces enough pure water to meet the needs of 500,000 people.

An additional benefit of the GWR System is that as the newly purified, low-mineral water mixes with existing groundwater, it reduces the average mineral content of the water in the Orange County groundwater basin. Lowering the level of minerals in the water, or reducing water hardness, decreases maintenance costs for residents and businesses by extending the life of water heaters, boilers, cooling towers, and plumbing fixtures.

---

*Source: Trends in Water & Wastewater, p. 136.

Today, the GWR System is the world's largest water purification plant for groundwater recharge. Orange County has established a blueprint for large-scale wastewater purification that is already being emulated in dry regions and nations, such as Singapore that leading the world in wastewater reuse technology.

Case Study 7.6: Water Reuse via Dual Membrane Technology—Water Company Supplies RO Quality Water from Treated Effluent

Challenge

Anglian Water supplies water in one of the driest part of the United Kingdom. Industrial development in the area is resulting in an ever-increasing demand for water.

Following the upgrading of one of their sewage treatment works, Anglian Water had a supply of a large amount of treated wastewater with a consistent quality within close proximity of one of its largest customers, Peterborough power station, which used up to 1,200 m³ of demineralized water every day. Anglian Water wanted to reduce the demands on the potable water supply by using a dual membrane treatment technology to produce water for the power station.

Furthermore, the current operation of the power station's demineralization plant was subject to seasonable variation of incoming surface water quality, with associated downtime and elevated operating costs. The variable organic loadings were also resulting in steam purity levels falling below that specified by the turbine equipment manufacturer's recommended levels.

Location:	Peterborough, Cambridgeshire, England
Commissioned:	August 2000
Application:	Municipal wastewater effluent reuse
Feed water type:	Secondary effluent
Feed water quality:	Cond 900 to 1500 µS, TSS 14 mg/L (mean)
Product quality:	Cond < 70 µS, Turb < 0.1 ntu
Capacity:	1.2 mld of product water
Technologies:	Microfiltration, reverse osmosis

Solution

By utilizing Pall Corporation's high-performance membrane microfiltration (MF) system, Alpheus Environmental Ltd. (a subsidiary of Anglian Water) was able to produce filtrate that was suitable for feeding to a reverse osmosis (RO) membrane plant.

The effluent being fed to the MF plant is prefiltered using a 150-µm rotating drum filter. The membranes are Microza Aria 0.1-µm polyvinylidene fluoride hollow fibers with an out to in flow path. At Flag Fen, the plant comprises two streams.

The MF plant operates in batch sequence filtration mode with automated air scour and dilute sodium hypochlorite reverse flush on a timed basis. Additionally, an automated alkaline sodium hypochlorite clean in place (CIP) at preset feed pressure (typically every 2 to 3 weeks) to remove organic contaminants from the membrane surface. Low pH CIP can also be carried out, if required, to control inorganic contaminants.

The two streams operate in duty/assist mode controlled by the level in the MF tank.

Water recycling plant schematic flow diagram

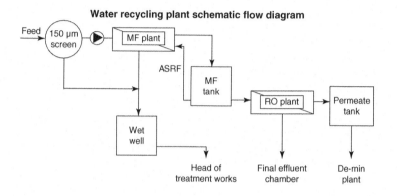

The thin-film composite polyamide RO membranes are configured in two streams of 4:2 arrays mounted on a single skid also operating in duty assist mode. The plant operates at 80% recovery, producing 1.2 mld, which is fed directly to the demineralization plant at the power station. CIP of the RO skid is performed with high and low pH regimes in sequence.

The whole process is automatic using a PC-based SCADA system, allowing remote monitoring and control. The site is unmanned and only requires a weekly visit to check on chemical levels.

Value Delivered

Since the system's commissioning in the year 2000, the power station has realized considerable savings. Performance of the demineralization plant has dramatically improved; regeneration intervals have increased from 8 h to 60 h. This equates to a run time volume increase from 530 m^3 to 7,000 m^3. An associated reduction in chemical consumption has been seen, with an overall decrease of 87% in total delivered chemical volumes from, 30 deliveries per year to three or four per year.

The availability of the demineralization plant has increased from 78% to 98%, enabling an increase of almost 20% in water produced. Trade effluent discharge has reduced by 87% by virtue of the reduced regeneration frequency. This equates to a reduction of 28,000 m^3/y, with the associated reduction in disposal costs.

Finally, reduced organic contaminant levels in the treated water have ensured that steam purity levels are now within the recommended levels, reducing the risk of corrosion and turbine failure.

Chemical Usage Data

Chemical consumption has been reduced by 87%, with a reduction of approximately 428 tons per year in usage of 36% hydrochloric acid and 255 tons per year in usage of 47% caustic soda. The chemicals use to clean the MF system are approximately 12 tons of 32% caustic soda, 9 tons of sodium hypochlorite, 0.2 tons of citric acid, and 0.05 tons of 60% nitric acid. The protection offered by the MF is such that the RO membranes need cleaning less than once per year.

Case Study 7.7: Membrane Design and Optimization for Treating Variable Wastewater Sources

Scott Caothien, Tony Wachinski, PhD, PE, and Ron Van Bemmel
Pall Water Processing, New York, USA

New advances in membrane design and flux maintenance protocols are presented to show the versatility of membranes in meeting variable, challenging wastewaters. With variable feed sources, different membranes will behave differently depending upon their ability to resist or handle fouling, which has a big impact on the economics of the membrane plant.

The configuration of a membrane system—pressurized membrane or immersed membrane—plays an important role in design flux and power requirement. The numbers of modules and treatment trains, spatial footprint, and operations and maintenance costs depend upon membrane flux and configuration. Therefore, standardizing membrane fluxes across all membrane platforms results in a less than optimum membrane system design and more expensive cost of treated water.

Examples of optimized design during pilot-scale testing and full-scale installations with membrane design fluxes ranging up to 120 lmh are presented for a various feed waters. One Australian recycled water installation is shown comparing low membrane flux protocol to a pilot-scale study with optimized flux protocol on the same feed water.

An optimized flux approach can differentiate membrane systems based on the life cycle costs, number of modules required, and replacement costs. It will deliver the lowest cost of treatment through savings on capital costs and operating expenses.

Keywords

Recycled water, membrane filtration, secondary effluent, pressurized membranes

Introduction

Demands for reliable and uninterrupted water supplies have been growing for years. Many industries as well as municipalities are driving the need for recycled water. Historically, conventional treatment technologies were used; but as regulations tighten, they become unreliable and expensive to operate, often requiring high doses of water treatment chemicals. Conventional treatments are rapidly being replaced by advanced membrane technologies.

State of the art membrane technologies are particularly well suited for use in variable wastewater and are providing a new source of water for many applications. Membranes are less sensitive to variations in feed water and can handle peak flows without deterioration in effluent quality. Modular membrane trains are easily expandable and have significantly come down in price. More and more affordable membrane solutions are offered in pre-engineered, packaged, and also mobile designs. The "plug and play" designs allow rapid deployment and assembly on-site. Regardless of treatment capacity, membrane systems continue to provide a product of water that often surpasses stringent water purity specifications. Membrane technologies are now the treatment of choice for wastewater reuse.

This case study describes packaged membrane systems for treatment capacities under 15 mld and custom-designed membrane systems for up to 350 mld. An overview of the membrane properties, characteristics, and regulatory claims is provided. Case histories and operating experiences of full-scale and demonstration-scale installations are reviewed specifically for irrigation, indirect potable use, river discharge, and supply to power industry. Recent advances in cleaning and flux maintenance protocols are presented to show the versatility and flexibility of membranes to meet changing and challenging wastewater sources. Finally, budgetary cost estimates for membrane plants are provided in relation to treatment capacity.

MF Membrane Technology

The Pall Microza hollow fiber MF membrane is made from a polyvinylidene fluoride polymer and has a uniform construction throughout its thickness. Figure CS 7-1 shows SEM photographs of the MF membrane at various sections.

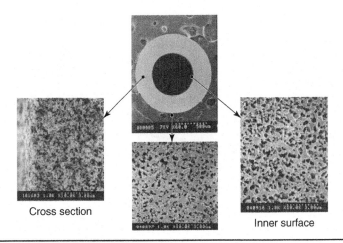

Cross section

Inner surface

Figure CS 7-1 SEM photos of Pall Microza MF membrane.

Because of its unique symmetrical configuration, there is no skin or coating which can deliminate or separate from the substrate under operating stress. The membrane is resistant to surface defects and pinholes.

The Pall MF module comes in a standard size of 2 m long × 0.2 m across in an ABS housing or shell. Each module contains 6,350 fibers. The hollow fiber membrane is rated at 0.1 µm and has a high porosity and high permeability, with a clean-water flux of 440 lmh at 100 kPa (20°C). For potable water treatment, the Pall membrane has been approved by the California Department of Health Services to operate at a flux of 204 lmh and a maximum transmembrane pressure (TMP) of 310 kPa. For wastewater recycling, the department does not limit the flux or the operating TMP.

Regulatory agencies in the United States have granted the Microza membrane log reduction values (LRVs) of more than 4 for *Cryptosporidium* oocysts and more than 6 for *Escherichia coli* without pretreatment. When direct coagulation is applied, the membrane has an LRV of 2.5 for MS-2 bacteriophages.

MF System Technology

The Pall Aria MF systems are engineered in prepackaged models with treatment capacities up to 4 mld (Fig. CS 7-2) or can be custom designed to treat 350 mld or higher. Figure CS 7-3 shows a 25-mld custom designed Aria system.

The modules-and-racks designs have typically lower installation costs and faster delivery schedules than submerged or vacuum membrane systems. Identification of fiber breakages and repair for modules-and-racks designs are simple and can be done without removing the module from the racks and without the assistance of a crane or

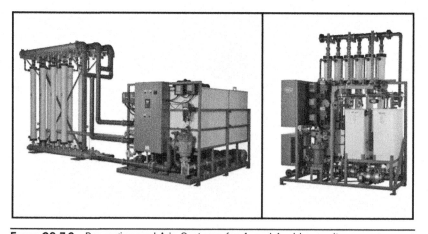

Figure CS 7-2   Pre-engineered Aria Systems for 4- and 1-mld capacity.

FIGURE **CS 7-3**   Custom-designed system for 25-mld capacity.

any other specialized equipment. Broken fibers can be identified and repaired in about 30 min.

Another feature of Aria systems is the ability to control fouling through an operating strategy called enhanced flux maintenance (EFM), in which a chlorinated solution at pH 11 is recirculated through the modules at set intervals (daily or weekly) for a period of 30 to 40 min. The EFM process is fully automated and does not affect the life of the membranes. EFM desensitizes the membranes from variations in water quality and allows the membranes to operate in an almost-clean condition and therefore at lower average operating TMP.

EFM is much more effective than chemically enhanced backwash because it has a longer exposure time but at the same time generates less chemical waste through recirculation and reuse of the chemicals.

Variable Wastewater Sources and Usage

Wastewater for Indirect Potable Use

The Groundwater Replenishment System in Orange County, California, treats secondary effluent to very stringent levels that meet both state and federal standards for drinking water. A Pall Aria MF demonstration-scale system (3.8 mld), consisting of two racks of 50 modules each, was operated from the end of 2000 to the

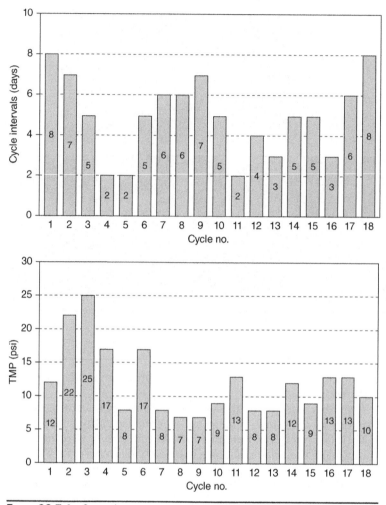

Figure CS 7-4 Operating parameters for groundwater replenishment system secondary effluent.

beginning of 2001 at a flux of 60.5 lmh with an average EFM interval of 5 days and a clean in place (CIP) interval of more than 3 months (Fig. CS 7-4). The average TMP throughout the test was around 117 kPa.

Peak flows of 50% increase, simulating wet weather conditions, were demonstrated for a 48-h period without shutting down the MF system. Long-term test data indicated LRVs of 3 for MS-2 bacteriophage viruses and 5 for coliforms. The high virus LRV was attributed to the formation of a biofilm and the adsorption of viruses on solid matter present in the feed.

Coagulated Effluent for Irrigation and Industrial Uses

Denver Metro, Colorado, wanted to reclaim secondary effluent for irrigation purposes. A major water quality objective was total phosphorus of less than 0.5 mg/L. To that end, direct coagulation with ferric chloride at a dose of 15 mg/L was employed to assist in phosphorus removal. A Pall Aria MF pilot plant was evaluated for 5 months at various membrane design fluxes. It was observed that direct coagulation with ferric chloride significantly improved the performance of pressurized membrane systems. A membrane design flux of 85 to 98 lmh without EFM was submitted. Water quality objectives for phosphorus, turbidity, and coliforms were met irrespective of design fluxes.

However, at that period of time about 5 years ago, a conventional treatment plant was deemed less expensive and more conservative than a membrane plant.

Secondary Effluent for Refinery Boiler Feed

To provide boiler feed water to a neighboring refinery, West Basin Municipal Water in California further treated Title 22 wastewater with MF and RO. The high quality Title 22 wastewater resulted in less fouling for the MF system, which could operate at higher membrane design flux of 85 lmh with a lower average operating TMP of 100 kPa. CIP interval was greater than 3 months, and no EFM was required.

The pilot study at West Basin provided the design basis for Alamitos Barrier, where Title 22 water was further treated with Pall MF, RO, and UV for reinjection into wells to prevent seawater intrusion.

Recycled Wastewater for University Use

The University of California at Davis plans to recycle secondary effluent to meet various water needs at the university. A research and development program was initiated to evaluate the use of pressurized MF on secondary effluent which has been pretreated with a cloth filter media to Title 22 standards. The prefiltered secondary effluent allows the MF system to operate at very high membrane flux of 128 lmh using a daily EFM regime and CIP interval of a month. Significant economic savings can be realized by pretreating secondary effluent to Title 22 standards.

Secondary Effluent for River Discharge

Sacramento Regional Wastewater Treatment Plant in California considered MF to supplement treatment capacity for existing sand filtration. Water quality objectives included meeting Title 22 standards and removing phosphorus. To reduce the costs of membrane installation, the beneficial effects of conventional pretreatment using coagulation and dissolved air flotation clarification were evaluated.

On secondary effluent, the Pall Aria MF system was operated at 85 lmh with daily EFM. To remove phosphorus, a direct coagulation process with alum was used. Furthermore, a dissolved air flotation clarifier was installed to lower the solids prior to MF. On coagulated

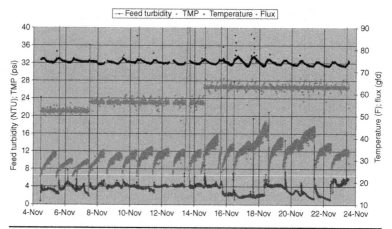

FIGURE **CS 7-5** MF performance with coagulated and clarified secondary effluent.

and clarified effluent, the MF performance substantially improved to 107 lmh with an average operating TMP of 100 kPa (Fig. CS 7-5).

Australian Perspective on Recycled Water

Brisbane City constructed a Pall Aria MF membrane plant to treat 15 mld of BNR secondary effluent. Currently, the majority of the wastewater from Luggage Point Wastewater Treatment Plant goes out into Moreton Bay, but a small proportion (15 mld) is "recovered" and further treated with a Pall MF and Filmtec RO and sent to a BP refinery for cooling water and boiler feed.

The Pall MF system was designed at a flux of 45 lmh with six racks (including one standby) and 66 modules per rack. Since 2000, the system has operated with an average TMP of 60 kPa and CIP intervals of 4 to 6 weeks. The plant was not designed with the EFM feature.

During the latter part of 2006, a Pall pilot plant was installed side by side with the full-scale plant to demonstrate performance using new operating protocols with EFM. The pilot has been operating at 80 lmh for more than 6 years without the need for a CIP. A daily EFM with 500 ppm NaOCl and weekly EFM with 1,000 ppm citric acid were set for this run. The spikes in TMP observed were due to a faulty chemical pump, so EFM was performed without the benefit of chlorine. Unfortunately, these incidents occurred on two separate occasions. Once the EFM resumed, the recovered TMP was significantly reduced.

On average, the TMP is around 70 kPa (Fig. CS7-6). Since the installation of the Pall MF plant 6 years ago at Luggage Point, there have been advances in operating protocols that can mitigate fouling and promote a better performance by 80% higher flux and similar TMP range. Running the pilot plant side by side provided some beneficial new parameters for improvements on the existing installation.

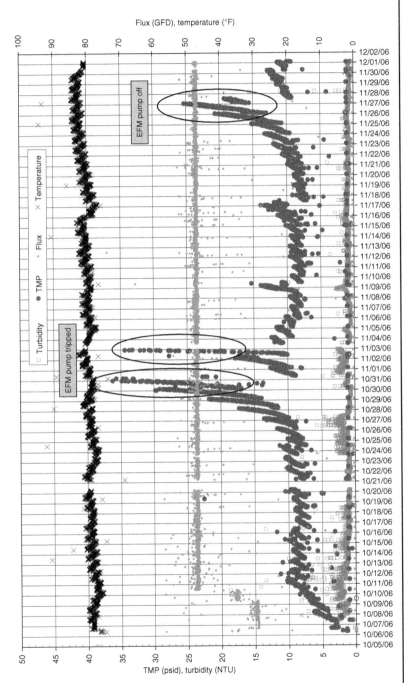

Flux (GFD), temperature (°F)

TMP (psid), turbidity (NTU)

□ Turbidity   ● TMP   ◆ Flux   × Temperature

EFM pump off

EFM pump tripped

Figure CS 7-6 Luggage point pilot plant performance.

227

Future optimization would include the use of coagulants such as alum to further control biofouling and at the same time improve treated water quality by removing total phosphorus.

Cost of Recycled Water vs. Treatment Capacity

This section provides some quick guidelines with respect to the cost of recycled water as a function of treatment capacity. Capital costs including one set of replacement modules are considered, without redundancy built into the systems. The evaluation is calculated based on a membrane flux of 62 lmh, a 20-year plant life, and a membrane warranty of 10 years.

As the treatment capacity increases, the costs of recycled water decrease. For example, a 2-mld plant will provide treated water at $0.25/L, while a 20-mld plant will provide treated water at $0.12/L (Fig. CS 7-7).

Conclusion

Membrane technologies can treat variable wastewaters into new sources of supply for indirect potable, industrial, irrigation, discharge into rivers, and other general reuses. Variable feed waters have an impact on membrane design flux. However, a strategy to mitigate fouling can be used to maximize the potential of these membranes in terms of design flux. The use of pre-engineered systems with optimized membrane designs has driven down the cost of wastewater treatment. As treatment sizes become increasingly greater, savings will be realized with economy of scale, thereby making custom-designed plants more and more common.

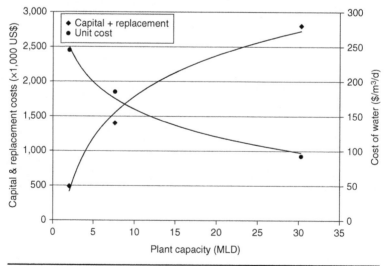

Figure CS 7-7   Cost of recycled water and treatment size.

There has been significant experience in fouling control with EFM and also feed pretreatment by either direct coagulation or prefiltration with cloth media, which has increased membrane fluxes and at the same time lowered energy costs and replacement module costs. An example is seen with side by side comparative tests at Luggage Point, Queensland, Australia, which indicated that significant flux improvement can be achieved by using the latest operating protocols while maintaining a low energy demand.

Due to the variety of feed sources and their characteristics, each membrane must be optimized to derive the highest membrane flux with least energy required. This approach will yield the lowest cost of ownership and treatment.

References

Carollo Engineers. 2005. "Submerged Versus Module Membrane Systems: Two Systems Design for Evaluation." Texas Water.

Dwyer, P. L., and R. Collins. 2001. "Report of *Cryptosporidium parvum* Oocysts and MS2 Bacteriophage Virus Removals by the Pall Microfiltration System Plant at NE Bakersfield (CA) Water Treatment Plant." *UNH* (December).

Dwyer, P., M. Smith, and R. Collins. 2004. "Final Report on the Challenges of Pall Microza USV-6203, UNA-520A and XUSV-5203." *UNH* (June).

Leslie, G. L., W. R. Mills, T. M. Dawes, J. C. Kennedy, D. F. McIntyre, and B. P. Anderson. 1999. "Meeting the Demand of Potable Water in Orange County in the 21st Century: The Role of Membrane Processes," Presented at AWWA Membrane Technology Conference, March 1999.

Marshall, T., and L. Don. "Delivery and Performance of the Luggage Point Wastewater Treatment Plant Water Reclamation Project."

Sakaji, R. H. 2002. "Pall Microza Microfiltration System: Conditional Acceptance of Higher Flux with Ferric Chloride Addition." California Department of Health Services, May 2002.

Sakaji, R. H. 2004. "Conditional Acceptance of Flux Increase for Pall USV-6203; Conditional Acceptance of Pall Microza UNA-620A as Alternative Filtration Technology." California Department of Health Services letter, July 2004.

Schimmoller, L., B. McEwen, and J. Fisher. 2002. "Ferric Chloride Pretreatment Improves Membrane Performance: Pilot Results for Denver's New 30 MGD Water Reclamation Plant." Presented at AWWA Membrane Technology Conference, March 2001.

Tamada, R. K., S. Hayes, S. L. Crawford, and T. L. Long. 2006. "Weatherford's 6-MGD Membrane Expansion: A Design/Build Case Study." Texas Water, April 2006.

Thompson, D. M., and M. C. White. 2005. "Price Sensitivity for Evaluated Bid Membrane Procurements." Presented at AWWA Membrane Technology Conference, March 2005.

Vickers, J. C. 2005. "Comparing Membranes and Conventional Treatment Options for Reuse." AWWA ACE, June 2005.

Wachinski, A. M., and C. Liu. 2007. "Design Considerations for Small Drinking Water Membrane Systems." *Water Conditioning & Purification* (March).

Wehner, M. 2006. "Barriers the Solution to Recycle Water." *WME magazine* (May).

Case Study 7.8: High Purity Water from Tidal Canal—Water Company Supplies High Purity Boiler Feed Water with Membrane/Membrane Technology

Challenge

A large chemical manufacturer needed to supply its Antwerp plant with high quality water at a low cost. Rather than purchasing municipal water, the company opted to process brackish water with high salinity from a tidal dock using an advanced membrane system. Pall Corporation was selected to design, build, and operate a 100 m^3/h system providing water with a final salt content of less than 3 mg/L TDS. The challenge was to find a cost-effective means of converting this raw water into high quality water.

Solution

A two-stage integrated membrane system—Pall Corporation's microfilter upstream of GE reverse osmosis (RO)—was installed. The water entering the plant is a mixture of fresh water from inland canals and brackish water coming from the Scheldt tidal estuary.

The quality of the raw water varies with the season and tide but is generally poor. Table 1 shows raw water parameters and their ranges of values.

The raw water is passed through two 300-μm automatic backwashing prefilters before entering Pall Corporation's 0.1-μm polyvinylidene fluoride membrane system. The filtrate is then treated by a two stage GE SWRO system, producing the final required water quality.

Energy recovery is applied via a GE Fedco HPB Energy recovery turbine.

Parameter	Average	Minimum	Maximum
Temperature (°C)	15	7	30
pH	8.0	6.5	9.5
TDS (mg/L)	10,000	—	14,000
Suspended solids (mg/L)	10	—	18
Turbidity (ntu)	3	—	10
Oils/hydrocarbons (mg/L)		—	4.0
TOC (mg/L)	12	—	25
Cl⁻ (mg/L)	5,500		7,300
Fe (mg/L)	0.5		0.9

TABLE CS 8-1   Raw Water Quality

Value Delivered

The system is designed for full automatic operation with minimal operator attendance.

An integrated membrane system is needed to process the poor quality water. The prefiltration membrane stage has produced water with very low fouling potential (SDI < 2). The membrane filters are highly effective at removing algae, bacteria, cysts, and other fine particulate matter. Consequently, RO membrane performance has remained stable and water quality has remained consistent regardless of variations in season, salinity, or temperature.

The goal of producing low-cost high quality water has been met. Since the project began, an integrated membrane system has reliably produced filtered water with a turbidity of < 0.2 ntu, total dissolved solids of < 3 mg/L, and conductivity of <2 mS/cm.

This innovative, efficient filtration has supplied finished water that has surpassed customer requirements. Table 2 illustrates system performance by comparing actual finished water quality to that required by the customer. Actual finished water quality is also compared with actual raw water quality in this table.

All parameters are remotely monitored.

This system has affirmed that water treatment using this innovative two-stage filtration process is cost effective and produces water that meets the customer's process standards.

Conclusion

A large chemical manufacturer in the port of Antwerp contracted with Pall Corporation for a design-build-operate project to produce water from the tidal dock. Both Pall and GE technology was used.

Both supplier and customer have found the contracting arrangements advantageous. Pall has provided evaluation of customer needs and goals, technical assessment, design, and fabrication of the water treatment system.

Parameter	Raw Water Measurement	Finished Water Requirement	Finished Water Measurement
Flow rate (m³/h)	140	100	100 (at 98% availability)
Temperature (°C)	≥ 10		
Turbidity (ntu)	12 (average)		
TDS (mg/L)	12,000 (average)	< 3.0	
TOC (mg/L)	12		0.1
Conductivity	> 14 μS/cm	< 10 μS/cm	< 2 μS/cm

TABLE CS 8-2   System Performance Related to Water Quality

The project has demonstrated the following benefits:

- Low cost to customer for process water
- Mutually beneficial contract arrangements
- Water quality and flux rate that exceed customer expectations
- Simple installation using modular skids
- Streamlined process for desalinating water
- Operator-friendly controls
- Innovative pretreatment process that does not require sand filters
- Economical upgrade to a larger system

CHAPTER 8

Future Trends and Challenges

Introduction

Rising water scarcity, combined with booming population growth, is placing unprecedented demands on the world's freshwater resources, creating a greater need for water conservation and new sources of potable supply. The stakes are even higher and the need more imminent in dryer regions of the world, including metropolitan areas at risk of interruptions in supply due to water shortages and drought. Securing clean and reliable water for the future, both for people and for natural systems, is becoming increasingly more urgent as communities, industries, and agriculture all vie for limited resources.

Globally, the availability of water is variable, with freshwater sources and population density unevenly distributed. Many countries, including the United States and China, have sufficient water, but because water supply and demand are disproportionately allocated, water shortages ensue.

These mounting water shortage concerns are driving initiatives in alternative water supply solutions, including expansions in the water reuse industry—opening up more opportunities for water reuse projects, spurring higher demand in water reuse treatment technologies, and growing the market for reclaimed water.

Membranes are factoring into this trend considerably and will continue to play a major role as more water treatment for reuse is needed. Currently in the United States, the vast majority of water reuse is limited to California, Florida, Arizona, and Texas; states where acute water shortage problems and the consistent threat of drought are driving new project initiatives. However, across the nation, water reuse is on the rise with more states pursuing projects that utilize recycled or reclaimed water on some level.

The same trend is occurring globally. Demand for treatment technologies and reuse is increasing, with total global reuse as a percentage of treated water expected to rise from 4% in 2010 to 33% by 2025.

As of 2010, Kuwait was the global leader in water reuse, with 91% reuse of total treated wastewater, followed by Israel (85%), Singapore (35%), and Egypt (32%), with the United States trailing in seventh, at just 11% (Global Water Summit, 2010).

Target Opportunities for Water Reuse

A growing number of opportunities for implementing water reuse initiatives are becoming available. Key target areas for projects that will help continue to expand the water reuse market include agricultural and golf course irrigation, landscape watering, recreational use, toilet flushing, and fire suppression. Projects incorporating more advanced treatment processes will also continue to play a vital role in producing purified water for supplementing potable supplies, including reservoir augmentation and groundwater recharge.

Water reuse strategies are also being utilized in greater frequency across a growing number of industrial sectors as a safeguard against disruptions in supply. Water scarcity and declining water availability can destabilize industrial and manufacturing operations where high quality process water is needed for a variety of purposes, including production, material processing, and cooling. These risks can be amplified for operations located in regions experiencing water scarcity (Ceres, 2009).

Industries that can mitigate water-related risks by implementing water reuse and recycling measures include high tech and electronics companies, bottling plants, the biotechnology and pharmaceutical industries, and other facilities dependent on consistent, purified water flow. Greater industrial capitalization on water reuse will also effectively reduce industry's reliance on potable supplies, thus easing pressures on local water reserves.

With maturation in the water reuse industry, treatment is expected to become increasingly more targeted towards specific end users, producing water qualities that are custom-designed to match the needs of planned end uses. This approach will result in greater energy savings and more economical water reuse. As an example, using highly treated, purified water for golf course irrigation or agriculture would not be the most optimum or cost-effective solution, considering that these applications do not require the same level of robust treatment as is needed for potable or high-end industry processes. Aggressive nutrient removal would also not necessarily be needed if the planned reuse were irrigation related, as these end uses could potentially benefit from the nitrogen and phosphorus present in lower, yet adequately treated wastewater.

Newer, cutting-edge applications of wastewater reuse are also expected to come into greater use, including the evolution of wastewater reuse to the more innovative practice of wastewater mining—a strategy incorporating the progressive view that effluent presents an opportunity and is considered more of a resource than a waste. When wastewater is mined, water is treated for reuse, but additionally, useful resources are extracted from the waste stream for recycling and beneficial purposes. These resources can include inorganic compounds, energy, metals, salts, and nutrients such as phosphorus and nitrogen which in turn can be used as high quality fertilizers.

Technologies

Industry-wide, research and development efforts are focused towards improving membrane technologies, producing products that offer improved separation performances at lower energies, thus leading to higher membrane efficiencies and more cost-effective treatment. As a result of material innovations—including ceramic and Teflon membranes, and polymers with nanoengineered particles—membrane options are becoming more durable, offering improved mechanical strength and higher resilience to chemical degradation and fouling. Membranes with improved scouring efficiencies and lower life cycle costs are further focus areas of the industry.

With increasing demand for high quality purified water for potable supply and industry-related needs, advanced membrane processes will continue to develop, including membranes for microfiltration (MF), ultrafiltration (UF), nanofiltration (NF), and reverse osmosis (RO). Additionally, as the market for water reuse grows, an ever-growing number of treatment trains are becoming available, including membrane technologies that can be coupled in a plug-and-play fashion with other treatment processes—such as advanced oxidation, ultraviolet light (UV) treatment, and granular activated carbon—to produce a range of water qualities suitable for a variety of end uses. Of all available treatment options, hollow fiber MF is regarded as the kingpin of membrane technologies, with the ability to manage relatively higher contaminated waters at higher flows and reduced pressure requirements. However, the market for hollow fiber UF is also developing; that process could become more of a factor moving forward, especially as a pretreatment option.

Future trends in membranes include technological advancements in the area of integrity testing, including the development of online direct integrity tests that incorporate the same sensitivity as traditional off-line procedures. Expanding pretreatment technologies will also increase the use of MF in treating wastewater, and the marriage of MF with posttreatment technologies such as UV holds significant promise in reducing chemical and chlorine requirements.

More membrane-related advancements include improvements in membrane packaging, which are contributing to lower energy consumption and more consistent effluent water quality. Newer membrane technologies incorporating more innovative fiber bundling are also leading to higher flux potential and increased membrane surface areas.

Membrane treatment facilities are trending towards larger capacity designs, driven in large part by MF and UF retrofits. Larger capacity MF and UF facilities are more inclined towards submerged systems, while smaller capacity facilities are tending to employ encased, skid-mounted packaged systems.

In the near term, the reuse industry is addressing challenges associated with waste volumes generated from low pressure membranes by increasing recoveries, which effectively reduces the amount of waste to be disposed. Research efforts are also underway towards developing the next generation of water reuse treatment technologies for dealing with highly contaminated, complex, and difficult-to-treat waters such as those associated with flue gas desulfurization (FGD) or generated from heavier industrial processes, including mining activities, unconventional gas production, and oil refining. Treatment solutions which combine membranes with advanced chemical processes are capable of achieving recoveries in the range of 75% to 80%. However, for extremely high efficiency recoveries of up to 98%, progressive treatment applications such as thermal evaporation, zero liquid discharge, crystallization, and biopolishing hold significant promise and value for the future. Additionally, high recoveries of up to 99.6% are possible when MF reject is treated by further MF processes. As a comparison, only 80% recoveries are possible when RO membranes treat MF processed wastewater.

Public Perception Challenges—Indirect Potable Reuse

Water reuse projects that involve treating secondary effluent for indirect potable use present a particular challenge because of the public perception element of utilizing reclaimed water for these purposes. The public's view towards water reuse in this regard can be a substantial obstacle that would otherwise not necessarily be encountered for other forms of water reuse projects, such as those that utilize reclaimed water for agriculture or industry. Indeed, the cumulative impact of the public's sensitivity towards using treated wastewater for replenishing drinking water supplies—heightened significantly by political opposition and sensationalized media coverage—can be powerful enough to effectively shut down water reuse projects.

An example is the East Valley Water Reclamation and Recycling Project, a $55 million indirect potable reuse project constructed in the 1990s in Los Angeles, which would have involved utilizing highly treated effluent for recharging underground aquifers in parts of the San Fernando Valley. Following completion, the project encountered stiff political resistance, fueled by negative media coverage, which led to heavy public opposition and the eventual closing of the plant.

Media reporting of the project included use of the term *toilet to tap*—a negative and misleading connotation, as this phrase detracts attention from the important wastewater treatment processes responsible for purifying effluent to a level suitable for potable reuse. However, public opposition can be contained and effectively managed by implementing strategic public outreach and education programs, efforts which have resulted in higher success rates with indirect potable water reuse projects. Outreach strategies that have proven effective include early public involvement, straightforward communications, and active engagement with key stakeholders. The outreach process should also be tailored and custom-designed to the individual community, with a focus on building support and credibility through transparency and openness in the planning process.

Imagery and language are also very influential factors in terms of shaping public perception towards water reuse projects. Research conducted by Linda Macpherson, reuse principal technologist at CH2M Hill, and Paul Slovic of Decision Research found that certain terminology used to describe water reuse projects—words such as *sewage* and *treated wastewater*—can have the effect of stigmatizing the concept of water reuse and deterring people from acceptance.

According to Macpherson (2011), outreach should emphasize positive-based language that is geared towards helping the public better understand the water cycle and how treatment can be an integral part of a water management system. This includes relaying the message that in reality, all water users are downstream in some way from another user. Macpherson and Slovic also concluded that the public's knowledge of water science is low, particularly in regard to how substances are removed from water. However, with increased knowledge and higher quality information, the public's understanding of water reuse will improve, leading to higher acceptance.

The public outreach campaign jointly launched and executed by the Orange County Water District and the Orange County Sanitation District in support of the $481 million Groundwater Replenishment (GWR) System is an example of how strategic outreach can lead to high approval and successful project implementation. A major indirect potable water reuse project that purifies highly treated wastewater for replenishing a large groundwater basin, the GWR System, supplies approximately 70% of the potable water demand for more than 2.3 million people in Orange County, California.

The carefully planned, community-tailored public outreach and education program associated with the GWR System included extensive stakeholder involvement, consistent and transparent public communications, and a public educational effort that was able to successfully convey the safety and purity of the water treatment technology.

The program was especially beneficial in its effectiveness at establishing leadership backing by gathering hundreds of letters of support from community leaders, and at building public support across a diverse cultural population through extensive presentations and educational workshops. After gaining credibility and trust, the water reuse project was successfully built and has been operational since early 2008.

Challenges Associated with the Cost of Water

Demand for water is escalating quickly, but the current cost of water does not reflect this demand, as prices are being kept artificially low. The costs associated with meeting higher water demands by increasingly supply is rising quickly, as lower cost water supply options such as groundwater pumping are giving way to higher cost solutions such as desalination. This is causing an increasing price mismatch between the cost of bringing new water supplies online and the price that those new water supplies will fetch in the marketplace.

The anticipated outcome is that over the next decade, governments will realize that they can no longer finance this growing discrepancy, resulting in rising water tariffs and stricter limits on "free" water in agriculture, followed by the introduction of higher irrigation fees. This in turn will help provide financing for infrastructure and will also greatly incentivize new water conservation measures. The water "crisis" will eventually solve itself in the long term, and we will see accelerated growth in the revenues earned in the water sector.

Jar Test Procedures

Introduction

Jar testing is most often used to determine coagulant doses for conventional water treatment. However, it can also be used with bench scale microfiltration (MF) and ultrafiltration (UF) modules, or vacuum filtration with the pore size of the MF to be considered, for a reuse application, e.g., to determine MF/UF performance on a given water with various pretreatments.

One important difference is that when coagulation is used with low pressure membranes, only a pin floc is required.

Jar testing—bench scale membrane testing—provides valuable data including candidate coagulants, optimum dosing, and required detention time. It can also be used to select candidate oxidants, concentration, and detention time.

Jar testing on actual wastewater provides "good" design data in a few hours with minimum expense.

Background

Coagulation is a process in which a metal salt (i.e., coagulant) destabilizes mineral colloids. The destabilized colloids aggregate to form flocs that are large and dense enough to settle. In meantime, metal salts themselves also form flocs of metal hydroxides and settle. During the formation and aggregation of flocs, small particles can be adsorbed or embodied into the flocs and removed by subsequent liquid/solid separation processes such as settling, conventional filtration, and membrane filtration.

The majority of natural organic matter (NOM) in water is considered dissolved, meaning it ends up in the filtrate of 0.45-μm membrane discs. The typical range of molecular weight of NOM is 500 to 20,000 Da. NOM may carry negative charge from dissociation of hydrogen ions from its carboxyl functional groups. Coagulation can be very effective for NOM removal via charge neutralization and hydrophobic adsorption under certain conditions. Alkalinity and pH are two important water quality factors that affect NOM removal by coagulation. At lower pH values, NOM removal is more effective. Coagulant dose is probably the most important operating factor that affects NOM removal. This is easy to understand when picturing NOM removal by coagulation as an adsorption process: the higher the coagulant dose, the more flocs available to adsorb NOM. In addition, a high dose of coagulant will cause more pH downshift, which facilitates NOM removal.

General Description of a Jar Test

A jar test is a simple and rapid way to learn process chemistry and the impact of different conditions on coagulation operation. It is conducted with a jar tester, a six-station stirring batch reactor with identical mixing conditions. Figure A-1 is a schematic drawing of a jar tester.

Each of the six stations has a 2-L reaction vessel equipped with a mixer. Each vessel has a draining port (not shown) located in the front, about two thirds from the top of the vessel. Each mixer can be disengaged or engaged independently. Once they are engaged, the rotation speed of all mixers is synchronized and adjustable by turning the knob on the left side of the machine. Important factors affecting chemical reaction—such as dose, pH, and mixing intensity—can be studied conveniently with a jar tester.

Equipment and Apparatus You Need to Conduct Jar Test

1. Jar tester
2. Stopwatch
3. Thermometer
4. pH meter

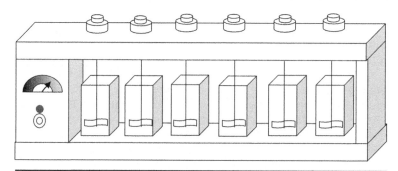

Figure A-1   Schematic drawing of jar tester.

5. Small graduated cylinder or pipette

6. Small beakers or test tubes for containing coagulant solution for each station. If test tubes are used, you also need a rack for keeping them in an upright position.

7. Membrane filter discs (0.2-µm rating and 47-mm diameter) and filtration kit (equipped with a vacuum pump)

8. Squeeze bottle for deionized water

9. Sample vials

10. Coagulants (alum, $FeCl_3$, etc.)

11. Deionized water (for dilution of coagulant)

12. Volumetric flask (for dilution of coagulant)

Before You Start

1. Prepare Stock Solution of Coagulant with Known Concentration

    In order to have more accurate control over the appropriate coagulant dose, you probably need to prepare or dilute a coagulant solution to the appropriate concentration. For most applications, it is convenient to make a coagulant solution at a concentration equivalent of 5 mg/mL (5,000 mg/L). Thus, if you need to conduct the jar test at coagulant doses of, say, 0, 10, 20, 30, 40, and 60 mg/L, you can add 0, 2, 4, 6, 8, and 12 mL (respectively) into six containers. (It is always a good practice to include a zero dose sample in your test plan.) If a higher dose range is needed, you may double the concentration of stock solution.

2. Get Water Sample Ready

    To conduct a jar test, you will need at least $6 \times 2 = 12$ L of water. Therefore, you may need to have a 5-gal jug sample. If you want to conduct a jar test to include more variables than just coagulant dose—e.g., pH—then you may need more water sample. It is always a good practice to plan to have more water samples than you actually need, so that if you need to repeat the test or test the effect of more variables, you will not run out of water sample. Another thing to remember is to have water samples at the temperature they would be at in actual operation. If the sample is refrigerated but the test is to be conducted at ambient temperature, then take the water sample out of the refrigerator sufficiently ahead of the time for the test. It takes quite a while to bring the temperature up to ambient temperature without heating.

Getting Started

The machine is very simple and straightforward to set up and operate. The procedure is as follows:

1. After finishing setup, learn how to engage and disengage the mixers.

2. Put the mixers in engaged position. Plug into the power source and slowly turn up the rotation speed.

3. Be familiar with adjustment of the rotation speed. Two settings are typically used: high speed (at 300 rpm) for rapid mixing and low speed (at 30 rpm) for flocculation. Some jar testers have a toggle or button to switch between high and low speed settings. Be sure you are familiar with operating the machine, and mark the approximate positions of the speed-adjusting knob for the desired settings.

Now you are ready to conduct the test with the machine.

Test Procedure

1. Put a 2-L water sample in each of the six jars.

2. Prepare the correct dose for each jar by using a pipette to transfer the required volume of stock solution of coagulant into small beaker or test tubes. You may wish to start with test doses of 0, 10, 20, 30, 40, and 60 mg/L.

3. Measure and record the temperature and pH of each sample.

4. If the temperature of the samples is close to the desired temperature condition, start the mixers and set the speed at 300 rpm.

5. Start the stopwatch.

6. At 10 s, pour the stock solution of coagulant into the corresponding jars *at the same time*. You will need someone to help you to do this, because the pouring needs to be simultaneous. Ideally, three people will participate, with each pouring two.

7. At 40 (70) s (i.e., 30 [60] s after pouring the coagulants), set the mixers at the low-speed setting of 30 rpm.

8. At the desired flocculation time (typically 10 to 15 min) plus 40 (70) s, stop the mixers and measure and record temperature and pH.

9. Filter the samples through 0.2-μm membrane discs and collect filtrate samples for total organic carbon analysis. Please note that some membrane discs may have significant organics

leaching out during the filtration. Be sure to select membrane discs without this problem and rinse them thoroughly before use. Or run a blank to determine the impact of the filter on the test.

Data Collection

A sample data record is shown here. This is merely a guide. The data required depends upon the objective of the test.

Sample Source:_____Date of Collection:_____

Recorded by:_____Date of Test:_____

Sample ID	Volume of Stock (mL)	Coagulant Dose (mg/L)	Before Temp. (°C)	pH	After Temp. (°C)	pH	Observation of Flocs (Floc Size, Settling, etc.)

APPENDIX B

Tables and Conversion Factors

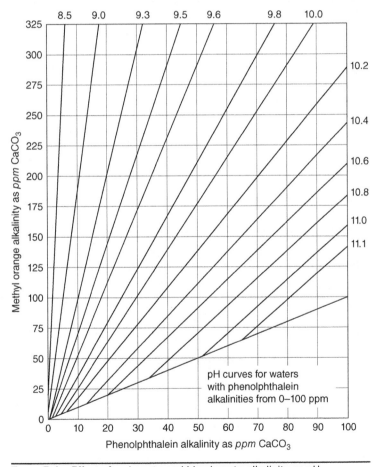

FIGURE B-1 Effect of carbonate and bicarbonate alkalinity on pH.

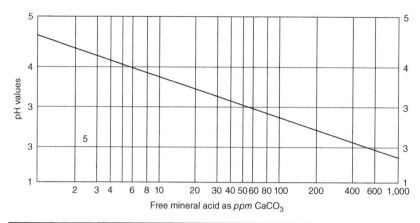

FIGURE B-2 Effect of mineral acidity on pH.

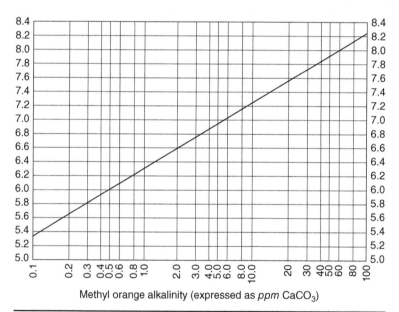

Methyl orange alkalinity (expressed as *ppm* CaCO₃)

FIGURE B-3 Effect of carbonate and bicarbonate alkalinity pH.

Specific Gravity	Baumé	% HCl	Normality	g/L	lb/ft³	lb/gal
1.0032	0.5	1	0.2750	10.03	0.6263	0.0837
1.0082	1.2	2	0.5528	20.16	1.259	0.1683
1.0181	2.6	4	1.117	40.72	2.542	0.3399
1.0279	3.9	6	1.691	61.67	3.850	0.5147
1.0376	5.3	8	2.276	83.01	5.182	0.6927
1.0474	6.6	10	2.871	104.7	6.539	0.8741
1.0574	7.9	12	3.480	126.9	7.921	1.059
1.0675	9.2	14	4.100	149.5	9.330	1.247
1.0776	10.4	16	4.728	172.4	10.76	1.439
1.0878	11.7	18	5.370	195.8	12.22	1.634
1.0980	12.9	20	6.022	219.6	13.71	1.833
1.1083	14.2	22	6.686	243.8	15.22	2.035
1.1187	15.4	24	7.363	268.5	16.76	2.241
1.1290	16.6	26	8.049	293.5	18.32	2.450
1.1392	17.7	28	8.748	319.0	19.91	2.662
1.1493	18.8	30	9.456	344.8	21.52	2.887
1.1593	19.9	32	10.17	371.0	23.16	3.096
1.1691	21.0	34	10.90	397.5	24.81	3.317
1.1789	22.0	36	11.64	424.4	26.49	3.542
1.1885	23.0	38	12.38	451.6	28.19	3.769
1.1980	24.0	40	13.14	479.2	29.92	3.999

TABLE **B-1**   Hydrochloric Acid (Aqueous Hydrochloric Acid Solutions)

Specific Gravity	Baumé	% NaOH	Normality	g/L	lb/ft³	lb/gal
1.0095	1.4	1	0.2524	10.10	0.6302	0.0842
1.0207	2.9	2	0.5101	20.41	1.274	0.1704
1.0318	4.5	3	0.7814	30.95	1.932	0.2583
1.0428	6.0	4	1.042	41.71	2.604	0.3841
1.0538	7.4	5	1.317	52.69	3.289	0.4397
1.0648	8.8	6	1.597	63.89	3.988	0.5332
1.0758	10.2	7	1.902	75.31	4.701	0.6284
1.0869	11.6	8	2.175	86.95	5.428	0.7256
1.0979	12.9	9	2.470	98.81	6.168	0.8246
1.1089	14.2	10	3.772	110.9	6.923	0.9254
1.1309	16.8	12	3.932	135.7	8.472	1.133
1.1530	19.2	14	4.034	161.4	10.08	1.347
1.1751	21.6	16	4.699	188.0	11.74	1.569
1.1972	23.9	18	5.387	215.5	13.45	1.798
1.2191	26.1	20	6.094	243.8	15.22	2.035
1.2411	28.6	22	6.824	273.0	17.05	2.279
1.2629	30.2	24	7.577	303.1	18.92	2.529
1.2848	32.1	26	8.349	334.0	20.85	2.788
1.3064	34.0	28	9.145	365.8	22.84	3.053
1.3279	35.8	30	9.960	398.4	24.87	3.324
1.3490	37.5	32	10.79	431.7	26.95	3.602
1.3696	39.1	34	11.64	465.7	29.07	3.886
1.3900	40.7	36	12.51	500.4	31.24	4.176
1.4101	42.2	38	13.39	535.8	33.45	4.472
1.4300	43.6	40	14.30	572.0	35.71	4.773
1.4494	45.0	42	15.22	608.7	38.00	5.080
1.4685	46.3	44	16.15	646.1	40.34	5.392
1.4873	47.5	46	17.10	684.2	42.71	5.709
1.5065	48.8	48	18.08	723.1	45.14	6.035
1.5253	49.9	50	19.07	762.7	47.61	6.364

TABLE B-2   Sodium Hydroxide (Aqueous Sodium Hydroxide Solutions)

Specific Gravity	Baumé	% H_2SO_4	g/L	lb/ft³	lb/gal
1.0051	0.7	1	10.05	0.6275	0.0839
1.0118	1.7	2	20.24	1.263	0.1689
1.0184	2.6	3	30.55	1.907	0.2550
1.0250	3.5	4	41.00	2.560	0.3422
1.0317	4.5	5	51.59	3.220	0.4305
1.0385	5.4	6	62.31	3.890	0.5200
1.0453	6.3	7	73.17	4.568	0.6106
1.0522	7.2	8	84.18	5.255	0.7025
1.0591	8.1	9	95.32	5.950	0.7955
1.0661	9.0	10	106.6	6.655	0.8897
1.0731	9.9	11	118.0	7.369	0.9851
1.0802	10.8	12	129.6	8.092	1.082
1.0874	11.7	13	141.4	8.825	1.180
1.0947	12.5	14	153.3	9.567	1.279
1.1020	13.4	15	165.3	10.32	1.379
1.1094	14.3	16	177.5	11.08	1.481
1.1168	15.2	17	189.9	11.85	1.584
1.1243	16.0	18	202.4	12.63	1.689
1.1318	16.9	19	215.0	13.42	1.795
1.1394	17.7	20	227.9	14.23	1.902
1.1471	18.6	21	240.9	15.04	2.010
1.1548	19.4	22	254.1	15.86	2.120
1.1626	20.3	23	267.4	16.69	2.231
1.1704	21.1	24	280.9	17.54	2.344
1.1783	21.9	25	294.6	18.39	2.458
1.1832	22.8	26	308.4	19.25	2.574
1.1942	23.6	27	322.4	20.13	2.691
1.2023	24.4	28	336.6	21.02	2.809
1.2104	25.2	29	351.0	21.91	2.929
1.2185	26.0	30	365.6	22.82	3.051
1.2267	26.8	31	380.3	23.74	3.173
1.2349	27.6	32	395.2	24.67	3.298
1.2432	28.4	33	410.3	25.61	3.424
1.2515	29.1	34	425.5	26.56	3.551

TABLE B-3 Sulfuric Acid (Aqueous Sulfuric Acid Solutions)

Specific Gravity	Baumé	% H$_2$SO$_4$	g/L	lb/ft^3	lb/gal
1.2599	29.9	35	441.0	27.53	3.680
1.2684	30.7	36	456.6	28.51	3.811
1.2769	31.4	37	472.5	29.49	3.943
1.2855	32.2	38	488.5	30.49	4.077
1.2941	33.0	39	504.7	31.51	4.212
1.3028	33.7	40	521.1	32.53	4.349
1.3116	34.5	41	537.8	33.57	4.488
1.3205	35.2	42	554.6	34.62	4.628
1.3294	35.9	43	571.6	35.69	4.770
1.3384	36.7	44	588.9	36.76	4.914
1.3476	37.4	45	606.4	37.86	5.061
1.3569	38.1	46	624.2	38.97	5.209
1.3663	38.9	47	642.2	40.09	5.359
1.3758	39.6	48	660.4	41.23	5.511
1.3854	40.3	49	678.8	42.38	5.665
1.3951	41.1	50	697.6	43.55	5.821
1.4049	41.8	51	716.5	44.73	5.979
1.4148	42.5	52	736.7	45.93	6.140
1.4248	43.2	53	755.1	47.14	6.302
1.4350	44.0	54	774.9	48.37	6.467

TABLE B-3   Sulfuric Acid (Aqueous Sulfuric Acid Solutions) (*Continued*)

Specific Gravity	Baumé	% NaOH	g/L	lb/ft³	lb/gal
1.0095	1.4	1	10.10	0.6302	0.0842
1.0207	2.9	2	20.41	1.274	0.1704
1.0318	4.5	3	30.95	1.932	0.2583
1.0428	6.0	4	41.71	2.604	0.3481
1.0538	7.4	5	52.69	3.289	0.4397
1.0648	8.8	6	63.89	3.988	0.5332
1.0758	10.2	7	75.31	4.701	0.6284
1.0869	11.6	8	86.95	5.428	0.7256
1.0979	12.9	9	98.81	6.168	0.8246
1.1089	14.2	10	110.9	6.923	0.9254
1.1309	16.8	12	135.7	8.472	1.133
1.1530	19.2	14	161.4	10.08	1.347
1.1751	21.6	16	188.0	11.74	1.569
1.1972	23.9	18	215.5	13.45	1.798
1.2191	26.1	20	243.8	15.22	2.035
1.2411	28.2	22	273.0	17.05	2.279
1.2629	30.2	24	303.1	18.92	2.529
1.2848	32.1	26	334.0	20.85	2.788
1.3064	34.0	28	365.8	22.84	3.053
1.3279	35.8	30	398.4	24.87	3.324
1.3490	37.5	32	431.7	26.95	3.602
1.3696	39.1	34	465.7	29.07	3.886
1.3900	40.7	36	500.4	31.24	4.176
1.4101	42.2	38	535.8	33.45	4.472
1.4300	43.6	40	572.0	35.71	4.773
1.4494	45.0	42	608.7	38.00	5.080
1.4685	46.3	44	646.1	40.34	5.392
1.4873	47.5	46	684.2	42.71	5.709
1.5065	48.8	48	723.1	45.14	6.035
1.5253	49.9	50	762.7	47.61	6.364

TABLE B-4   Sodium Hydroxide (Aqueous Sodium Hydroxide Solutions)

APPENDIX C

Atomic Numbers and Atomic Weights

Element	Symbol	Atomic Number	Atomic Weight
Actinium	Ac	89	227.0278
Aluminum	Al	13	26.98154
Americium	Am	95	(243)*
Antimony	Sb	51	121.75
Argon	Ar	18	39.948
Arsenic	As	33	74.9216
Astatine	At	85	(210)
Barium	Ba	56	137.33
Berkelium	Bk	97	(247)
Beryllium	Be	4	9.01218
Bismuth	Bi	83	208.9804
Boron	B	5	10.9804
Bromine	Br	35	79.904
Cadmium	Cd	48	112.41
Calcium	Ca	20	40.08
Californium	Cf	98	(251)
Carbon	C	6	12.011
Cerium	Ce	58	140.12
Cesium	Cs	55	132.9054
Chlorine	Cl	17	35.453
Chromium	Cr	24	51.996
Cobalt	Co	27	58.9332
Copper	Cu	29	6,63.546
Curium	Cm	96	(247)
Dysprosium	Dy	66	162.50
Einsteinium	Es	99	(254)
Erbium	Er	68	167.26

(Continued)

Element	Symbol	Atomic Number	Atomic Weight
Europium	Eu	63	151.96
Fermium	Fm	100	(257)
Fluorine	F	9	18.99840
Francium	Fr	87	(223)
Gadolinium	Gd	64	157.25
Gallium	Ga	31	69.72
Germanium	Ge	32	72.59
Gold	Au	79	196.9665
Hafnium	Hf	72	178.49
Helium	He	2	4.00260
Holmium	Ho	67	164.9304
Hydrogen	H	1	1.0079
Indium	In	49	114.82
Iodine	I	53	6,126.9045
Iridium	Ir	77	192.22
Iron	Fe	26	55.847
Krypton	Kr	36	83.80
Lanthanum	La	57	138.9055
Lawrencium	Lr	103	(260)
Lead	Pb	82	207.2
Lithium	Li	3	6.941
Lutetium	Lu	71	174.97
Magnesium	Mg	12	24.305
Manganese	Mn	25	54.9380
Mercury	Hg	80	200.59
Molybdenum	Mo	42	95.94
Neodymium	Nd	60	144.24
Neon	Ne	10	20.179
Neptunium	Np	93	237.0482
Nickel	Ni	28	58.70
Niobium	Nb	41	92.9064
Nitrogen	N	7	14.0067
Nobelium	No	102	(259)
Osmium	Os	76	190.2
Oxygen	O	8	15.9994
Palladium	Pd	46	106.4
Phosphorus	P	15	30.97376
Platinum	Pt	78	195.09

Element	Symbol	Atomic Number	Atomic Weight
Plutonium	Pu	94	(244)
Polonium	Po	84	(209)
Potassium	K	19	39.0983
Praseodymium	Pr	59	140.9077
Promethium	Pm	61	(145)
Protactinium	Pa	91	231.03589
Radium	Ra	88	226.0254
Radon	Rn	86	(222)
Rhenium	Re	75	186.207
Rhodium	Rh	45	102.9055
Rubidium	Rb	37	85.4678
Ruthenium	Ru	44	101.07
Samarium	Sm	62	150.4
Scandium	Sc	21	44.9559
Selenium	Se	34	78.96
Silicon	Si	14	28.0855
Silver	Ag	47	107.868
Sodium	Na	11	22.98977
Strontium	Sr	38	87.62
Sulfur	S	16	32.06
Tantalum	Ta	73	180.9479
Technetium	Tc	43	(97)
Tellurium	Te	52	127.60
Terbium	Tb	65	158.9254
Thallium	Tl	81	204.37
Thorium	Th	90	232.0381
Thulium	Tm	69	168.9342
Tin	Sn	50	118.69
Titanium	Ti	22	47.90
Tungsten	W	74	183.85
Uranium	U	92	238.029
Vanadium	V	23	50.9414
Xenon	Xe	54	131.30
Ytterbium	Yb	70	173.04
Yttrium	Y	39	88.9059
Zinc	Zn	30	65.38
Zirconium	Zr	40	91.22

*Value in parenthesis is the mass number of the longest lived isotope of the element.

APPENDIX D

Examples of State Water Reuse Criteria for Selected Nonpotable Applications[*]

---

*Source: National Academy of Sciences (2012).

State	Fodder Crop Irrigation[a] Quality Limits	Treatment Required	Processed Food Crop Irrigation[b] Quality Limits	Treatment Required	Food Crop Irrigation[c,d] Quality Limits	Treatment Required	Restricted Recreational Impoundments[e] Quality Limits	Treatment Required
Arizona	• 1,000 fecal coliforms/100 mL	• Secondary (stabilization ponds)	Not covered	Not covered	• No detectable fecal coliforms/100 mL • 2 ntu	• Secondary • Filtration • Disinfection	• No detectable fecal coliforms/100 mL • 2 ntu	• Secondary • Filtration • Disinfection
California	Not specified	Secondary	Not specified	Secondary	• 2.2 total coliforms/100 mL • 2 ntu	• Secondary • Coagulation[f] • Filtration • Disinfection	2.2 total coliforms/100 mL	• Secondary • Disinfection
Colorado	Not covered	Not covered	Not covered	Not covered	Not covered	Not covered	Not covered	Not covered
Florida	• 200 fecal coliforms/100 mL • 20 mg/L CBOD • 20 mg/L TSS	• Secondary • Disinfection	• No detectable fecal coliforms/100 mL • 20 mg/L CBOD • 5 mg/L TSS	• Secondary • Filtration • Disinfection	• No detectable fecal coliforms/100 mL • 20 mg/L CBOD • 5 mg/L TSS	• Secondary • Filtration • Disinfection	• No detectable fecal coliforms/100 mL • 20 mg/L CBOD • 5 mg/L TSS	• Secondary • Filtration • Disinfection
Texas	• 200 fecal coliforms or E. coli/100 mL • 35 enterococci/100 mL • 20 mg/L BOD • 15 mg/L CBOD	Not specified	• 200 fecal coliforms or E. coli/100 mL • 35 enterococci/100 mL • 20 mg/L BOD • 15 mg/L CBOD	Not specified	• 20 fecal coliforms or E. coli/100 mL • 4 enterococci/100 mL • 3 ntu • 5 mg/L BOD or CBOD	Not specified	• 20 fecal coliforms or E. coli/100 mL • 4 enterococci/100 mL • 3 ntu • 5 mg/L BOD or CBOD	Not specified

State					
Utah	• 200 fecal coliforms/100 mL • 25 mg/L BOD • 25 mg/L TSS • Secondary • Disinfection	• No detectable fecal coliforms/100 mL • 10 mg/L BOD • 2 ntu • Secondary • Filtration • Disinfection	• No detectable fecal coliforms/100 mL • 10 mg/L BOD • 2 ntu • Secondary • Filtration • Disinfection	• 200 fecal coliforms/100 mL • 25 mg/L BOD • 25 mg/L TSS • Secondary • Filtration • Disinfection	• 200 fecal coliforms/100 mL • 25 mg/L BOD • 25 mg/L TSS • Secondary • Disinfection
Washington	• 240 total coliforms/100 mL • Secondary • Disinfection	• 240 total coliforms/100 mL • Secondary • Disinfection	• 2.2 total coliforms/100 mL • 2 ntu • Secondary • Coagulation • Filtration • Disinfection	• 2.2 total coliforms/100 mL • Secondary • Disinfection	• 200 fecal coliforms/100 mL • 25 mg/L BOD • 25 mg/L TSS • Secondary • Filtration • Disinfection

[a] In some states, more restrictive requirements apply where milking animals are allowed to graze on pasture irrigated with reclaimed water.

[b] Physical or chemical processing sufficient to destroy pathogenic microorganisms. Less restrictive requirements may apply where there is no direct contact between reclaimed water and the edible portion of the crop.

[c] Food crops eaten raw where there is direct contact between reclaimed water and the edible portion of the crop.

[d] In Florida and Texas, "Irrigation of edible crops that will not be peeled, skinned, cooked, or thermally processed before consumption is allowed if an indirect application method is used which will preclude direct contact with the reclaimed water" (such as ridge and furrow irrigation, drip irrigation, or a subsurface distribution system) (30 Texas Administrative Code §210.24).

[e] Recreation is limited to fishing, boating, and other non-body-contact activities.

[f] Not needed if filter effluent turbidity does not exceed 2 ntu; turbidity of the influent to the filters is continually measured; influent turbidity does not exceed 5 ntu for more than 15 min and never exceeds 10 ntu; and there is capability to automatically activate chemical addition or divert the wastewater should the filter influent turbidity exceed 5 ntu for more than 15 min.

Adapted from Arizona Department of Environmental Quality (2011), California Department of Environmental Quality (2011), California Department of Public Health (2009), Colorado Department of Health and Environment (2007), Florida Department of Environmental Protection (2007), Texas Commission on Environmental Quality (2010), Utah Department of Environmental Quality (2011), Washington Department of Health and Washington Department of Ecology (1997).

TABLE D-1 Some Examples of State Water Reuse Criteria for Selected Nonpotable Applications

State	Restricted Access Irrigation[a] Quality Limits	Treatment Required	Unrestricted Access Irrigation[b] Quality Limits	Treatment Required	Toilet Flushing[c] Quality Limits	Treatment Required	Industrial Cooling Water[d] Quality Limits	Treatment Required
Arizona	200 fecal coliforms/100 mL	• Secondary • Disinfection	• No detectable fecal coliforms/ 100 mL • 2 ntu	• Secondary • Filtration • Disinfection	• No detectable fecal coliforms/ 100 mL • 2 ntu	• Secondary • Filtration • Disinfection	Not covered	Not covered
California	23 total coliforms/100 mL	• Secondary • Disinfection	• 2.2 total coliforms/ 100 mL • 2 ntu	• Secondary • Coagulation[e] • Filtration • Disinfection	• 2.2 total coliforms/ 100 mL • 2 ntu	• Secondary • Filtration • Disinfection	• 2.2 total coliforms/ 100 mL • 2 ntu	• Secondary • Coagulation[e] • Filtration • Disinfection
Colorado	• 126 E. coli/ 100 mL • 30 mg/L TSS	• Secondary • Disinfection	• No detectable E. coli/100 mL • 3 ntu	• Secondary • Filtration • Disinfection	Not covered	Not covered	• 126 E. coli/ 100 mL • 30 mg/L TSS	• Secondary • Disinfection
Florida	• 200 fecal coliforms/ 100 mL • 20 mg/L CBOD • 20 mg/L TSS	• Secondary • Disinfection	• No detectable fecal coliforms/ 100 mL • 20 mg/L CBOD • 5 mg/L TSS	• Secondary • Filtration • Disinfection	• No detectable fecal coliforms/ 100 mL • 20 mg/L CBOD • 5 mg/L TSS	• Secondary • Filtration • Disinfection	• No detectable fecal coliforms/ 100 mL • 20 mg/L CBOD • 5 mg/L TSS	• Secondary • Filtration • Disinfection
Texas	• 200 fecal coliforms or E. coli/100 mL • 35 enterococci/ 100 mL • 20 mg/L BOD • 15 mg/L CBOD	Not specified	• 20 fecal coliforms or E. coli/100 mL • 4 enterococci/ 100 mL • 3 ntu • 5 mg/L BOD or CBOD	Not specified	• 20 fecal coliforms or E. coli/100 mL • 4 enterococci/ 100 mL • 3 ntu • 5 mg/L BOD or CBOD	Not specified	• 200 fecal coliforms or E. coli/100 mL • 35 enterococci/ 100 mL • 20 mg/L BOD • 15 mg/L CBOD	Not specified

TABLE D-2 Further Examples of State Water Reuse Criteria for Selected Nonpotable Applications

State					
Utah	• 200 fecal coliforms/100 mL • 25 mg/L BOD • 25 mg/L TSS • Secondary • Disinfection	• No detectable fecal coliforms/100 mL • 10 mg/L BOD • 2 ntu • Secondary • Disinfection	• No detectable fecal coliforms/100 mL • 10 mg/L BOD • 2 ntu • Secondary • Filtration • Disinfection	• 200 fecal coliforms/100 mL • 25 mg/L BOD • 25 mg/TSS • Secondary • Filtration • Disinfection	• 200 fecal coliforms/100 mL • 25 mg/L BOD • 25 mg/TSS • Secondary • Disinfection
Washington	23 total coliforms/100 mL • Secondary • Disinfection	• 2.2 total coliforms/100 mL • 2 ntu • Secondary • Disinfection	• 2.2 total coliforms/100 mL • 2 ntu • Secondary • Coagulation • Filtration • Disinfection	• 2.2 total coliforms/100 mL • 2 ntu • Secondary • Coagulation • Filtration • Disinfection	• 2.2 total coliforms/100 mL • 2 ntu • Secondary • Coagulation • Filtration • Disinfection

[a] Classification varies by state; generally includes irrigation of cemeteries, freeway medians, restricted access golf courses, and similar restricted access areas.

[b] Includes irrigation of parks, playgrounds, schoolyards, residential lawns, and similar unrestricted access areas.

[c] Not allowed in single-family residential dwelling units.

[d] Cooling towers where a mist is created that may reach populated areas.

[e] Not needed if filter effluent turbidity does not exceed 2 ntu; turbidity of the influent to the filters is continually measured; influent turbidity does not exceed 5 ntu for more than 15 min and never exceeds 10 ntu; and there is capability to automatically activate chemical addition or divert the wastewater should the filter influent turbidity exceed 5 ntu for more than 15 min.

Adapted from Arizona Department of Environmental Quality (2011), California Department of Environmental Quality (2011), California Department of Public Health (2009), Colorado Department of Health and Environment (2007), Florida Department of Environmental Protection (2007), Texas Commission on Environmental Quality (2007), Utah Department of Environmental Quality (2010), Washington Department of Health and Washington Department of Ecology (1997).

APPENDIX E

The National Pretreatment Program and Expanding Source Control*

Τhe Clean Water Act (CWA), passed in 1972, was designed to eliminate the discharge of pollutants into the nation's waters and to achieve fishable and swimmable water quality levels. The United States Environmental Protection Agency's (USEPA) National Pollutant Discharge Elimination System (NPDES), one of the CWA's key components, requires that all direct discharges to the nation's waters comply with an NPDES permit, but many industries discharge through municipal wastewater treatment plants. Consequently, the USEPA established the National Pretreatment Program, which requires industrial and commercial dischargers to treat or control pollutants in their wastewater prior to discharge to municipal wastewater treatment plants.

Generally, wastewater treatment plants are designed to treat domestic wastewater only. Under the National Pretreatment Program, local governments must implement pretreatment standards requiring that pollutants be removed from any industrial or commercial discharge to a wastewater collection system. The current objectives of the program are to

- prevent the discharge of pollutants that may pass through the municipal wastewater treatment plant untreated,

---

*Source: National Academy of Sciences.

- protect wastewater treatment plants from hazards posed by untreated industrial wastewater, and

- improve the quality of effluents and biosolids so that they can be used for beneficial purposes (Alan Plummer Associates, 2010).

Under this program, wastewater authorities must adopt ordinances, issue permits, monitor compliance, and take enforcement action when violations occur. The USEPA has established numeric effluent guidelines for 56 categories of industry, and the Clean Water Act requires that the USEPA annually review its effluent guidelines and pretreatment standards to identify new categories for standards.

A summary of the National Pretreatment Program's achievements (USEPA, 2003b) demonstrates that it has resulted in significant reductions in the discharge of toxic chemicals to the environment. Most standards have been based on the 129 priority pollutants, which were included in the 1977 Amendments to the Clean Water Act. Recently, an update has been proposed to the Universal Wastes Rule to incorporate pharmaceuticals and thereby streamline disposal of hazardous pharmaceutical wastes and reduce the amount of these chemicals in wastewater (73 Fed. Reg. 73520, Dec. 2, 2008), although no subsequent action has been taken.

In *Issues in Potable Reuse* the National Research Council (NRC, 1998), recommended that the USEPA develop a priority list of contaminants of public health significance that are known or anticipated to occur in wastewater, and that individual communities institute stringent industrial pretreatment and pollutant source control programs based on this guidance. The USEPA has not developed such a list, but some utilities have taken actions on their own. For example, the Orange County Sanitation District, which supplies reclaimed water for the Orange County Water District's Groundwater Replenishment System, has expanded the agency's source control program to include pollutant prioritization, enhanced outreach to industry and the public, and a geographic information-system-based toxics inventory. Through its source control program, the Orange County Sanitation District was able to reduce the industrial discharge of 1,4-dioxane and N-nitrosodimethylamine (NDMA) into the wastewater collection system. Oregon is developing rules that will require municipal wastewater treatment plants to develop plans for reducing listed priority persistent pollutants. The Oregon list includes well-studied pollutants as well as some for which little information exists (Alan Plummer Associates, 2010). Other programs have been developed to reduce the introduction of pharmaceutical products into wastewater systems.

Under existing federal regulations, class V injection wells do not require a federal permit if they do not endanger underground sources of drinking water and comply with other UIC program requirements (49 CFR § 144.82). However, states may include additional requirements with regard to treatment, well construction, and water quality monitoring standards prior to permitting any injection of reclaimed water into aquifers that are currently being, or could be, used for potable supply.

U.S. Drinking Water Regulations: The Safe Drinking Water Act

The U.S. drinking water regulations set standards that all drinking water treatment plants are required to meet, whether they use pristine water supply sources, supply water from potable reuse projects, or practice de facto reuse. This section provides a review of the regulatory framework and an evaluation of its adequacy for potable reuse.

In 1974, Congress passed the Safe Drinking Water Act (SDWA), which provides authority to the USEPA to establish and enforce national standards for drinking water to protect public health. For priority contaminants, the USEPA determines a maximum contaminant level goal, the level below which there is no known or expected risk to human health. A maximum contaminant level (MCL) is the highest concentration of a contaminant that is allowed in drinking water through an enforceable primary standard. MCLs are set as close to MCL goals as possible, considering best available treatment technology and costs versus benefits. Regular testing and reporting is required to ensure that contaminants do not exceed MCLs. For some contaminants, including microorganisms, the USEPA requires that specific treatment techniques (TTs) be used in the drinking water treatment process in lieu of an MCL. Individual states are allowed to adopt more stringent standards, if desired. In 2009, the USEPA National Primary Drinking Water Regulations included three MCLs for disinfectants, four MCLs for radionuclides, five MCLs or TTs for microorganisms, 16 MCLs or TTs for inorganic chemicals, and 53 MCLs or TTs for organic chemicals (EPA, 2009b).

Consideration of De Facto Water Reuse in U.S. Drinking Water Standards

The U.S. Public Health Service published drinking water standards in 1962 (U.S. Public Health Service, 1962) which provide some insight into concerns regarding de facto (or unplanned) water reuse. Although the standards specifically state that "the water supply should be obtained from the most desirable source which is feasible," the document goes on to state: "If the source is not adequately protected by natural means, the supply shall be adequately protected by treatment." The 1962 standards included alkylbenzene sulfonate (ABS), an anionic surfactant that was commonly used in detergent. The statement is made that "waters containing ABS are likely to be at least 10 percent of sewage origin for each mg ABS/liter present." Also of pertinent interest was the use of carbon chloroform extract (CCE) in the 1962 standards as an indicator of anthropogenic organic compounds in water. A standard of 200 µg/L CCE was established to "represent an exceptional and unwarranted dosage of the water consumer with ill-defined chemicals," whether from wastewater or other sources. The ABS and CCE standards promulgated in 1962 demonstrate that the federal government understood that de facto water reuse was occurring and that the contamination of drinking water from a diversity of synthetic organic contaminants was possible.

NDMA was likely present in reclaimed and potable waters for quite some time at concentrations far greater than 0.7 ng/L, a USEPA-established groundwater cleanup level (USEPA, 2010b). Although nitrosamines were known to occur in potable water systems as early as the 1970s, NDMA did not gain widespread attention until the 1990s, when it was discovered in elevated levels in California reuse systems (Najm and Trussell, 2001). NDMA was added to the CCL in 2009 and was included in the UCMR2.

Protection Against Greater Microbial Risks

Under the Safe Drinking Water Act (SDWA), viruses and protozoa are regulated by treatment techniques rather than MCL. Under the original Surface Water Treatment Rule (SWTR [42 USCA 300g-1 (b)(2)(c)]), all surface water treatment plants (unless exempt by waiver) had to have treatment sufficient to achieve 99.9% reduction in *Giardia* and 99.99% reduction in viruses, and the operational characteristics of treatment steps needed to achieve this were defined in guidance manuals. Bacterial pathogens are also presumed to be reduced. Under the Long Term 2 SWTR (LT2SWTR), utilities have been required to take measurements of the source water concentrations of *Cryptosporidium* to determine if further reductions of *Cryptosporidium* are required. This additional reduction (either by additional processes or by more intensive application of existing processes) would also result in increased reduction of bacteria, viruses, and *Giardia*. It is uncertain whether this regulatory framework is sufficient when source waters contain a high proportion of wastewater.

Failure of any of the treatment processes used to control pathogens would carry a risk of sporadic breakthrough of pathogens. To the degree that high levels of pathogen reduction are achieved by engineered processes rather than use of a protected watershed (with lower levels of pathogens), it becomes more critical to maintain multiple barriers designed to improve reliability, whether in a planned reuse situation or in a conventional water system treating impaired surface waters.

Assessment of the Existing Federal Regulatory Framework for Potable Reuse

Reclaimed water used for potable reuse ultimately is required to meet all physical, chemical, radiological, and microbiological standards for drinking water. The SDWA provides a measure of human health protection in terms of discrete chemicals based upon standards established and enforced by the USEPA (whether in the form of a numerical MCL or a treatment technique). However, the SDWA does not yet establish standards for all potentially harmful constituents that may be present in wastewater. At present, the rules promulgated under the CWA and SDWA do not sufficiently address the public health concerns associated with reclaimed water for potable reuse. Also, the data sets used to develop the universe of contaminants considered for regulation are not yet sufficient to capture the range of contaminants that may be present in reclaimed water for potable reuse applications. More detailed reuse regulations exist in some states to address some, but not all, of these concerns (discussed later). A discussion of potential advantages and disadvantages of federal reuse regulations follows the discussion of state reuse regulations. However, it is critical to understand that many drinking water systems in the United States utilize source waters with significant contributions from treated wastewater. Therefore, a revised regulatory paradigm that provides greater protection for potable reuse applications would need to consider the extent of de facto reuse to provide equivalent protection for all consumers.

Water Reuse Regulations and Guidelines

There are no federal regulations specifically governing water reclamation and reuse in the United States; hence, the regulation of water reuse rests with the individual states. However, the federal government does provide guidance to states via the USEPA's *Guidelines for Water Reuse*, which "presents and summarizes recommended water reuse guidelines for the benefit of the water and wastewater utilities and regulatory agencies" (USEPA, 2004). Regulations differ from guidelines in that regulations are legally adopted, enforceable, and mandatory, whereas guidelines are advisory and compliance is voluntary. Guidelines sometimes become enforceable requirements if they are incorporated into state regulations or water reuse permits.

Water reuse regulations and guidelines can be based on a variety of considerations but are directed principally at public health protection. For nonpotable reclaimed water applications, criteria generally address only microbiological and environmental concerns; existing regulations and guidelines for nonpotable reuse generally are not risk based. For potable reuse applications, health risks associated with pathogenic microorganisms and chemical constituents are both addressed. Reuse guidelines also generally address proper controls and safety precautions implemented at areas where nonpotable reclaimed water is used (e.g., warning signs, color-coded pipes, cross-connection control provisions). Additionally, guidelines may include water quality considerations that are unrelated to public health or environmental protection but are important to the success of specific nonpotable reuse applications (e.g., irrigation or industrial cooling). The following sections summarize the federal reuse guidelines and state guidelines or regulations for nonpotable and potable reuse.

USEPA Guidelines for Water Reuse

The USEPA's *Guidelines for Water Reuse* (USEPA, 2004), which covers both potable and nonpotable reuse, is intended to provide reasonable guidance, with supporting information, for utilities and regulatory agencies in the United States. The guidelines are particularly useful for states that have not developed their own water reuse regulations or are revising or expanding existing regulations. The guidelines contain a plethora of information on various aspects of water reuse, including treatment technology, public health concerns, legal and institutional issues, public involvement programs, and suggested water quality treatment and quality requirements for different reuse applications. The remainder of this section focuses on the suggested water treatment and quality requirements included in the guidelines.

The guidelines summarize the treatment processes and water quality limits for a variety of nonpotable and potable reclaimed water applications. Also included are monitoring frequencies, setback distances, and other controls for each water reuse application. The suggested guidelines pertaining to treatment and water quality are based primarily on wastewater reclamation and reuse data from the United States. The guidelines apply to the reclamation of domestic wastewater from treatment plants with limited industrial waste inputs and "are not intended to be used as definitive water reclamation and reuse criteria" (USEPA, 2004).

Nonpotable Reuse

The USEPA guidelines (2004) recommend two different levels of disinfection for nonpotable uses of reclaimed water. For applications where direct or indirect reclaimed water contact is probable or expected, and for dual-water systems where cross-connections are always possible, disinfection to a level of no detectable fecal coliform organisms per 100 mL is advised (based on the median value of the last 7 days for which analyses have been completed). In any given sample, the USEPA also recommends that fecal coliforms not exceed 14 per 100 mL (2004). For applications where no direct public or worker contact with reclaimed water occurs, the guidelines recommend disinfection to achieve a fecal coliform concentration not exceeding 200 per 100 mL (based on the median value of the last 7 days of analyses). It is noteworthy that the USEPA guidelines for nonpotable reuse applications are not based on rigorous health risk assessment methodology.

Additional recommendations for nonpotable reuse applications include

- clear, colorless, odorless, and nontoxic water;
- a setback distance of 50 ft between areas irrigated with reclaimed water and potable water supply wells;
- maintenance of a chlorine residual of greater than or equal to 0.5 mg/L in the distribution system;
- reliable treatment and emergency storage or disposal alternatives for inadequately treated water;
- cross-connection control devices; and
- color-coded nonpotable water lines and appurtenances.

The guidelines include similar design and operational recommendations for the other reclaimed water applications.

Potable Reuse

The USEPA's guidelines provide some specific wastewater treatment and reclaimed water quality recommendations for potable reuse via groundwater recharge and surface water augmentation. The guidelines outline the extensive treatment, water quality, and monitoring requirements that are likely to be imposed for potable reuse projects and are based principally on California's draft groundwater recharge regulations and Florida's potable reuse regulations in place at the time the guidelines were written. The guidelines recommend that potable reuse projects meet drinking water standards and also monitor for hazardous compounds (or classes of compounds) that are not included in the drinking water standards (USEPA, 2004). The USEPA guidelines' focus is on end-point water quality.

APPENDIX F
State Websites

State	Type	Agency	Rules	Website
Alabama	Guidelines	Department of Environmental Management	Guidelines and Minimum Requirements for Municipal, Semi-Public and Private Land Treatment Facilities	www.adem.state.al.us/ http://209.192.62.106/ Land treatment guidelines not found on website
Alaska	Regulations	Department of Environmental Conservation	Alaska Administrative Code, Title 18—Environmental Conservation, Chapter 72, Article 2, Section 275—Disposal Systems	www.state.ak.us/local/akpages/ ENV.CONSERV/home.htm www.state.ak.us/local/akpages/ ENV.CONSERV/title18/aac72ndx.htm
Arizona	Regulations	Department of Environmental Quality	Arizona Administrative Code, Title 18—Environmental Quality, Chapter 11, Article 3—Reclaimed Water Quality Standards and Chapter 9, Article 7—Direct Reuse of Reclaimed Water	www.sos.state.az.us/ www.sos.state.az.us/public_services/Table_of_Contents.htm
Arkansas	Guidelines	Department of Environmental Quality	Arkansas Land Application Guidelines for Domestic Wastewater	www.adeq.state.ar.us/default.htm www.adeq.state.ar.us/water/branch_permits/default.htm Land application guidelines not found on website
California	Regulations	Department of Health Services	California Department of Health Services Regulations and Guidance for Recycled Water (The Purple Book) California Code of Regulations, Titles 17 and 22	www.dhs.cahwnet.gov www.dhs.ca.gov/ps/ddwem/publications/waterrecycling/waterrecyclingindex.htm http://ccr.oal.ca.gov/

Colorado	Regulations	Department of Public Health and Environment	Water Quality Control Commission Regulation 84—Reclaimed Domestic Wastewater Control Regulation	www.cdphe.state.co.us/ cdphehom.asp www.cdphe.state.co.us/op/regs/ waterregs/100284.pdf
Connecticut	Neither	Department of Environmental Protection	None	http://dep.state.ct.us/
Delaware	Regulations	Department of Natural Resources and Environmental Control	Guidance and Regulations Governing the Land Treatment of Wastes	www.dnrec.state.de.us/dnrec2000/ www.dnrec.state.de.us/ water2000/Sections/GroundWat/ GWDSRegulations.htm
Florida	Regulations	Department of Environmental Protection	Florida Administrative Code, Chapter 62-610—Reuse of Reclaimed Water and Land Application	www.dep.state.fl.us/ www.dep.state.fl.us/water/reuse/ index.htm http://fac.dos.state.fl.us/
Georgia	Guidelines	Department of Natural Resources	Environmental Protection Division Guidelines for Water Reclamation and Urban Water Reuse	www.dnr.state.ga.us/dnr/environ/ www.ganet.org/dnr/environ/ techguide_files/wpb/reuse.pdf
Hawaii	Guidelines	Department of Health	Guidelines for the Treatment and Use of Recycled Water	www.state.hi.us/doh/ www.state.hi.us/doh/eh/wwb/reuse-final.pdf
Idaho	Regulations	Department of Environmental Quality	58.01.17 Wastewater Land Application Permit Rules	www2.state.id.us/adm/index.htm www2.state.id.us/adm/adminrules/ rules/idapa58/58index.htm

(Continued)

State	Type	Agency	Rules	Website
Illinois	Regulations	Environmental Protection Agency	Illinois Administrative Code, Title 35, Subtitle C, Part 372—Illinois Design Standards for Slow Rate Land Application of Treated Wastewater	www.ipcb.state.il.us/ www.ipcb.state.il.us/SLR/IPCBandIE PAEnvironmentalRegulations-Title35. asp
Indiana	Regulations	Department of Environmental Management	Indiana Administrative Code, Title 327, Article 6.1—Land Application of Biosolid, Industrial Waste Product, and Pollutant-Bearing Water	www.in.gov/idem/ www.in.gov/legislative/iac/ title327.html
Iowa	Regulations	Department of Natural Resources	Environmental Protection Division Iowa Wastewater Design Standards, Chapter 21—Land Application of Wastewater	www.state.ia.us/epd/ www.state.ia.us/epd/wastewtr/ design.htm
Kansas	Guidelines	Department of Health and Environment	KDHE Administrative Rules and Regulations, 28-16—Water Pollution Control	www.kdhe.state.ks.us/ www.kdhe.state.ks.us/regs/
Kentucky	Neither	None	None	http://kentucky.gov/Default.html
Louisiana	Neither	None	None	www.state.la.us/
Maine	Neither	None	None	www.state.me.us/
Maryland	Guidelines	Department of the Environment	Department of the Environment Guidelines for Land Treatment of Municipal Wastewaters, Title 26	www.mde.state.md.us/index.asp www.dsd.state.md.us/comar/ subtitle_chapters/26_Chapters.htm
Massachusetts	Guidelines	Massachusetts Department of Environmental Protection	Interim Guidelines on Reclaimed Water (Revised)	www.state.ma.us/dep/dephome.htm www.state.ma.us/dep/brp/wwm/ t5regs.htm

State			
Michigan	Regulations	Department of Environmental Quality	Part 22 Rules of Part 31 Groundwater Quality Rules Part 22 Guidesheet II Irrigation Management Plan Rule 2215 Various Aboveground Disposal Systems	www.michigan.gov/deq www.michigan.gov/deq/0,1607, 7-135-3313_3682-14902-,00.html www.michigan.gov/deq/0,1607, 7-135-3312_4117-9782-,00.html www.deq.state.mi.us/documents/ deq-wmd-gwp-Rule2215VariousAbove GroundDisposalSystems-
Minnesota	Neither	None	None	www.state.mn.us/cgi-bin/portal/mn/ jsp/home.do?agency=NorthStar
Mississippi	Neither	None	None	www.mississippi.gov/
Missouri	Regulations	Department of Natural Resources	Code of State Regulations, Title 10, Division 20, Chapter 8—Design Guides	www.sos.mo.gov/ www.sos.mo.gov/adrules/csr/ current/10csr/10csr.asp
Montana	Guidelines	Department of Environmental Quality	Design Standards for Wastewater Facilities, Standards for the Spray Irrigation of Wastewater*	www.deq.state.mt.us/ www.deq.state.mt.us/wqinfo/ Circulars/DEQ2.PDF
Nebraska	Regulations	Department of Environmental Quality	Title 119, Chapter 9—Disposal of Sewage Sludge and Land Application of Effluent—Regulations refer to the use of Guidelines for Treated Wastewater Irrigation Systems, February 1986	www.deq.state.ne.us/

(Continued)

State	Type	Agency	Rules	Website
Nevada	Regulations	Department of Conservation and Natural Resources	Division of Environmental Protection, Nevada Administrative Code 445A.275—Use of Treated Effluent for Irrigation: General Design Criteria for Reclaimed Water Irrigation Use	http://ndep.nv.gov/ http://ndep.nv.gov/nac/445a-226.pdf http://ndep.nv.gov/bwpc/wts1a.pdf
New Hampshire	Neither	None	None	www.state.nh.us/
New Jersey	Guidelines	Department of Environmental Protection, Division of Water Quality	Technical Manual for Reclaimed Water for Beneficial Reuse	www.state.nj.us/dep/dwq/techman.htm
New Mexico	Guidelines	Environment Department	Use of Domestic Wastewater Effluent for Irrigation	www.nmenv.state.nm.us/ Guidelines not found on website
New York	Guidelines	Department of Environmental Conservation	State Guidelines for the Use of Land Treatment of Wastewater	www.dec.state.ny.us/ Guidelines not found on website
North Carolina	Regulations	Department of Environment and Natural Resources	Administrative Rules, Title 15A, Chapter 02, Subchapter H, .0200—Waste Not Discharged to Surface Waters	www.oah.state.nc.us/rules/ http://ncrules.state.nc.us/ncadministrativ_/title15aenviron_/chapter02enviro_/default.htm

North Dakota	Guidelines	Department of Health	Division of Water Quality Criteria for Irrigation with Treated Wastewater Recommended Criteria for Land Disposal of Effluent	www.health.state.nd.us/wq/
Ohio	Guidelines	Environmental Protection Agency	The Ohio State University Extension Bulletin 860—Reuse of Reclaimed Wastewater Through Irrigation	www.epa.state.oh.us/ http://ohioline.osu.edij/b860/
Oklahoma	Regulations	Department of Environmental Quality	Title 252, Chapters 621 and 656	www.deq.state.ok.us/mainlinks/ deqrules.htm

*Standards for the Spray Irrigation of Wastewater in Design Standards for Wastewater Facilities (1999), Circular DEQ2, Montana Department of Environmental Quality, Helena, MT.

APPENDIX G

California Code of Regulations, Title 17

Division 1. State Department of Health Services

Chapter 5. Sanitation (Environmental)

Group 4. Drinking Water Supplies

Article 1. General

7583. Definitions In addition to the definitions in Section 4010.1 of the Health and Safety Code, the following terms are defined for the purpose of this Chapter:

(a) "Approved Water Supply" is a water supply whose potability is regulated by a State or local health agency.

(b) "Auxiliary Water Supply" is any water supply other than that received from a public water system.

(c) "Air-Gap Separation (AG)" is a physical break between the supply line and a receiving vessel.

(d) "AWWA Standard" is an official standard developed and approved by the American Water Works Association (AWWA).

(e) "Cross-Connection" is an unprotected actual or potential connection between a potable water system used to supply water for drinking purposes and any source or system containing unapproved water or a substance that is not or cannot be approved as safe, wholesome, and potable. By-pass arrangements, jumper connections, removable sections, swivel or changeover devices, or other devices through which backflow could occur, shall be considered to be cross-connections.

(f) "Double Check Valve Assembly (DC)" is an assembly of at least two independently acting check valves including tightly closing shut-off valves on each side of the check valve assembly and test cocks available for testing the watertightness of each check valve.

(g) "Health Agency" means the California Department of Health Services, or the local health officer with respect to a small water system.

(h) "Local Health Agency" means the county or city health authority.

7585. Evaluation of Hazard The water supplier shall evaluate the degree of potential health hazard to the public water supply which may be created as a result of conditions existing on a user's premises. The water supplier, however, shall not be responsible for abatement of cross-connections which may exist within a user's premises. As a minimum, the evaluation should consider: the existence of cross-connections, the nature of materials handled on the property, the probability of a backflow occurring, the degree of piping system complexity and the potential for piping system modification. Special consideration shall be given to the premises of the following types of water users:

(a) Premises where substances harmful to health are handled under pressure in a manner which could permit their entry into the public water system. This includes chemical or biological process waters and water from public water supplies which have deteriorated in sanitary quality.

(b) Premises having an auxiliary water supply, unless the auxiliary supply is accepted as an additional source by the water supplier and is approved by the health agency.

(c) Premises that have internal cross-connections that are not abated to the satisfaction of the water supplier or the health agency.

(d) Premises where cross-connections are likely to occur and entry is restricted so that cross-connection inspections cannot be made with sufficient frequency or at sufficiently short notice to assure that cross-connections do not exist.

(e) Premises having a repeated history of cross-connections being established or re-established.

7586. User Supervisor The health agency and water supplier may, at their discretion, require an industrial water user to designate a user supervisor when the water user's premises has a multipiping system that conveys various types of fluids, some of which may be hazardous, and where changes in the piping system are frequently made. The user supervisor shall be responsible for the avoidance of cross-connections during the installation, operation and maintenance of the water user's pipelines and equipment.

California Health Laws Related to Recycled Water   *June 2001*
Title 17   *Edition*

TABLE 1
TYPE OF BACKFLOW PROTECTION REQUIRED

Degree of Hazard	Minimum Type of Backflow Prevention
(a) Sewage and Hazardous Substances	
(1) Premises where there are waste water pumping and/or treatment plants and there is no interconnection with the potable water system. This does not include a single-family residence that has a sewage lift pump. A RP be provided in lieu of an AG if approved by the health agency and water supplier.	AG
(2) Premises where hazardous substances are handled in any manner in which the substances may enter the potable water system. This does not include a single-family residence that has a sewage lift pump. A RP may be provided in lieu of an AG if approved by the health agency and water supplier.	AG
(3) Premises where there are irrigation systems into which fertilizers, herbicides, or pesticides are, or can be, injected.	RP
(b) Auxiliary Water Supplies	
(1) Premises where there is an unapproved auxiliary water supply which is interconnected with the public water system. A RP or DC may be provided in lieu of an AG if approved by the health agency and water supplier.	AG
(2) Premises where there is an unapproved auxiliary RP water supply and there are no interconnections with the public water system. A DC may be provided in lieu of a RP if approved by the health agency and water supplier.	RP

7604. Type of protection required The type of protection that shall be provided to prevent backflow into the public water supply shall be commensurate with the degree of hazard that exists on the consumer's premises. The type of protective device that may be required (listed in an increasing level of protection) includes: Double Check Valve Assembly—(DC), Reduced Pressure Principle Backflow Prevention Device—(RP) and an Air-Gap Separation—(AG). The water user may choose a higher level of protection than required by the water supplier. The minimum types of backflow protection required to protect the public water supply, at the water user's connection to premises with various degrees of hazard, are given in Table 1. Situations not covered in Table 1 shall be evaluated on a case-by-case basis and the appropriate backflow protection shall be determined by the water supplier or health agency.

(c) Recycled water

(1) Premises where the public water system is used to supplement the recycled water supply.

(2) Premises where recycled water is used, other than as allowed in paragraph (3), and there is no interconnection with the potable water system.

(3) Residences using recycled water for landscape irrigation as part of an approved dual plumbed use area established pursuant to sections 60313 through 60316 unless the recycled water supplier obtains approval of the local public water supplier, or the Department if the water supplier is also the supplier of the recycled water, to utilize an alternative backflow protection plan that includes an annual inspection and annual shutdown test of the recycled water and potable water systems pursuant to subsection 60316(a).

(d) Fire Protection Systems

(1) Premises where the fire system is directly supplied from the public water system and there is an unapproved auxiliary water supply on or to the premises (not interconnected).

(2) Premises where the fire system is supplied from the public water system and interconnected with an unapproved auxiliary water supply. An RP may be provided in lieu of an AG if approved by the health agency and water supplier.

(3) Premises where the fire system is supplied from the public water system and where either elevated storage tanks or fire pumps which take suction from private reservoirs or tanks are used.

(4) Premises where the fire system is supplied from the public water system and where recycled water is used in a separate piping system within the same building.

APPENDIX H

California Code of Regulations, Title 22

There are very stringent water quality laws that apply to recycled water. The California Department of Health Services standards for recycled water are referred to as Title 22, because they are incorporated in Title 22, Chapter 3, Division 4 of the California Code of Regulations, with stipulations applying to various types of reuse and levels of required treatment. The Regional Water Quality Control Board is involved with respect to the use and application of recycled water and any associated runoff.

Title 22 allows for many uses of recycled water. In San Diego the uses for recycled water include irrigation of food crops, parks, playgrounds, school yards, residential landscaping, cemeteries, freeway landscaping, golf courses, ornamental nurseries, pasture for animals, orchards, and vineyards. In addition, recycled water can be used for fishing or boating recreational impoundments, fish hatcheries, cooling towers, and decorative fountains. Other allowable uses include flushing toilets and urinals, industrial process water, commercial laundries, making artificial snow, soil compaction, mixing concrete, and flushing sanitary sewers.

Division 4. Environmental Health

Chapter 1. Introduction

Article 1. Definitions

60001. Department   Whenever the term "department" is used in this division, it means the State Department of Health Services, unless otherwise specified.

60003. Director   Whenever the term "director" is used in this division, it means the Director, State Department of Health Services, unless otherwise specified.

Chapter 2. Regulations for the Implementation of the California Environmental Quality Act

Article 1. General Requirements and Categorical Exemptions

60100. General Requirements   The Department of Health Services incorporates by reference the objectives, criteria, and procedures as delineated in Chapters 1, 2, 2.5, 2.6, 3, 4, 5, and 6, Division 13, Public Resources Code, Sections 21000 et seq., and the Guidelines for the Implementation of the California Environmental Quality Act, Title 14, Division 6, Chapter 3, California Administrative Code, Sections 15000 et seq.

60101. Specific Activities Within Categorical Exempt Classes   The following specific activities are determined by the Department to fall within the classes of categorical exemptions set forth in Sections 15300 et seq. of Title 14 of the California Administrative Code:

(a) Class 1: Existing Facilities.

   (1) Any interior or exterior alteration of water treatment units, water supply systems, and pump station buildings where the alteration involves the addition, deletion, or modification of mechanical, electrical, or hydraulic controls.

   (2) Maintenance, repair, replacement, or reconstruction to any water treatment process units, including structures, filters, pumps and chlorinators.

(b) Class 2: Replacement or Reconstruction.

   (1) Repair or replacement of any water service connections, meters, and valves for backflow prevention, air release, pressure regulating, shut-off and blow-off or flushing.

   (2) Replacement or reconstruction of any existing water supply distribution lines, storage tanks and reservoirs of substantially the same size.

   (3) Replacement or reconstruction of any water wells, pump stations and related appurtenances.

(c) Class 3: New Construction of Small Structures.

   (1) Construction of any water supply and distribution lines of less than sixteen inches in diameter, and related appurtenances.

   (2) Construction of any water storage tanks and reservoirs of less than 100,000 gallon capacity.

(d) Class 4: Minor Alterations to Land.

   (1) Minor alterations to land, water, or vegetation on any officially existing designated wildlife management areas

or fish production facilities for the purpose of reducing the environmental potential for nuisances or vector production.

(2) Any minor alterations to highway crossings for water supply and distribution lines.

Chapter 3. Water Recycling Criteria

Article 1. Definitions

60301. Definitions

60301.100. Approved Laboratory "Approved laboratory" means a laboratory that has been certified by the Department to perform microbiological analyses pursuant to section 116390, Health and Safety Code.

60301.160. Coagulated Wastewater "Coagulated wastewater" means oxidized wastewater in which colloidal and finely divided suspended matter have been destabilized and agglomerated upstream from a filter by the addition of suitable floc-forming chemicals.

60301.170. Conventional Treatment "Conventional treatment" means a treatment chain that utilizes a sedimentation unit process between the coagulation and filtration processes and produces an effluent that meets the definition for disinfected tertiary recycled water.

60301.200. Direct Beneficial Use "Direct beneficial use" means the use of recycled water that has been transported from the point of treatment or production to the point of use without an intervening discharge to waters of the State.

60301.220. Disinfected Secondary-2.2 Recycled Water "Disinfected secondary-2.2 recycled water" means recycled water that has been oxidized and disinfected so that the median concentration of total coliform bacteria in the disinfected effluent does not exceed a most probable number (MPN) of 2.2 per 100 milliliters utilizing the bacteriological results of the last seven days for which analyses have been completed, and the number of total coliform bacteria does not exceed an MPN of 23 per 100 milliliters in more than one sample in any 30 day period.

60301.225. Disinfected Secondary-23 Recycled Water "Disinfected secondary-23 recycled water" means recycled water that has been oxidized and disinfected so that the median concentration of total coliform bacteria in the disinfected effluent does not exceed a most probable number (MPN) of 23 per 100 milliliters utilizing the bacteriological results of the last seven days for which analyses have been completed, and the number of total coliform bacteria does not exceed an MPN of 240 per 100 milliliters in more than one sample in any 30 day period.

60301.230. Disinfected tertiary recycled water "Disinfected tertiary recycled water" means a filtered and subsequently disinfected wastewater that meets the following criteria:

(a) The filtered wastewater has been disinfected by either:

   (1) A chlorine disinfection process following filtration that provides a CT (the product of total chlorine residual and modal contact time measured at the same point) value of not less than 450 milligram-minutes per liter at all times with a modal contact time of at least 90 minutes, based on peak dry weather design flow;

or

   (2) A disinfection process that, when combined with the filtration process, has been demonstrated to inactivate and/or remove 99.999 percent of the plaque-forming units of F-specific bacteriophage MS2, or polio virus in the wastewater. A virus that is at least as resistant to disinfection as polio virus may be used for purposes of the demonstration.

(b) The median concentration of total coliform bacteria measured in the disinfected effluent does not exceed an MPN of 2.2 per 100 milliliters utilizing the bacteriological results of the last seven days for which analyses have been completed and the number of total coliform bacteria does not exceed an MPN of 23 per 100 milliliters in more than one sample in any 30 day period. No sample shall exceed an MPN of 240 total coliform bacteria per 100 milliliters.

60301.240. Drift "Drift" means the water that escapes to the atmosphere as water droplets from a cooling system.

60301.245. Drift Eliminator "Drift eliminator" means a feature of a cooling system that reduces to a minimum the generation of drift from the system.

60301.250. Dual Plumbed System "Dual plumbed system" or "dual plumbed" means a system that utilizes separate piping systems for recycled water and potable water within a facility and where the recycled water is used for either of the following purposes:

(a) To serve plumbing outlets (excluding fire suppression systems) within a building or

(b) Outdoor landscape irrigation at individual residences.

60301.300. F-specific Bacteriophage MS-2   "F-specific bacteriophage MS-2" means a strain of a specific type of virus that infects coliform bacteria that is traceable to the American Type Culture Collection (ATCC 15597B1) and is grown on lawns of *E. coli* (ATCC 15597).

60301.310. Facility   "Facility" means any type of building or structure, or a defined area of specific use that receives water for domestic use from a public water system as defined in section 116275 of the Health and Safety Code.

60301.320. Filtered Wastewater   "Filtered wastewater" means an oxidized wastewater that meets the criteria in subsection (a) or (b):

(a) Has been coagulated and passed through natural undisturbed soils or a bed of filter media pursuant to the following:

   (1) At a rate that does not exceed 5 gallons per minute per square foot of surface area in mono, dual or mixed media gravity, upflow or pressure filtration systems, or does not exceed 2 gallons per minute per square foot of surface area in traveling bridge automatic backwash filters; and

   (2) So that the turbidity of the filtered wastewater does not exceed any of the following:

      (A) An average of 2 NTU within a 24-hour period;

      (B) 5 NTU more than 5 percent of the time within a 24-hour period: and

      (C) 10 NTU at any time.

(b) Has been passed through a microfiltration, ultrafiltration, nanofiltration, or reverse osmosis membrane so that the turbidity of the filtered wastewater does not exceed any of the following:

   (1) 0.2 NTU more than 5 percent of the time within a 24-hour period; and

   (2) 0.5 NTU at any time.

60301.330. Food Crops   "Food crops" means any crops intended for human consumption.

60301.400. Hose Bibb   "Hose bibb" means a faucet or similar device to which a common garden hose can be readily attached.

60301.550. Landscape Impoundment   "Landscape impoundment" means an impoundment in which recycled water is stored or used for aesthetic enjoyment or landscape irrigation, or which otherwise serves a similar function and is not intended to include public contact.

60301.600. Modal Contact Time "Modal contact time" means the amount of time elapsed between the time that a tracer, such as salt or dye, is injected into the influent at the entrance to a chamber and the time that the highest concentration of the tracer is observed in the effluent from the chamber.

60301.620. Nonrestricted Recreational Impoundment "Nonrestricted recreational impoundment" means an impoundment of recycled water, in which no limitations are imposed on body-contact water recreational activities.

60301.630. NTU "NTU" (nephelometric turbidity unit) means a measurement of turbidity as determined by the ratio of the intensity of light scattered by the sample to the intensity of incident light as measured by method 2130 B in *Standard Methods for the Examination of Water and Wastewater*, 20th ed.; Eaton, A. D., Clesceri, L. S., and Greenberg, A. E., Eds; American Public Health Association: Washington, DC, 1995; p. 2–8.

60301.650. Oxidized Wastewater. "Oxidized wastewater" means wastewater in which the organic matter has been stabilized, is nonputrescible, and contains dissolved oxygen.

60301.660. Peak Dry Weather Design Flow "Peak dry weather design flow" means the arithmetic mean of the maximum peak flow rates sustained over some period of time (for example three hours) during the maximum 24-hour dry weather period. Dry weather period is defined as periods of little or no rainfall.

60301.700. Recycled Water Agency "Recycled water agency" means the public water system, or a publicly or privately owned or operated recycled water system, that delivers or proposes to deliver recycled water to a facility.

60301.710. Recycling Plant "Recycling plant" means an arrangement of devices, structures, equipment, processes and controls which produce recycled water.

60301.740. Regulatory Agency "Regulatory agency" means the California Regional Water Quality Control Board(s) that have jurisdiction over the recycling plant and use areas.

60301.750. Restricted Access Golf Course "Restricted access golf course" means a golf course where public access is controlled so that areas irrigated with recycled water cannot be used as if they were part of a park, playground, or school yard and where irrigation is conducted only in areas and during periods when the golf course is not being used by golfers.

60301.760. Restricted Recreational Impoundment   "Restricted recreational impoundment" means an impoundment of recycled water in which recreation is limited to fishing, boating, and other non-body-contact water recreational activities.

60301.800. Spray Irrigation   "Spray irrigation" means the application of recycled water to crops to maintain vegetation or support growth of vegetation by applying it from sprinklers.

60301.830. Standby Unit Process   "Standby unit process" means an alternate unit process or an equivalent alternative process which is maintained in operable condition and which is capable of providing comparable treatment of the actual flow through the unit for which it is a substitute.

60301.900. Undisinfected Secondary Recycled Water   "Undisinfected secondary recycled water" means oxidized wastewater.

60301.920. Use Area   "Use area" means an area of recycled water use with defined boundaries. A use area may contain one or more facilities.

Article 2. Sources of Recycled Water

60302. Source Specifications   The requirements in this chapter shall only apply to recycled water from sources that contain domestic waste, in whole or in part.

Article 3. Uses of Recycled Water

60303. Exceptions   The requirements set forth in this chapter shall not apply to the use of recycled water onsite at a water recycling plant, or wastewater treatment plant, provided access by the public to the area of onsite recycled water use is restricted.

60304. Use of Recycled Water for Irrigation

(a) Recycled water used for the surface irrigation of the following shall be a disinfected tertiary recycled water, except that for filtration pursuant to Section 60301.320(a) coagulation need not be used as part of the treatment process provided that the filter effluent turbidity does not exceed 2 NTU, the turbidity of the influent to the filters is continuously measured, the influent turbidity does not exceed 5 NTU for more than 15 minutes and never exceeds 10 NTU, and that there is the capability to automatically activate chemical addition or divert the wastewater should the filter influent turbidity exceed 5 NTU for more than 15 minutes:

(1) Food crops, including all edible root crops, where the recycled water comes into contact with the edible portion of the crop,

(2) Parks and playgrounds,

(3) School yards,

(4) Residential landscaping,

(5) Unrestricted access golf courses, and

(6) Any other irrigation use not specified in this section and not prohibited by other sections of the California Code of Regulations.

(b) Recycled water used for the surface irrigation of food crops where the edible portion is produced above ground and not contacted by the recycled water shall be at least disinfected secondary-2.2 recycled water.

(c) Recycled water used for the surface irrigation of the following shall be at least disinfected secondary-23 recycled water:

(1) Cemeteries,

(2) Freeway landscaping,

(3) Restricted access golf courses,

(4) Ornamental nursery stock and sod farms where access by the general public is not restricted,

(5) Pasture for animals producing milk for human consumption, and

(6) Any nonedible vegetation where access is controlled so that the irrigated area cannot be used as if it were part of a park, playground or school yard

(d) Recycled wastewater used for the surface irrigation of the following shall be at least undisinfected secondary recycled water:

(1) Orchards where the recycled water does not come into contact with the edible portion of the crop,

(2) Vineyards where the recycled water does not come into contact with the edible portion of the crop,

(3) Non food-bearing trees (Christmas tree farms are included in this category provided no irrigation with recycled water occurs for a period of 14 days prior to harvesting or allowing access by the general public),

(4) Fodder and fiber crops and pasture for animals not producing milk for human consumption,

(5) Seed crops not eaten by humans,

(6) Food crops that must undergo commercial pathogen-destroying processing before being consumed by humans, and

(7) Ornamental nursery stock and sod farms provided no irrigation with recycled water occurs for a period of 14 days prior to harvesting, retail sale, or allowing access by the general public.

(e) No recycled water used for irrigation, or soil that has been irrigated with recycled water, shall come into contact with the edible portion of food crops eaten raw by humans unless the recycled water complies with subsection (a).

60305. Use of Recycled Water for Impoundments

(a) Except as provided in subsection (b), recycled water used as a source of water supply for nonrestricted recreational impoundments shall be disinfected tertiary recycled water that has been subjected to conventional treatment.

(b) Disinfected tertiary recycled water that has not received conventional treatment may be used for nonrestricted recreational impoundments provided the recycled water is monitored for the presence of pathogenic organisms in accordance with the following:

(1) During the first 12 months of operation and use the recycled water shall be sampled and analyzed monthly for *Giardia*, enteric viruses, and *Cryptosporidium*. Following the first 12 months of use, the recycled water shall be sampled and analyzed quarterly for *Giardia*, enteric viruses, and *Cryptosporidium*. The ongoing monitoring may be discontinued after the first two years of operation with the approval of the department. This monitoring shall be in addition to the monitoring set forth in section 60321.

(2) The samples shall be taken at a point following disinfection and prior to the point where the recycled water enters the use impoundment. The samples shall be analyzed by an approved laboratory and the results submitted quarterly to the regulatory agency.

(c) The total coliform bacteria concentrations in recycled water used for nonrestricted recreational impoundments, measured at a point between the disinfection process and the point of entry to the use impoundment, shall comply with the criteria specified in section 60301.230(b) for disinfected tertiary recycled water.

(d) Recycled water used as a source of supply for restricted recreational impoundments and for any publicly accessible impoundments at fish hatcheries shall be at least disinfected secondary-2.2 recycled water.

(e) Recycled water used as a source of supply for landscape impoundments that do not utilize decorative fountains shall be at least disinfected secondary-23 recycled water.

60306. Use of Recycled Water for Cooling

(a) Recycled water used for industrial or commercial cooling or air conditioning that involves the use of a cooling tower, evaporative condenser, spraying or any mechanism that creates a mist shall be disinfected tertiary recycled water.

(b) Use of recycled water for industrial or commercial cooling or air conditioning that does not involve the use of a cooling tower, evaporative condenser, spraying, or any mechanism that creates a mist shall be at least disinfected secondary-23 recycled water.

(c) Whenever a cooling system, using recycled water in conjunction with an air conditioning facility, utilizes a cooling tower or otherwise creates a mist that could come into contact with employees or members of the public, the cooling system shall comply with the following:

(1) A drift eliminator shall be used whenever the cooling system is in operation.

(2) A chlorine, or other, biocide shall be used to treat the cooling system recirculating water to minimize the growth of *Legionella* and other microorganisms.

60307. Use of Recycled Water for Other Purposes

(a) Recycled water used for the following shall be disinfected tertiary recycled water, except that for filtration being provided pursuant to section 60301.320(a) coagulation need not be used as part of the treatment process provided that the filter effluent turbidity does not exceed 2 NTU, the turbidity of the influent to the filters is continuously measured, the influent turbidity does not exceed 5 NTU for more than 15 minutes and never exceeds 10 NTU, and that there is the capability to automatically activate chemical addition or divert the wastewater should the filter influent turbidity exceed 5 NTU for more than 15 minutes:

(1) Flushing toilets and urinals,

(2) Priming drain traps,

(3) Industrial process water that may come into contact with workers,

(4) Structural fire fighting,

(5) Decorative fountains,

(6) Commercial laundries,

(7) Consolidation of backfill around potable water pipelines,

(8) Artificial snow making for commercial outdoor use, and

(9) Commercial car washes, including hand washes if the recycled water is not heated, where the general public is excluded from the washing process.

(b) Recycled water used for the following uses shall be at least disinfected secondary-23 recycled water:

(1) Industrial boiler feed,

(2) Nonstructural fire fighting,

(3) Backfill consolidation around nonpotable piping,

(4) Soil compaction,

(5) Mixing concrete,

(6) Dust control on roads and streets,

(7) Cleaning roads, sidewalks and outdoor work areas, and

(8) Industrial process water that will not come into contact with workers.

(c) Recycled water used for flushing sanitary sewers shall be at least undisinfected secondary recycled water.

Article 4. Use Area Requirements

60310. Use Area Requirements

(a) No irrigation with disinfected tertiary recycled water shall take place within 50 feet of any domestic water supply well unless all of the following conditions have been met:

(1) A geological investigation demonstrates that an aquitard exists at the well between the uppermost aquifer being drawn from and the ground surface.

(2) The well contains an annular seal that extends from the surface into the aquitard.

(3) The well is housed to prevent any recycled water spray from coming into contact with the wellhead facilities.

(4) The ground surface immediately around the wellhead is contoured to allow surface water to drain away from the well.

(5) The owner of the well approves of the elimination of the buffer zone requirement.

(b) No impoundment of disinfected tertiary recycled water shall occur within 100 feet of any domestic water supply well.

(c) No irrigation with, or impoundment of, disinfected second-ary-2.2 or disinfected secondary-23 recycled water shall take place within 100 feet of any domestic water supply well.

(d) No irrigation with, or impoundment of, undisinfected secondary recycled water shall take place within 150 feet of any domestic water supply well.

(e) Any use of recycled water shall comply with the following:

(1) Any irrigation runoff shall be confined to the recycled water use area, unless the runoff does not pose a public health threat and is authorized by the regulatory agency.

(2) Spray, mist or runoff shall not enter dwellings, designated outdoor eating areas or food handling facilities.

(3) Drinking water fountains shall be protected against contact with recycled water spray, mist or runoff.

(f) No spray irrigation of any recycled water, other than disinfected tertiary recycled water, shall take place within 100 feet of a residence or a place where public exposure could be similar to that of a park, playground or school yard.

(g) All use areas where recycled water is used that are accessible to the public shall be posted with signs that are visible to the public, in a size no less than 4 inches high by 8 inches wide, that include the following wording : "RECYCLED WATER - DO NOT DRINK". Each sign shall display an international symbol similar to that shown in Fig. 60310-A. The Department may accept alternative signage and wording, or an educational program, provided the applicant demonstrates to the Department that the alternative approach will assure an equivalent degree of public notification.

(h) Except as allowed under section 7604 of Title 17, California Code of Regulations, no physical connection shall be made or allowed to exist between any recycled water system and any separate system conveying potable water.

(i) The portions of the recycled water piping system that are in areas subject to access by the general public shall not include any hose bibbs. Only quick couplers that differ from those used on the potable water system shall be used on the portions of the recycled water piping system in areas subject to public access.

Article 5. Dual Plumbed Recycled Water Systems

60313. General Requirements

(a) No person other than a recycled water agency shall deliver recycled water to a dual-plumbed facility.

(b) No recycled water agency shall deliver recycled water for any internal use to any individually-owned residential units including free-standing structures, multiplexes or condominiums.

(c) No recycled water agency shall deliver recycled water for internal use except for fire suppression systems, to any facility that produces or processes food products or beverages. For purposes of this Subsection, cafeterias or snack bars in a facility whose primary function does not involve the production or processing of foods or beverages are not considered facilities that produce or process foods or beverages.

(d) No recycled water agency shall deliver recycled water to a facility using a dual plumbed system unless the report required pursuant to section 13522.5 of the Water Code, and which meets the requirements set forth in section 60314, has been submitted to, and approved by, the regulatory agency.

FIGURE H-1   Water Recycling Criteria.

60314. Report Submittal

(a) For dual-plumbed recycled water systems, the report submitted pursuant to section 13522.5 of the Water Code shall contain the following information in addition to the information required by section 60323:

  (1) A detailed description of the intended use area identifying the following:

    (A) The number, location, and type of facilities within the use area proposing to use dual plumbed systems,

    (B) The average number of persons estimated to be served by each facility on a daily basis,

    (C) The specific boundaries of the proposed use area including a map showing the location of each facility to be served,

    (D) The person or persons responsible for operation of the dual plumbed system at each facility, and

    (E) The specific use to be made of the recycled water at each facility.

  (2) Plans and specifications describing the following:

    (A) Proposed piping system to be used,

    (B) Pipe locations of both the recycled and potable systems,

    (C) Type and location of the outlets and plumbing fixtures that will be accessible to the public, and

    (D) The methods and devices to be used to prevent backflow of recycled water into the public water system.

  (3) The methods to be used by the recycled water agency to assure that the installation and operation of the dual plumbed system will not result in cross-connections between the recycled water piping system and the potable water piping system. This shall include a description of pressure, dye or other test methods to be used to test the system every four years.

(b) A master plan report that covers more than one facility or use site may be submitted provided the report includes the information required by this section. Plans and specifications for individual facilities covered by the report may be submitted at any time prior to the delivery of recycled water to the facility.

60315. Design Requirements

The public water supply shall not be used as a backup or supplemental source of water for a dual-plumbed recycled water system unless the connection between the two systems is protected by an air-gap separation which complies with the requirements of sections 7602(a) and 7603(a) of Title 17,

California Code of Regulations, and the approval of the public water system has been obtained.

60316. Operation Requirements

(a) Prior to the initial operation of the dual-plumbed recycled water system and annually thereafter, the Recycled Water Agency shall ensure that the dual plumbed system within each facility and use area is inspected for possible cross-connections with the potable water system. The recycled water system shall also be tested for possible cross-connections at least once every four years. The testing shall be conducted in accordance with the method described in the report submitted pursuant to section 60314. The inspections and the testing shall be performed by a cross-connection control specialist certified by the California-Nevada section of the American Water Works Association or an organization with equivalent certification requirements. A written report documenting the result of the inspection or testing for the prior year shall be submitted to the department within 30 days following completion of the inspection or testing.

(b) The recycled water agency shall notify the department of any incidence of backflow from the dual-plumbed recycled water system into the potable water system within 24 hours of the discovery of the incident.

(c) Any backflow prevention device installed to protect the public water system serving the dual-plumbed recycled water system shall be inspected and maintained in accordance with section 7605 of Title 17, California Code of Regulations.

Article 5.1. Groundwater Recharge

60320. Groundwater Recharge

(a) Reclaimed water used for groundwater recharge of domestic water supply aquifers by surface spreading shall be at all times of a quality that fully protects public health. The State Department of Health Services' recommendations to the Regional Water Quality Control Boards for proposed groundwater recharge projects and for expansion of existing projects will be made on an individual case basis where the use of reclaimed water involves a potential risk to public health.

(b) The State Department of Health Services' recommendations will be based on all relevant aspects of each project, including the following factors: treatment provided; effluent quality and quantity; spreading area operations; soil characteristics; hydrogeology; residence time; and distance to withdrawal.

(c) The State Department of Health Services will hold a public hearing prior to making the final determination regarding the public health aspects of each groundwater recharge project. Final recommendations will be submitted to the Regional Water Quality Control Board in an expeditious manner.

Article 5.5. Other Methods of Treatment

60320.5. Other Methods of Treatment   Methods of treatment other than those included in this chapter and their reliability features may be accepted if the applicant demonstrates to the satisfaction of the State Department of Health (Department of Health Services) that the methods of treatment and reliability features will assure an equal degree of treatment and reliability.

Article 6. Sampling and Analysis

60321. Sampling and Analysis

(a) Disinfected secondary-23, disinfected secondary-2.2, and disinfected tertiary recycled water shall be sampled at least once daily for total coliform bacteria. The samples shall be taken from the disinfected effluent and shall be analyzed by an approved laboratory.

(b) Disinfected tertiary recycled water shall be continuously sampled for turbidity using a continuous turbidity meter and recorder following filtration. Compliance with the daily average operating filter effluent turbidity shall be determined by averaging the levels of recorded turbidity taken at four-hour intervals over a 24-hour period. Compliance with turbidity pursuant to section 60301.320(a)(2)(B) and (b)(1) shall be determined using the levels of recorded turbidity taken at intervals of no more than 1-2-hours over a 24-hour period. Should the continuous turbidity meter and recorder fail, grab sampling at a minimum frequency of 1.2-hours may be substituted for a period of up to 24-hours. The results of the daily average turbidity determinations shall be reported quarterly to the regulatory agency.

(c) The producer or supplier of the recycled water shall conduct the sampling required in subsections (a) and (b).

Article 7. Engineering Report and Operational Requirements

60323. Engineering Report

(a) No person shall produce or supply reclaimed water for direct reuse from a proposed water reclamation plant unless he files an engineering report.

(b) The report shall be prepared by a properly qualified engineer registered in California and experienced in the field of wastewater treatment, and shall contain a description of the design of the proposed reclamation system. The report shall clearly indicate the means for compliance with these regulations and any other features specified by the regulatory agency.

(c) The report shall contain a contingency plan which will assure that no untreated or inadequately treated wastewater will be delivered to the use area.

60325. Personnel

(a) Each reclamation plant shall be provided with a sufficient number of qualified personnel to operate the facility effectively so as to achieve the required level of treatment at all times.

(b) Qualified personnel shall be those meeting requirements established pursuant to Chapter 9 (commencing with section 13625) of the Water Code.

60327. Maintenance   A preventive maintenance program shall be provided at each reclamation plant to ensure that all equipment is kept in a reliable operating condition.

60329. Operating Records and Reports

(a) Operating records shall be maintained at the reclamation plant or a central depository within the operating agency. These shall include: all analyses specified in the reclamation criteria; records of operational problems, plant and equipment breakdowns, and diversions to emergency storage or disposal; all corrective or preventive action taken.

(b) Process or equipment failures triggering an alarm shall be recorded and maintained as a separate record file. The recorded information shall include the time and cause of failure and corrective action taken.

(c) A monthly summary of operating records as specified under (a) of this section shall be filed monthly with the regulatory agency.

(d) Any discharge of untreated or partially treated wastewater to the use area, and the cessation of same, shall be reported immediately by telephone to the regulatory agency, the State Department of Health, and the local health officer.

60331. Bypass   There shall be no bypassing of untreated or partially treated wastewater from the reclamation plant or any intermediate unit processes to the point of use.

Article 8. General Requirements of Design

60333. Flexibility of Design   The design of process piping, equipment arrangement, and unit structures in the reclamation plant must allow for efficiency and convenience in operation and maintenance and provide flexibility of operation to permit the highest possible degree of treatment to be obtained under varying circumstances.

60335. Alarms

(a) Alarm devices required for various unit processes as specified in other sections of these regulations shall be installed to provide warning of:

    (1) Loss of power from the normal power supply.

    (2) Failure of a biological treatment process.

    (3) Failure of a disinfection process.

    (4) Failure of a coagulation process.

    (5) Failure of a filtration process.

    (6) Any other specific process failure for which warning is required by the regulatory agency.

(b) All required alarm devices shall be independent of the normal power supply of the reclamation plant.

(c) The person to be warned shall be the plant operator, superintendent or any other responsible person designated by the management of the reclamation plant and capable of taking prompt corrective action.

(d) Individual alarm devices may be connected to a master alarm to sound at a location where it can be conveniently observed by the attendant. In case the reclamation plant is not attended full time, the alarm(s) shall be connected to sound at a police station, fire station or other full time service unit with which arrangements have been made to alert the person in charge at times that the reclamation plant is unattended.

60337. Power Supply   The power supply shall be provided with one of the following reliability features:

(a) Alarm and standby power source.

(b) Alarm and automatically actuated short-term retention or disposal provisions as specified in section 60341.

(c) Automatically actuated long-term storage or disposal provisions as specified in section 60341.

Article 9. Reliability Requirements for Primary Effluent

60339. Primary Treatment   Reclamation plants producing reclaimed water exclusively for uses for which primary effluent is permitted shall be provided with one of the following reliability features:

(a) Multiple primary treatment units capable of producing primary effluent with one unit not in operation.

(b) Long-term storage or disposal provisions as specified in section 60341.

Note: Use of primary effluent for recycled water is no longer allowed [repeal of section 60309, effective December 2000].

Article 10. Reliability Requirements for Full Treatment

60341. Emergency Storage or Disposal

(a) Where short-term retention or disposal provisions are used as a reliability feature, these shall consist of facilities reserved for the purpose of storing or disposing of untreated or partially treated wastewater for at least a 24-hour period. The facilities shall include all the necessary diversion devices, provisions for odor control, conduits, and pumping and pump back equipment. All of the equipment other than the pump back equipment shall be either independent of the normal power supply or provided with a standby power source.

(b) Where long-term storage or disposal provisions are used as a reliability feature, these shall consist of ponds, reservoirs, percolation areas, downstream sewers leading to other treatment or disposal facilities or any other facilities reserved for the purpose of emergency storage or disposal of untreated or partially treated wastewater. These facilities shall be of sufficient capacity to provide disposal or storage of wastewater for at least 20 days, and shall include all the necessary diversion works, provisions for odor and nuisance control, conduits, and pumping and pump back equipment. All of the equipment other than the pump back equipment shall be either independent of the normal power supply or provided with a standby power source.

(c) Diversion to a less demanding reuse is an acceptable alternative to emergency disposal of partially treated wastewater provided that the quality of the partially treated wastewater is suitable for the less demanding reuse.

(d) Subject to prior approval by the regulatory agency, diversion to a discharge point which requires lesser quality of wastewater is an acceptable alternative to emergency disposal of partially treated wastewater.

(e) Automatically actuated short-term retention or disposal provisions and automatically actuated long-term storage or disposal provisions shall include, in addition to provisions of (a), (b), (c), or (d) of this section, all the necessary sensors, instruments, valves and other devices to enable fully automatic diversion of untreated or partially treated wastewater to approved emergency storage or disposal in the event of failure of a treatment process and a manual reset to prevent automatic restart until the failure is corrected.

60343. Primary Treatment   All primary treatment unit processes shall be provided with one of the following reliability features:

(a) Multiple primary treatment units capable of producing primary effluent with one unit not in operation.

(b) Standby primary treatment unit process.

(c) Long-term storage or disposal provisions.

60345. Biological Treatment   All biological treatment unit processes shall be provided with one of the following reliability features:

(a) Alarm and multiple biological treatment units capable of producing oxidized wastewater with one unit not in operation.

(b) Alarm, short-term retention or disposal provisions, and standby replacement equipment.

(c) Alarm and long-term storage or disposal provisions.

(d) Automatically actuated long-term storage or disposal provisions.

60347. Secondary Sedimentation   All secondary sedimentation unit processes shall be provided with one of the following reliability features:

(a) Multiple sedimentation units capable of treating the entire flow with one unit not in operation.

(b) Standby sedimentation unit process.

(c) Long-term storage or disposal provisions.

60349. Coagulation

(a) All coagulation unit processes shall be provided with the following mandatory features for uninterrupted coagulant feed:

(1) Standby feeders,

(2) Adequate chemical stowage and conveyance facilities,

(3) Adequate reserve chemical supply, and

(4) Automatic dosage control.

(b) All coagulation unit processes shall be provided with one of the following reliability features:

(1) Alarm and multiple coagulation units capable of treating the entire flow with one unit not in operation;

(2) Alarm, short-term retention or disposal provisions, and standby replacement equipment;

(3) Alarm and long-term storage or disposal provisions;

(4) Automatically actuated long-term storage or disposal provisions, or

(5) Alarm and standby coagulation process.

60351. Filtration   All filtration unit processes shall be provided with one of the following reliability features:

(a) Alarm and multiple filter units capable of treating the entire flow with one unit not in operation.

(b) Alarm, short-term retention or disposal provisions and standby replacement equipment.

(c) Alarm and long-term storage or disposal provisions.

(d) Automatically actuated long-term storage or disposal provisions.

(e) Alarm and standby filtration unit process.

60353. Disinfection

(a) All disinfection unit processes where chlorine is used as the disinfectant shall be provided with the following features for uninterrupted chlorine feed:

(1) Standby chlorine supply,

(2) Manifold systems to connect chlorine cylinders,

(3) Chlorine scales, and

(4) Automatic devices for switching to full chlorine cylinders.

Automatic residual control of chlorine dosage, automatic measuring and recording of chlorine residual, and hydraulic performance studies may also be required.

(b) All disinfection unit processes where chlorine is used as the disinfectant shall be provided with one of the following reliability features:

(1) Alarm and standby chlorinator;

(2) Alarm, short-term retention or disposal provisions, and standby replacement equipment;

(3) Alarm and long-term storage or disposal provisions;

(4) Automatically actuated long-term storage or disposal provisions; or

(5) Alarm and multiple point chlorination, each with independent power source, separate chlorinator and separate chlorine supply.

60355. Other Alternatives to Reliability Requirements   Other alternatives to reliability requirements set forth in Articles 8 to 10 may be accepted if the applicant demonstrates to the satisfaction of the State Department of Health that the proposed alternative will assure an equal degree of reliability.

Guidelines for Water Reuse Applications

Type of Use	Treatment	Reclaimed Water Quality
Urban uses,[a] food crops eaten raw, recreational impoundments[b]	• Secondary[c] • Filtration • Disinfection	• pH = 6–9 • ≤ 10 mg/L BOD • ≤ 2 ntu[d] • No detectable fecal coliforms/100 mL[e] • ≥ 1 mg/L Cl$_2$ residual[f]
Restricted access area irrigation,[g] surface irrigation of orchards and vineyards, processed food crops,[h] nonfood crops,[i] aesthetic impoundments,[j] construction uses,[k] industrial cooling,[l] environmental reuse[m]	• Secondary[c] • Disinfection	• pH = 6–9 • ≤ 30 mg/L BOD • ≤ 30 mg/L TSS • ≤ 200 fecal coliforms/100 mL[e] • ≥ 1 mg/L Cl$_2$ residual[f] (except for environmental reuse)
Groundwater recharge of nonpotable aquifers by spreading	• Site specific and use dependent • Primary (minimum)	Site specific and use dependent
Groundwater recharge of nonpotable aquifers by injection	• Site specific and use dependent • Secondary (minimum)	Site specific and use dependent
Groundwater recharge of potable aquifers by spreading	• Site specific • Secondary[c] and disinfection (minimum) • May also need filtration and/or advanced wastewater treatment	• Site specific • Meet drinking water standards after percolation through vadose zone

(Continued)

Type of Use	Treatment	Reclaimed Water Quality
Groundwater recharge of potable aquifers by injection	• Secondary[c] • Filtration • Disinfection • Advanced wastewater treatment	Includes, but not limited to, the following: • pH = 6.5–8.5 • ≤ 2 ntu[d] • No detectable total coliforms/100 mL[e] • ≥ 1 mg/L Cl_2 residual[f] • ≤ 3 mg/L TOC • ≤ 0.2 mg/L TOX (total organic halogen) • Meet drinking water standards
Groundwater recharge of potable aquifers by augmentation of surface supplies	• Secondary[c] • Filtration • Disinfection • Advanced wastewater treatment	Includes, but not limited to, the following: • pH = 6.5–8.5 • ≤ 2 ntu[d] • No detectable total coliforms/100 mL[e] • ≥ 1 mg/L Cl_2 residual[f] • ≤ 3 mg/L TOC • Meet drinking water standards

[a]All types of landscape irrigation, toilet and urinal flushing, vehicle washing, use in fire protection systems and commercial air conditioner systems, and other uses with similar access or exposure to the water.

[b]Fishing, boating, and full body contact allowed.

[c]Secondary treatment should produce effluent in which both the BOD and TSS do not exceed 30 mg/L.

[d]Should be met prior to disinfection. Average based on a 24-h time period. Turbidity should not exceed 5 ntu at any time. If TSS is used in lieu of turbidity, the TSS should not exceed 5 mg/L.

[e]Based on the median value of the last 7 days for which analyses have been completed.

[f]After a minimum contact time of 30 min.

[g]Sod farms, silviculture sites, and other areas where public access is prohibited, restricted, or infrequent.

[h]Undergo chemical or physical processing sufficient to destroy pathogens prior to sale to the public or others.

[i]Pasture for milking animals; fodder, fiber, and seed crops.

[j]Public contact with reclaimed water is not allowed.

[k]Includes soil compaction, dust control, aggregate washing, and making concrete.

[l]Once-through cooling. Reclaimed water for recirculating cooling towers may need additional treatment.

[m]Wetlands, marshes, wildlife habitat, and stream augmentation.

Adapted from USEPA (2004).

TABLE I-1   USEPA Suggested Guidelines for Water Reuse Applications

State Water Reuse Regulations and Guidelines

States generally develop water reuse regulations or guidelines in response to a need to regulate water reuse activities that are occurring or expected to occur in the near future. Water reuse criteria vary among the states that have developed regulations, and some states have no regulations or guidelines. Some states have regulations or guidelines directed at land treatment of wastewater or land application as a means of wastewater disposal rather than regulations oriented to the intentional beneficial use of reclaimed water. Water reuse regulations typically include wastewater treatment process requirements, treatment reliability requirements, reclaimed water quality criteria, reclaimed water conveyance and distribution system requirements, and area use controls. No state's regulations cover all potential applications of reclaimed water, and few states have regulations that address potable reuse. When state regulations do not address specific reuse applications, those applications are not necessarily prohibited; instead, they may be evaluated and permitted on a case-by-case basis. The following sections provide an overview of state approaches to nonpotable and potable reuse regulations.

State Guidelines and Regulations for Nonpotable Reuse

Examples of state regulations for various nonpotable applications are summarized in Appendix D, Table D-1. The table includes water quality limits and, where imposed, treatment process requirements. Water quality requirements usually include maximum limits based on averages or geometric means over a specific time period, or median values for a specific number of consecutively collected samples. They also usually include maximum values (particularly for microbial indicator organisms) that cannot be exceeded at any time, although these limits are not included in the table.

Table D-1 shows clear variations in the treatment and quality requirements among the states for the types of uses listed. Key areas of significant variation are discussed here.

Microbial Indicator Organisms

Some states use total coliforms as the indicator organism, whereas others use fecal coliforms, *Escherichia coli*, or enterococci. Total coliforms represent a more conservative measure of the microbial water quality and include fecal coliforms and some nonfecal bacteria, such as soil bacteria. Some states have based their requirements on the USEPA guidelines (USEPA, 2004), which suggest using fecal coliforms as the indicator organism. Regulatory decisions regarding the selection of indicator organism to use are somewhat subjective, as is the acceptable limit. The rationale regarding the selection of which indicator organism to use and the methods used to determine whether acceptable microbial limits have been met are not consistent in all states. For example, in California the total coliform reporting limit is based on a running median of the last 7 days for which analyses have been completed, whereas in Florida the fecal coliform limit must be met in at least 75% of the samples over a 30-day period. Daily sampling is required in both states.

Turbidity Versus Total Suspended Solids (TSS)

For uses where human contact with the reclaimed water is expected or likely, some states specify turbidity limits, whereas others specify TSS limits. The removal of suspended matter is related to health protection. Particulate matter can reduce the effectiveness of disinfection processes, such as chlorine and UV radiation (see Chapter 2), by either exerting a chlorine demand or by absorbing the UV.

APPENDIX J

Development of a Comprehensive Integrity Verification Manual*

Introduction

The ability to maintain system integrity is one of the most important operational concerns associated with any membrane filtration facility, whether applied for compliance with the Long Term 2 Enhanced Surface Water Treatment Rule (LT2ESWTR) or for any other treatment objective. Because a membrane represents a physical barrier to pathogens and other drinking water contaminants, the means to ensure that this barrier remains uncompromised is critical for the ongoing protection of public health. Moreover, the number and variety of integrity-related compliance requirements for membrane filtration under the LT2ESWTR—ranging from various forms of testing to repair to data collection and reporting, as specified in Appnexdix L and Chapter 5 of the Membrane Filtration Guidance Manual—illustrate the potential complexity of the process of verifying and maintaining system integrity. As a result, this appendix has been prepared to serve as a tool to guide utilities in the development of a comprehensive, effective, and efficient integrity verification program (IVP). The various sections of this appendix are organized into a series of introductory questions that would be posed in the preparation of an IVP and corresponding discussions to elaborate on how these questions would be addressed in the context of an IVP.

---

*Source: *Membrane Filtration Guidance Manual* (USEPA, 2005).

What Is a Comprehensive IVP?

A comprehensive IVP is a customized site- and system-specific program that details all operating procedures associated with maintaining membrane filtration system integrity, including both federal and state requirements and any additional practices that are voluntarily implemented at the discretion of the utility. In the broadest terms, an IVP should serve as a master plan for preserving system integrity.

What Is the Purpose of an IVP?

The primary purpose of an IVP is to provide a utility with a rational and systematic blueprint for applying appropriate tools and techniques to efficiently conduct the following procedures:

- verifying integrity on an ongoing basis,
- identifying and correcting any integrity problems,
- recording and analyzing integrity test data, and
- preparing any required compliance reporting.

The successful execution of these procedures, in turn, allows a utility to track system performance and determine whether or not it is consistent with either that established by challenge testing or other requirements that may be applicable.

Why Is an IVP Important?

Because the process of maintaining system integrity has many aspects, an IVP is critical for organizing all of the various operating procedures relating to system integrity into a comprehensive plan. As an organizational tool, an IVP is also important for its ability to provide a framework to assist operators with conducting the proper procedures in the correct sequence both under normal operating conditions and when an integrity breach is either suspected or confirmed. As a whole, these IVP functions help to both ensure the production of safe drinking water and facilitate regulatory compliance.

What Are the Regulatory Requirements Associated with an IVP?

Although an IVP is not required for membrane filtration systems under the LT2ESWTR, the development of IVP can facilitate regulatory compliance by organizing the rule requirements into a comprehensive operational program. However, whether membrane filtration is applied for LT2ESWTR compliance or to meet any other treatment objectives, any IVP should be consistent with all USEPA and state requirements governing the operation of membrane treatment facilities. In addition, any requirements relating to maintaining system integrity should be incorporated into an IVP. Thus, while the development of an IVP is not required under the LT2ESWTR, it is strongly recommended for all utilities that utilize membrane filtration, particularly for disinfection applications.

What Are the Components of an IVP?

An IVP should incorporate all of the operational procedures associated with maintaining system integrity, including as a minimum the following major program elements:

- direct integrity testing,
- continuous indirect integrity monitoring,
- diagnostic testing,
- membrane repair and replacement,
- data collection and analysis, and
- reporting.

Other site-specific procedures or compliance requirements applicable to system integrity, but not specifically addressed under any of these program elements, should also be included as components of an IVP.

How Is IVP Guidance Presented in This Appendix?

This appendix is organized into sections according to the major program elements just listed. Each of these sections provides an overview of what the IVP should address with respect to that particular element and the various associated considerations that should be taken into account in the process of developing and implementing an IVP. Because compliance with the LT2ESWTR is critically related to maintaining system integrity, requirements from the rule are used as examples to illustrate program development, as well as how the various IVP components fit together into an integrated program.

The order of the sections in this appendix also illustrates the tiered approached to an IVP, ranging from the fundamental direct integrity test to successive levels of monitoring, testing, and repair that may be necessary. Note that this appendix is applicable to all types of membrane filtration addressed under the LT2ESWTR and consequently covered in the *Membrane Filtration Guidance Manual* (USEPA, 2005), including microfiltration (MF), ultrafiltration (UF), nanofiltration (NF), reverse osmosis (RO), and membrane cartridge filtration (MCF). Significant technology-specific nuances associated with one or more particular types of membrane filtration are addressed in context whenever practical.

Direct Integrity Testing

Direct integrity testing is the primary means of verifying integrity in membrane filtration systems and thus represents a fundamental component of an IVP. The series of questions into which this section is organized represents important aspects of direct integrity testing that should be addressed in an IVP. In addition, the questions are presented in a logical progression designed to parallel the step-by-step process of formulating a direct integrity testing strategy.

What Is the Purpose of Direct Integrity Testing?

Under the LT2ESWTR, a direct integrity test is defined as a physical test applied to a membrane unit in order to identify and isolate integrity breaches. Because direct integrity testing is the most accurate and precise means of determining whether or not a breach has occurred, it has historically been the primary means used to assess membrane integrity in potable water treatment applications in which pathogen removal is a principal concern. In addition, the test parameters and results can be correlated to the desired treatment objectives (e.g., log removal values) to yield a quantifiable assessment of system performance. In terms of LT2ESWTR compliance, as discussed in Chapter 4, requirements for direct integrity test *resolution* and *sensitivity* are specified for ensuring the necessary level of *Cryptosporidium* removal at a particular facility.

For example, the LT2ESWTR *resolution* requirement specifies that the direct integrity test parameters must be fixed such that a breach on the order of the smallest *Cryptosporidium* oocyst (i.e., 3 μm) is physically capable of contributing to a test response (40 CFR 141.719(b)(3)(ii)). Thus, for pressure-based direct integrity tests, the applied pressure (or vacuum) must be great enough to overcome the capillary static forces that hold water in a breach (i.e., the bubble point) of 3 μm in diameter in a fully wetted membrane, thereby allowing air to escape through a *Cryptosporidium*-sized hole and consequently enabling this loss of air to be potentially detected. If the applied pressure (or vacuum) were insufficient to overcome the bubble point, then the direct integrity test would be physically incapable of detecting any number of *Cryptosporidium*-sized breaches that would allow the passage of pathogens to the filtrate. Similarly, with marker-based tests, the marker used must be smaller than 3 μm to ensure that it would pass through a *Cryptosporidium*-sized hole, thus enabling a breach of this size to be potentially detected by the instrumentation that measures the concentration of the marker in the filtrate. Note that the 3 μm resolution requirement applies to all facilities utilizing membrane filtration for compliance with the LT2ESWTR.

However, unlike the resolution, the required *sensitivity*—the maximum log removal value (LRV) that can be reliably verified by the direct integrity test—may vary among different facilities under the LT2ESWTR, as the rule stipulates only that the test used have a sensitivity that exceeds the *Cryptosporidium* log removal credit awarded by the state (40 CFR 141.719(b)(3)(iii)). In some cases, the sensitivity of the test may be determined from the threshold test result that signifies the smallest detectable integrity breach, information that can be provided by the membrane filtration system supplier. This result, which represents the test sensitivity, may then be converted into a LRV using the methodology described in Chapter 4 or an alternative methodology approved by the state.

Although the compliance framework developed for direct integrity testing under the LT2ESWTR applies only to those utilities that are subject to the *Cryptosporidium* removal requirements of the rule, the methodology could be applied for other pathogens either by mandate of the state or at the discretion of the utility. If this methodology is applied to two or more different pathogens simultaneously at the same treatment facility, then the direct integrity test used would be required to have a resolution corresponding to the smallest of the pathogens; in addition, the test would need to have a sensitivity greater than the removal credit awarded by the state for each pathogen to which the methodology was applied. Even if a compliance framework similar to that for the LT2ESWTR is not applicable to a particular utility's membrane filtration system, a direct integrity test is still the most reliable means of determining whether or not a breach has occurred. For any system to which a compliance framework similar to that for the LT2ESWTR is applied, the resolution and sensitivity requirements for direct integrity testing should be established prior to placing the facility into service and specified in the IVP. If a different method is used to determine threshold integrity test results for applications other than LT2ESWTR compliance (e.g., fiber-cutting studies), these critical values should also be incorporated into the IVP. As a minimum, the IVP should include the direct integrity test result threshold specified by the manufacturer as indicative of a potential integrity breach.

What Type of Direct Integrity Test Should Be Used?

Currently, there are two general types of direct integrity tests that are commercially available for use with membrane filtration facilities: pressure-based tests and marker-based tests. The various specific tests that fall under each of these two general categories are described in Chapter 4. The particular test utilized in conjunction with a given system may depend on regulatory requirements, the type of membrane filtration system, the target organism (i.e., in terms of the required test resolution), test sensitivity, or the utility's preference.

The LT2ESWTR does not mandate the use of a particular direct integrity test for regulatory compliance; any test utilized must simply meet the criteria specified under the rule for resolution, sensitivity, and frequency. However, some states do stipulate the use of a specific test. In some cases, even if no particular test is specified, the required sensitivity relative to that of each potential direct integrity test method may govern which test(s) may be used. If no specific test is required by regulatory mandate, then any type of direct integrity test approved for use by the state may be used at the utility's discretion.

The direct integrity test used also depends to some extent on the type of membrane filtration system utilized, as some tests are not compatible with certain types of systems. For example, while a particulate marker test may be used with an MCF, MF, or UF system, a molecular marker test would not be a feasible means of assessing MCF, MF, or UF integrity, since the molecular marker may not be sufficiently removed by these membrane systems to demonstrate a reasonable LRV (e.g., 3 log). Conversely, a particulate marker test would probably not be used with NF or RO systems, since the particles could not easily be flushed from a spiral wound membrane module and would potentially foul or otherwise damage the membranes. Thus, for NF and RO systems, a molecular marker would be a more appropriate marker-based direct integrity test.

Pressure (or vacuum) decay tests are compatible with all the various types of membrane filtration as defined under the LT2ESWTR (i.e., MF, UF, NF, RO, and MCF), and the equipment necessary to conduct this type of test is typically included with most of the currently available proprietary MF and UF systems. Some types of direct integrity tests may not be available from a proprietary membrane filtration system supplier. If several types of tests are available and the utility is not otherwise constrained by regulatory requirements, the selection of a direct integrity test should take into account any site- or system-specific considerations that would either favor or preclude certain tests.

The type of direct integrity test used for a particular system and the justifying rationale should both be included in the facility IVP, as well as specific procedures for conducting the test. If the test is automated (as is common), the procedures specified in the IVP should include how the test works in terms of the automatic sequencing that the system undergoes. In addition, the procedure for manually conducting the test whenever necessary should also be specified, as well as any other responsibilities that system operators may have with respect to direct integrity testing. One particular consideration for systems with automated direct integrity testing is whether or not an operator must be present during the test. Although the presence of an operator may be beneficial, particularly if an integrity breach is detected, the ability of an operator to respond quickly to an automated alarm and notification system may render this unnecessary. If the state does not require direct operator supervision during direct integrity testing, the utility should exercise its own discretion in determining whether the presence of an operator is critical based on site- and system-specific considerations and degree of comfort with unsupervised testing.

How Frequently Should Direct Integrity Testing Be Conducted?

Because none of the various types of direct integrity tests currently available can feasibly assess integrity on a continuous basis while the system is online and producing filtered water, a membrane unit to be tested must be taken off-line and out of production during the testing process. Thus, more frequent direct integrity testing results in increased downtime for each membrane unit and consequently decreased system productivity. As a result, the frequency of direct integrity testing is an important parameter that hinges on striking an acceptable balance between the competing desires to maximize both confirmation of system integrity and production of treated water.

A number of states have established minimum test frequency requirements for membrane filtration systems; currently, these requirements range from as often as every 4 h to as relatively infrequent as once per week, depending on the state. If a utility applies membrane filtration for compliance with the LT2ESWTR, the rule requires that direct integrity testing be conducted on each membrane unit once per day at a minimum, unless the state approves less frequent testing based on demonstrated process reliability, the use of multiple barriers effective for *Cryptosporidium*, or reliable process safeguards (40 CFR 141.719(b)(3)(vi)). The state may also require testing on a more frequent basis at its discretion.

Even if subject to minimum test frequency requirements by federal or state regulations, a utility could opt to conduct direct integrity testing more frequently. In addition, a utility that is not otherwise constrained by any regulatory requirements should use its discretion to determine an appropriate direct integrity test frequency. In establishing a test frequency (regulatory requirements permitting), careful consideration should be given to the factors that may influence this decision. For example, testing less frequently may increase overall facility production and minimize the mechanical stress on the membrane module(s) that repeated testing might induce. However, increased testing provides more frequent assurance of system integrity, or in the case of an integrity breach, less operating time under compromised conditions that may allow the passage of pathogens or other undesirable prefiltered water constituents. The most important concern should be the maximum length of time that a utility feels comfortable potentially operating with an integrity breach of any magnitude, which would dictate the minimum acceptable integrity test frequency. Although the use of continuous indirect integrity monitoring (see Chapter 5 and later in this appendix) provides some measure of integrity confirmation between direct test applications, currently available indirect integrity monitoring techniques may not be able to detect a breach of potentially significant magnitude. It is important to note that the frequency of direct integrity testing should be based on public health considerations rather than the observed or anticipated frequency of the occurrence of integrity breaches.

When Should Direct Integrity Testing Be Conducted?

In the process of developing an IVP, some consideration should be given to the timing of conducting direct integrity testing relative to the normal operating cycle(s) of the system. For example, with systems that utilize some type of regular reverse flow process to remove foulants from the membrane surface (e.g., backwashing), the membrane will be the least obstructed immediately after this reverse flow process is complete. Implementing a direct integrity test at this point in the operational cycle would provide the most conservative estimate of system integrity, since the accumulated foulants that could plug any breaches, and thus effectively mask potential integrity problems, would be minimized. However, using this same example, for systems that employ a pressure-based direct integrity test, it is important that the test not be conducted too soon after the reverse flow processes (particularly if air is used), since the membrane must be fully-wetted for the integrity test to be effective.

It is also recommended that a direct integrity test be conducted after chemical cleaning or any other routine or emergency maintenance, in order to ensure that the system integrity has not been compromised by the procedure(s). The affected membrane module(s) or unit(s) should be returned to service only after system integrity has been confirmed by a direct integrity test. In cases in which a membrane unit has been taken out of service for diagnostic testing and/or repair, systems applying membrane filtration for LT2ESWTR compliance are required to conduct direct integrity testing on the affected unit to demonstrate system integrity prior to returning the unit to service (40 CFR 141.719(b)(3)(v)).

Any special circumstances that might call for direct integrity testing (i.e., in addition to the regularly scheduled periodic testing)—such as subsequent to chemical cleaning or membrane repair—whether in accordance with federal or state requirements or at the discretion of the utility, should be specified in the IVP.

How Should the Direct Integrity Test Results Be Interpreted?

For a given resolution, a marker- or pressure-based direct integrity test indicates whether or not an integrity breach has occurred by comparing the results of the test to the threshold result known to represent a breach. This threshold represents the test sensitivity and can be determined either by information provided by the manufacturer or through an on-site, system-specific assessment such as a fiber-cutting study.

Under the compliance framework of the LT2ESWTR, the ongoing test results obtained during facility operation for each membrane unit (as well as the threshold result representing the test sensitivity) can be converted into LRVs using the methodology described in Chapter 4 or another method approved by the state. Each successive test result (or LRV, as converted) is then compared to the log removal credit allocated to the system by the state (either as per the requirements of the

LT2ESWTR or for other treatment objectives) to determine compliance on an ongoing basis. If the LRV yielded by the direct integrity test is greater than the regulatory allocation, the system remains in compliance. However, if the LRV is below the required log removal, the membrane unit must be taken off-line for diagnostic testing and repair (see sections A.4 and A.5, respectively).

Thus, under the LT2ESWTR compliance framework, the direct integrity test results yield two important pieces of information: 1) whether or not an integrity breach has occurred, and 2) the maximum LRV that can be verified at the time of the test (when converted using the techniques described in Chapter 4 or an alternate methodology approved by the state). Because the test sensitivity may often be greater (i.e., a higher LRV) than the required log removal for any particular pathogen, it is possible for the direct integrity test to indicate that an integrity breach of some magnitude has occurred without the system being out of compliance. Thus, under the LT2ESWTR framework, a system could knowingly continue to operate with some level of integrity breach and still meet regulatory requirements. However, the USEPA recommends (and the state may require) that a membrane unit with a detectable integrity breach of any magnitude be immediately taken out of service for diagnostic testing and repair. At a minimum, utilities should conduct diagnostic testing and potential repair when the compromised unit is taken off-line for chemical cleaning, previously scheduled maintenance, or some other routine purpose.

The LT2ESWTR defines a control limit as an integrity test response which, if exceeded, indicates a potential integrity problem and triggers subsequent action. For the purposes of LT2ESWTR compliance, a direct integrity test control limit would be established as the test result that translates into an LRV equal to the *Cryptosporidium* removal credit awarded by the state. Any membrane unit for which the test results exceed the control limit (i.e., the LRV drops below the awarded removal credit) must be immediately taken off-line for diagnostic testing and repair (40 CFR 141.719(b)(3)(v)). Other control limits could also be established if the LT2ESWTR compliance framework is simultaneously applied for the removal of other pathogens with membrane filtration. However, because the system would be out of compliance if either of the two (or more) control limits were exceeded, the most stringent control limit (i.e., the lowest permissible integrity test response) would always represent the governing limit.

In addition to the critical or upper control limit (UCL) that governs regulatory compliance for a membrane filtration facility under the LT2ESWTR framework, lower control limits (LCLs) may be established as performance benchmarks either at the discretion of the utility or by mandate of the state. One or more LCLs may be identified between the integrity test result that indicates the smallest detectable breach and a breach that reduces the performance of the system such that it is just capable of meeting the required log removal (i.e., the UCL). LCLs may be useful for alerting system operators to the presence of an integrity breach, even if the detected breach is not sufficient to bring the system out of compliance. Rather than triggering unit shutdown and subsequent diagnostic testing, as with an UCL, exceeding an LCL might trigger increased operator attention or an investigation that can be conducted by operators while the system is still online in an attempt to determine the cause of the integrity problem and prevent the breach from expanding, if possible. Alternatively, in cases in which there is a particularly large gap between the test sensitivity and the UCL a utility may choose to voluntarily take a membrane unit off-line for diagnostic testing if a preferred LCL is exceeded, in order to minimize the risk to public health via the potential for pathogens to pass through a barrier known to be compromised, even if only to a small degree that is within regulatory tolerances.

As an example, even with the smallest detectable breach, the direct integrity test might still be able to verify 6-log removal (i.e., the test sensitivity), although the state only awards 2.5-log *Cryptosporidium* removal credit (i.e., the UCL) for the membrane filtration process. Under this scenario, a direct integrity test result yielding an LRV of 4 would indicate the presence of a significant integrity breach, even though the membrane filtration system would still be capable of achieving the required log removal. Thus, a utility might opt to establish an LCL at an LRV of 4 if an integrity breach of this magnitude represents an unacceptable public health risk independent of the ability of the system to maintain regulatory compliance.

Fiber-cutting studies that allow integrity test results to be quantified and correlated to a certain number of fiber breaks or a particular reduction in log removal capacity may be used to determine a single appropriate LCL or a series of tiered LCLs. Setting an LCL equal to the test sensitivity (i.e., the maximum log removal value that can be reliably verified by the direct integrity test) would represent the most conservative scenario, as in this case any detectable integrity breach would, at a minimum, alert an operator and potentially trigger some responsive action. All control limits, and the action triggered by exceeding each limit, should be clearly specified in the IVP.

Another issue that should be addressed in the IVP is the possibility of false positive and false negative direct integrity test results. For example, if a false positive result (i.e., a result incorrectly indicating a breach in a fully integral system) is suspected, the operator should check isolation valves and fittings on the system that are associated with the direct integrity test. In addition, a follow-up direct integrity test should be conducted both to confirm the results of the first test and to closely monitor the test for any system malfunctions. In any case, LT2ESWTR compliance requires that any membrane unit for which a direct integrity test result exceeds a UCL be taken off-line for diagnostic testing and repair, independent of whether or not a false positive result is suspected (40 CFR 141.719(b)(3)(v)).

False negative results (i.e., results that indicate either a fully integral system in the presence of an integrity breach or that significantly underestimate a substantial integrity breach) of direct integrity tests may be more difficult to identify; since the methods of continuous indirect integrity monitoring are less sensitive to integrity breaches, indirect monitoring data may not be capable of detecting a breach that is masked by a false negative direct test result. However, if the continuous indirect integrity monitoring data do suggest an integrity breach in contradiction of the direct test results, the membrane unit should be taken out of service to investigate the source of the discrepancy. One potential scenario that might result in a false negative result is an integrity breach that occurs in a membrane that is partially fouled. The accumulated foulants may obscure the breach, thus masking the integrity problem until after the next backwash or chemical cleaning. Nevertheless, in this case the false negative result may not represent a significant concern; if the direct integrity test is functioning properly, the false negative result would suggest that the system is functionally integral even with an integrity problem as a result of the breach being plugged. In this case, the breach would likely be detected after the next backwash or chemical cleaning that successfully removes the foulants from blocking the breach. A more problematic scenario involving false negative results might occur if the integrity test equipment is malfunctioning. Thus, is it important to incorporate a routine maintenance program for the integrity test system as part of the IVP.

Any particular strategies for minimizing the potential for false positive and false negative results should be specified in IVP documentation. A maintenance schedule for direct integrity monitoring equipment should also be specified. It is recommended that the direct integrity monitoring system receive a thorough check-up on at least an annual basis.

Continuous Indirect Integrity Monitoring

Continuous indirect integrity monitoring is a secondary means of verifying membrane filtration system integrity, intended to detect significant breaches between direct test applications. Thus, in the absence of a continuously applied direct test, continuous indirect integrity monitoring is critical to an IVP. As with the previous section, this section is organized into a series of questions that parallel the step-by-step process of formulating strategy for continuous indirect integrity monitoring. Each of these questions represents an important aspect of that strategy that should be included to some extent in an IVP.

What Is the Purpose of Indirect Integrity Monitoring?

For the purposes of LT2ESWTR compliance, continuous indirect integrity monitoring is defined as monitoring some aspect of filtrate water quality that is indicative of the removal of particulate matter at a frequency of at least once every 15 min. Although the various indirect monitoring methods are less sensitive techniques for assessing membrane integrity than the direct integrity tests, the value of utilizing the indirect methods is that they can be applied to assess integrity continuously while the system is online and producing water. In fact, since by definition indirect monitoring is applied to the filtrate, these techniques require that the membrane unit be in continuous production to assess membrane integrity.

Because currently available methods of direct integrity testing cannot be applied continuously, a successful direct test only indicates that no breach has occurred since the previous application of the test. Consequently, if the system were to become compromised immediately after a successful direct integrity test, the breach might not be detected until the next regularly scheduled direct test, which is typically as long as 1 day for the purposes of LT2ESWTR compliance. During this interval, a potentially significant breach could occur, allowing pathogens or other particulate matter to contaminate the filtrate for a period as long as an entire day. For applications other than LT2ESWTR compliance, this interval may be as short as 4 h or as long as one week or more, depending on state-specific regulatory policy. Thus, although continuous indirect integrity monitoring may not be able to detect small compromises in integrity, these techniques do provide the ability to identify larger breaches on an ongoing basis during production. As a result, periodic direct integrity testing and continuous indirect integrity monitoring are complementary tools for assessing system integrity, and both are critical for a comprehensive IVP.

Under the LT2ESWTR, continuous indirect integrity monitoring is required in the absence of a direct integrity test that can be applied continuously and that meets the resolution and sensitivity requirements of the rule (40 CFR 141.719(b)(4)). States may also have regulations requiring some form of continuous indirect integrity monitoring. In the absence of any applicable requirements, a utility may opt to employ some form of continuous indirect integrity monitoring at its discretion; however, note that turbidity monitoring, which is required under the various federal surface water treatment regulations as a measure of overall system performance, may also serve the dual purpose of continuous indirect integrity monitoring.

Unlike direct testing, there are no specific resolution or sensitivity requirements for continuous indirect integrity monitoring under the LT2ESWTR. However, these concepts, where applicable, may be useful tools for optimizing the ability of the various continuous indirect integrity monitoring methods to yield meaningful information about potential integrity breaches, as described in Chapter 5.

What Type of Indirect Integrity Monitoring Method Should Be Used?

There are a number of different methods and associated devices that may be used for continuous indirect integrity monitoring, including particle counting, particle monitoring, turbidimetry, laser turbidimetry, and conductivity monitoring. In general, any method that measures particulate matter in the filtrate as an indirect means of assessing integrity (such as particle counting, turbidimetry, etc.) is applicable to any of the various types of membrane filtration systems. Other methods that may measure dissolved constituents in the filtrate, such as conductivity monitoring, would only be applicable to NF or RO systems. The particular method of continuous indirect integrity monitoring employed by a utility for its system may be a function of regulatory requirements, test resolution or sensitivity, cost, confidence in the technology, or simply preference based on prior experience or other subjective criteria.

The LT2ESWTR requires the use of turbidity monitoring on each membrane unit as the default method of continuous indirect integrity monitoring for compliance, unless an alternate method is approved by the state (40 CFR 141.719(b)(4)(i)). Because the various federal surface water treatment regulations—the Surface Water Treatment Rule (SWTR), the Long-Term 1 Enhanced Surface Water Treatment Rule (LT1ESWTR), and the Interim Enhanced Surface Water Treatment Rule (IESWTR)—require turbidity monitoring as a means of assessing filtration performance, surface water facilities implementing membrane filtration for LT2ESWTR compliance may use turbidity monitoring to satisfy both requirements. States may have other specific requirements for continuous indirect integrity monitoring independent of the LT2ESWTR or may approve other methods for LT2ESWTR compliance. All federal and state requirements must be satisfied in a utility's IVP.

If not otherwise constrained by regulatory requirements, a utility's decision to use a specific type of indirect integrity monitoring technique may be influenced by the ability of a particular method to provide sufficient resolution or sensitivity. For example, because particle counters have been demonstrated to be more sensitive to breaches than are particle monitors or conventional turbidimeters, a utility may choose to use particle counting to maximize the ability to detect compromises in system integrity between periodic direct integrity test events (Jacangelo et al., 1997). Other utilities may select laser turbidimetry, which has been shown in some studies to perform comparably to particle counting in terms of sensitivity to integrity breaches (Banerjee et al., 2000; Colvin et al., 2001). In this case, if laser turbidimetry is approved by the state for both purposes, a utility may use laser turbidimeters for compliance with the applicable surface water treatment rules with the additional benefit of improved sensitivity for detecting integrity breaches over conventional turbidimeters.

Sensitivity may also be improved for any given continuous indirect integrity monitoring method by applying instrumentation to smaller groupings of membrane modules, such that any integrity breach would have a greater impact on filtrate quality. In this case, the benefits of the increased sensitivity should be weighed against the cost of additional instruments. This approach may be advantageous if the gain in sensitivity from using a greater number of less expensive instruments is justified by the comparable cost of fewer, more expensive instruments. Note that for LT2ESWTR compliance, an instrument for continuous indirect integrity monitoring must be applied to each membrane unit (40 CFR 141.719(b)(4)). A utility may apply instruments to smaller groupings of membrane modules at its discretion. For other applications, state requirements for monitoring various groupings of membrane modules in an overall system must be satisfied in a utility's IVP.

If test resolution is an important criterion, a utility might strongly consider particle counting. Because particle counting is the only method that assesses the size of particulate matter, it is the only method of indirect integrity monitoring to which the concept of resolution applies. For example, if membranes are applied specifically to remove *Cryptosporidium*, the particle counters should be well calibrated to accurately detect particles approximately 3 μm in size or larger. The target resolution may vary depending on the particular contaminant of concern.

What Constitutes Continuous Indirect Integrity Monitoring?

Under the LT2ESWTR, continuous monitoring is defined as monitoring conducted at a frequency of no less than once every 15 min. However, because the instrumentation used for the various methods of indirect integrity monitoring may allow data collection at much more frequent intervals, the state may have more stringent data collection requirements. In the absence of specific regulatory requirements, a utility may collect data at a frequency it determines to be appropriate. Nevertheless, since data acquisition can be automated, it is recommended that data be collected no less frequently than every 15 min, even if no other regulatory requirements apply.

While more frequent data collection provides both increased integrity monitoring and additional data to track system performance, there may be some potential complications from collecting data too frequently. For example, data could be collected frequently enough that a backwash (for applicable systems) cannot be completed between readings. In this case, a utility must program the system to cease data collection during the backwash cycle and for any length of time afterwards that the data remain artificially high, such that a direct integrity test is not triggered. (Data collection and analysis are further discussed later.) The implications of collecting indirect integrity monitoring data at different intervals should be considered in developing an IVP, and the frequency and any other associated qualifying guidelines or restrictions should be specified in the facility IVP documentation.

How Should Indirect Integrity Monitoring Results Be Interpreted?

Continuous indirect integrity monitoring is primarily intended to provide some indication of system integrity between direct integrity test applications. The indirect monitoring results are continuously compared to an established performance threshold that is known to represent a potential integrity breach. If this threshold level is exceeded, some type of response is triggered to further investigate the problem.

Under the LT2ESWTR, this performance threshold represents the UCL for continuous indirect integrity monitoring. If the UCL is exceeded, direct integrity testing is automatically triggered as a means to assess system integrity using a more sensitive technique. Unlike that for direct testing, the UCL for continuous indirect integrity monitoring does not necessarily correspond to a specific and quantifiable integrity loss. For example, the UCL established by the LT2ESWTR for turbidity monitoring (i.e., the default method in the absence of another technique specified by the state) is 0.15 ntu independent of site- or system-specific considerations; filtrate turbidity readings exceeding 0.15 ntu for a period of 15 min (or two consecutive 15 min readings exceeding 0.15 ntu) would immediately trigger direct testing (40 CFR 141.719(b)(4)(iv)). The 0.15 ntu threshold was selected because it is significantly below the 0.3 ntu threshold for filter performance required by the IESWTR for 95% of all turbidity samples yet, because membrane filtration systems are well documented to consistently produce filtered water below 0.05 ntu, a sustained filtrate turbidity exceeding 0.15 ntu strongly suggests a potential integrity problem. Note that if turbidity monitoring with the default UCL of 0.15 ntu is used as a means of continuous indirect integrity monitoring, a utility could simultaneously be in compliance with the IESWTR but out of compliance with the LT2ESWTR.

Although the LT2ESWTR specifies a UCL of 0.15 ntu with the default method of turbidity monitoring, the state may establish a more stringent standard at its discretion. In addition, for any approved method of continuous indirect integrity monitoring, the state may require a site- or system-specific performance-based UCL that is linked to a certain level of integrity loss (in terms of a specific number of broken fibers or LRV) as determined by a fiber-cutting study. These studies may also serve as the basis for establishing LCLs that trigger a particular response at a threshold prior to the point at which direct integrity testing would be required, either mandated by the state or voluntarily implemented by the utility. Voluntary LCLs could also be implemented after the membrane filtration system has been in operation for a certain period of time; this staggered implementation would allow sufficient baseline data to be collected such that operators could identify threshold levels that represent elevated or otherwise unusual results that are still below the UCL and do not necessarily indicate an integrity breach but nevertheless warrant observation. Some examples of actions associated with LCLs might be increased operator attention and diagnostic checks of the continuous indirect integrity monitoring instrumentation. All control limits (CLs)—both mandated by the state and voluntarily implemented by the utility—should be documented in an IVP, along with the rationale supporting the establishment of the CLs and the respective subsequent action associated with exceeding each.

As with direct integrity tests, false negative and false positive results are also possible with indirect integrity monitoring. For example, some indirect integrity monitoring instruments may indicate elevated levels of a parameter (e.g., turbidity, particle counts, etc.) after a routine maintenance event such as a backwash, particularly if air is employed in the process to scour or pulse the membrane surface. If significant, this air entrainment error could cause a CL to be exceeded, generating a false positive result (i.e., a result incorrectly suggesting a breach in a fully integral system); in the case of the UCL, this exceedance would inappropriately trigger direct integrity testing (and, under the LT2ESWTR, consequent reporting). This type of false positive result can be minimized by first characterizing typical system performance under a variety of operating conditions (such as after a backwash) and subsequently programming the data acquisition system to account for regularly occurring data aberrations of previously quantified magnitude and duration which are known not to represent an integrity problem, even if CLs are exceeded. (The importance of collecting and analyzing baseline data for IVP optimization is further discussed later.) In some cases, devices such as bubble traps (e.g., in the case of air entrainment) may be utilized to minimize modes of error that might generate false positive results.

False negative results (i.e., results that either indicate a fully integral system in the presence of an integrity breach or significantly underestimate a substantial integrity breach) may be more common with indirect integrity monitoring methods. Because currently available indirect integrity monitoring techniques are less sensitive to breaches than are direct integrity tests, it is possible that small breaches may occur that could be detectable via direct but not indirect methods. This potential may be minimized by utilizing a more sensitive method of continuous indirect integrity monitoring (such as the use of laser vs. conventional turbidimeters), if a utility is permitted such flexibility under state regulations. Alternatively, a utility could increase the sensitivity of an indirect method by utilizing a greater number of instruments (i.e., decreasing the number of membrane modules monitored per instrument). However, a utility considering these options should evaluate whether the cost of increasing indirect method sensitivity (via purchasing a greater number of instruments, more sensitive instruments, or both) is justified by the consequent level of heightened integrity monitoring ability between direct test events.

Any particular strategies for minimizing the potential for false positive and false negative results should be specified in IVP documentation. A calibration schedule for continuous indirect integrity monitoring instrumentation should also be specified. It is recommended that this instrumentation be calibrated on at least an annual basis.

Diagnostic Testing

Diagnostic testing is a process of identifying and isolating integrity breaches that have already been detected and confirmed using other methods, and is thus a critical component of an IVP. As with previous sections in this appendix, this section is organized into a series of questions that parallel the step-by-step process of formulating a diagnostic testing strategy for an IVP. Each of the questions presented in this section represents some aspect of diagnostic testing that should be clearly addressed and documented in an IVP.

What Is the Purpose of Diagnostic Testing?

The purpose of diagnostic testing is to identify and isolate integrity breaches in a membrane module that are detected via other methods. Because direct integrity testing and continuous indirect integrity monitoring techniques are only intended to detect the existence of a breach, diagnostic testing complements these methods, serving as a tool to pinpoint the exact location of a breach. In this way, diagnostic testing also serves as a critical link between identifying an integrity problem during the course of operation and repairing the breach. (Membrane repair and replacement are further discussed later.) Thus if an integrity breach is known or suspected in a membrane unit, that unit should be taken out of service for diagnostic testing in order to facilitate repairs. IVP documentation should clearly note the purpose of diagnostic testing so as to distinguish it from other forms of testing.

Under What Circumstances Should Diagnostic Testing Be Applied?

The LT2ESWTR requires that a membrane unit be taken out of service if the established direct integrity test UCL (i.e., that associated with the log removal credit awarded to the membrane process) is exceeded (40 CFR 141.719(b)(3)(v)). Under these conditions, diagnostic testing may be used to identify the location of an integrity breach. The use of diagnostic testing under these circumstances is also suggested for those utilities that do not use their respective membrane filtration systems for LT2ESWTR compliance. Thus, in essence, diagnostic testing is recommended any time an integrity breach is detected from the results of a direct integrity test.

Note that the LT2ESWTR does not link diagnostic testing directly to continuous indirect integrity monitoring. Even if indirect monitoring results clearly indicate an integrity problem, it is advisable to confirm the existence of a breach using more sensitive direct integrity testing methods. In any case, under the compliance framework for the LT2ESWTR, any continuous indirect integrity monitoring results that would clearly indicate an integrity breach would almost certainly exceed the indirect integrity monitoring UCL and thus trigger direct integrity testing, as required. However, a utility that is not otherwise constrained by regulatory requirements could voluntarily take a membrane unit out of service for diagnostic testing based on indirect monitoring results alone.

An IVP should clearly identify the specific circumstances under which diagnostic testing should be applied, including both those conditions that require diagnostic testing under a regulatory framework and those that might trigger the use of diagnostic testing on a voluntary basis by the utility.

What Type(s) of Diagnostic Testing Should Be Used?

By definition, most diagnostic tests are categorized as types of direct integrity tests; however, diagnostic tests are distinguished from other types of direct tests by their ability to not just detect an integrity breach in a membrane unit but help locate the specific module or fiber containing the breach, as well. In addition, most methods considered to be diagnostic tests are designed to be applied to specific membrane units on an as-needed basis, and thus are impractical for implementation on a scale that would satisfy the direct integrity testing requirements of the LT2ESWTR. For example, sonic testing—one type of diagnostic testing—requires an operator to manually apply an accelerometer to various locations on the membrane module to listen for vibrations caused by leaking air. While this technique fits the definition of direct integrity test, it would be infeasible to use such a test to check every module in a membrane filtration system for integrity breaches every day.

In addition to sonic testing, other methods of diagnostic testing include bubble testing, conductivity profiling, and simple visual inspection (where applicable). Each of these methods is described in further detail in Section 4.8. A pressure (or vacuum) decay test applied to a smaller number of modules (i.e., a subset of a full membrane unit) in order to identify a particular breached module may also be used as a form of diagnostic test. Single module testing represents the smallest form of incremental membrane unit testing. This type of diagnostic test generally involves removing the individual modules from the membrane unit and testing each on a specially designed single module apparatus. Other types of direct integrity tests may also have the potential to serve as diagnostic tests in a scaled-down form.

In some cases, a battery of diagnostic tests may be used to identify an integrity breach. For example, if an MF membrane unit fails a direct integrity test (i.e., the results exceed the associated UCL), after the unit is taken out of service a sonic test might be applied to each module in the unit in turn to identify the affected modules. (Note that although the membrane unit is taken out of service, it must still remain in operation in order to facilitate some types of diagnostic tests, such as the sonic test referenced in this example and described in Section 4.8.3. Therefore, in such cases, the unit must be operating in filter-to-waste mode.) The modules may then be removed from the unit to conduct a bubble test (see Section 4.8.2) in order to isolate particular fibers that may be subsequently removed from service permanently by pinning or sealing (as described in the next section). Thus, just as diagnostic testing complements direct integrity testing or continuous indirect integrity monitoring, different diagnostic tests can also complement each other.

A utility should develop its own system-specific protocol for diagnostic testing and document the procedures in its IVP. The IVP should specify which diagnostic tests are prescribed, the particular purpose of each test (e.g., to isolate a module or identify a particular fiber), the circumstances under which each test should be conducted, a list of necessary testing equipment, and detailed instructions for conducting the test. Some membrane filtration system manufacturers may provide guidance in developing an appropriate diagnostic testing plan. Although diagnostic testing may not be commonly employed, a utility should still keep all the equipment necessary to conduct each diagnostic test specified in its system IVP both on-site and in proper working order. Note that some diagnostic tests (such as sonic testing) require a greater degree of skill in both conducting the test and interpreting the results; operators who will have responsibility for conducting these tests should be designated and trained in advance in order to minimize membrane unit downtime.

Membrane Repair and Replacement

In the context of this guidance, "membrane repair and replacement" does not necessarily apply to just the membranes themselves, but to any component of a membrane filtration system that might allow an integrity breach if it were to fail. This section is organized into a series of questions presented in a logical order intended to parallel the step-by-step process of considering membrane repair and replacement in the context of an IVP. Each of the questions presented in this section should be addressed in a utility's IVP to an appropriate degree.

What Is the Purpose of Membrane Repair and Replacement?

The purpose of conducting repairs on a membrane filtration system or replacing irreparably damaged components is to correct any integrity breaches that have been detected, thus restoring and maintaining a fully integral system. In cases in which membrane filtration is applied for the removal of one or more specific pathogens of interest (e.g., for compliance with the LT2ESWTR), a more specific objective is to enable the system to maintain the full removal credit allocated by the state. As previously noted, although other system components may be essential for overall system operation, in this context repair and replacement are discussed in regard to only those system components that are critical to system integrity.

When Should Membrane Repair and Replacement Be Conducted?

In the simplest terms, repair or replacement should be conducted whenever an integrity breach is detected (e.g., using direct integrity testing or continuous indirect integrity monitoring). After the source of the breach has been isolated (e.g., via diagnostic testing), the problem should be corrected via component repair or replacement, as appropriate.

Under the LT2ESWTR, if the results of a direct integrity test exceed the upper (i.e., mandated) control limit, the affected membrane units must be immediately taken out of service (40 CFR 141.719(b)(3)(v)). Because the membrane units cannot be returned to service until a direct integrity test confirms that the UCL is no longer exceeded, some type of repair must be conducted to correct the integrity problem. If a utility has voluntarily implemented one or more tiered LCLs, it may in some cases be able to detect an integrity problem without exceeding the UCL. In this case the utility could opt to take corrective action, conducting diagnostic testing and subsequent repair immediately, or instead keep the affected membrane units in continued operation under increased observation until the next scheduled maintenance event. Although repair of any breach is recommended as soon as possible, a utility may exercise discretion in determining whether or not to implement any repairs based on the severity of the breach (assuming the UCL has not been exceeded). For example, if a utility is required to achieve only 2-log *Cryptosporidium* removal credit for LT2ESWTR compliance, it is possible that a significant integrity breach could occur without jeopardizing the ability of the membrane filtration system to obtain this credit. However, continued operation with a known integrity breach of any magnitude may not be permitted by the state. Even if operation under some compromised conditions is not explicitly prohibited, a utility should carefully consider the risk associated with the potential for pathogen passage before engaging in continued operation.

If a utility does not apply its membrane filtration system for compliance with the LT2ESWTR, it may have more flexibility with respect to the timing and necessity of conducting immediate repairs if an integrity breach is detected, unless otherwise constrained by state requirements. In the absence of any regulations prescribing system repair requirements, it is recommended that utilities adopt a conservative approach in order to help ensure the integrity of the membrane barrier against pathogens. The system IVP should clearly specify any regulatory requirements relating to membrane system repair, as well as other circumstances under which the utility would conduct repairs.

Membrane module repair—as opposed to replacement—is often advisable, if possible, since new membranes are typically expensive. However, if a membrane module is subject to repeated integrity breaches, a utility should consider replacing the module. Chronic repairs adversely affect treated water production and prevent operators from carrying out routine responsibilities. Also, if it is critical that an integrity breach be repaired as quickly as possible, for some types of membrane filtration systems it may be more efficient to replace the module with a new one from the utility's supply of spare components maintained on-site. Thus in some cases, the most expedient and cost-effective system repair could actually be membrane replacement.

Although the LT2ESWTR requires the use of direct integrity testing to confirm the success of any system repairs concerning integrity problems before returning the affected membrane units to service, this practice is recommended for all utilities, even if membrane filtration is not conducted for LT2ESWTR compliance (40 CFR 141.719(b) (3)(v)). This postrepair direct integrity testing confirms the success not only of the repair process but also of the proper reinstallation of any modules that may have been removed for repair. After any repair or replacement measures have been completed, both direct integrity test and continuous indirect integrity monitoring results for the repaired units should be closely tracked for an extended period, to gauge the long-term success of the repair and perhaps whether the problem is likely to recur.

What Are Some Common Modes of Integrity Breaches?

The types of integrity breaches to which a particular membrane filtration system is most susceptible can depend on both the type of system (i.e., MCF, MF, UF, NF, or RO) and the manufacturer. For example, while many NF and RO membranes are subject to chemical degradation by oxidants such as chlorine, only some MF and UF membranes are vulnerable to oxidation, depending on the membrane material used. This same example of chemical oxidation also illustrates potential causes of integrity breaches that are specific to a particular treatment application of membrane filtration. A treatment process that utilizes chlorine as a disinfectant upstream of RO membranes has an inherent potential source of chemical degradation, even though dechlorination may be implemented prior to the membranes. If the dechlorination system fails or is miscalibrated, the membranes could be subject to chlorine exposure. By contrast, an identical RO system used for a different application in which upstream disinfection is not required to meet treatment objectives would not have this additional element of risk. It is important to note that this example should not be interpreted as a recommendation against utilizing oxidants upstream of oxidant intolerant membranes. It is not uncommon to use upstream oxidants effectively, particularly for disinfection and bio-fouling control. However, a utility should be aware of the potential for chemical degradation of membrane material and take appropriate measures to prevent integrity breaches and protect treatment equipment.

In addition to chemical degradation, the most common causes of integrity breaches with NF and RO membranes are associated with O-rings and seals, which can be cracked, rolled, and/or improperly sized. Each of these defects can result in an integrity breach, as can foreign matter such as hairs or other fibrous material underneath O-rings. Other mechanisms for integrity breaches may be related to membrane defects, such as failures that may occur along glue lines or at weak spots in the membrane (i.e., creases or thin areas). Fiber breaks and potting problems are the most common types of integrity breaches associated with MF and UF membranes, although chemical degradation may also result in membrane failure. MCF systems are most likely to be compromised by improperly seated or sealed membrane cartridges or tears or punctures in the membrane material. MCF membranes may also exhibit integrity breaches along folds and creases, depending on the construction of the filter.

Although it is possible for breaches to occur at any time during operation, it is most common for integrity problems to occur during system start-up. These breaches typically result from either manufacturing defects or improper installation. As a result, it is important that an initial shakedown period be included as a part of the start-up process to enable these initial problems to be identified and corrected before the system is put into service.

An IVP should identify and document the most probable types of integrity breaches for a utility's specific membrane filtration system. These modes may be identified initially by consultation with the membrane manufacturer and consideration of site-specific circumstances (e.g., the use of upstream oxidants). Subsequently, the shakedown period and ongoing operational experience may yield important information about the most frequently occurring types of integrity problems. Also, the experience of other utilities using the same or similar filtration equipment may yield valuable information about the types of integrity failures common to a particular system.

How Should Membrane Repair and Replacement Be Conducted?

The types of repairs that may be conducted to correct integrity breaches vary significantly with the type of membrane filtration system. For example, for MF and UF systems, fiber breaks (the most common mode of failure) are not technically repaired, but rather isolated by inserting small pins or epoxy in the ends of the broken fiber, effectively removing them from service permanently and thus eliminating the system integrity breach. By contrast, it is generally not possible to repair compromised NF or RO membranes, although problems with O-rings and other seals may often be corrected by replacing the seals and making sure the membrane is properly seated in its housing. Likewise, because MCF systems utilize cartridges that are designed to be disposable, unless an integrity breach is the result of a seal problem, a damaged membrane cartridge would generally not be repaired. Thus, any integrity problems associated directly with either NF, RO, or MCF membranes themselves generally necessitate membrane replacement.

Prior to placing the system into operation, it is important to consult with the membrane manufacturer to determine what types of repairs can be made and which types of integrity breaches require membrane replacement. The manufacturer should also be able to provide both instructions and training for all applicable repair procedures. Because it is somewhat common for some types of integrity problems to occur during the system shakedown phase, it is recommended that this period be used for practicing repair procedures under the directions of a qualified manufacturer's representative.

When integrity problems occur during operation, it is important to identify the cause of the integrity problem, as well as its source. While some fiber breaks in MF and UF systems may be expected due to wear or mechanical stress over time, other breaches may have a specific cause that can be isolated and corrected to avoid further integrity problems. If an integrity breach occurs, it is important to check both the membrane filtration system and any upstream treatment processes to ensure that these are operating properly. For example, if a chemical that is incompatible with the membrane material is added upstream and not properly removed or quenched, some membrane damage and loss of integrity may occur.

Any necessary repair equipment or spare parts, including replacement modules that may be required in the event of an integrity breach, should be kept on-site. The system IVP should specify a list of these components, along with any applicable instructions. The IVP should also include suggestions for troubleshooting integrity problems.

Data Collection and Analysis

Diligent and rigorous collection and analysis of integrity testing and monitoring data are important components of any IVP. Careful data collection and analysis can serve as useful tools for preventing integrity problems, as well as for optimizing system performance and troubleshooting problems. Like other sections in this appendix, the following discussion is organized by addressing a series of critical questions in regard to data collection and analysis, each of which should be encompassed in a comprehensive IVP.

What Is the Purpose of Data Collection and Analysis?

Although the primary purpose of data collection and analysis may be to demonstrate regulatory compliance, a thorough and well-planned program can result in a number of other important benefits for preventing integrity problems and optimizing system performance. For example, a consistently maintained record of membrane unit performance during both direct integrity testing and continuous indirect integrity monitoring may help determine when some membranes are approaching the end of their useful lives. In addition, because the resistance or permeability of some membranes changes after an initial "setting" period, a careful record of integrity test results may indicate that either the UCL (subject to regulatory approval) or any voluntarily implemented LCLs should be adjusted.

Continuous indirect integrity monitoring data also have a number of advantages in addition to helping gauge membrane integrity. These data may be used to identify performance trends, such as those that may occur between backwashing or chemical cleaning events or over the entire life of the membranes. A well-documented data record can also identify any systematic or periodic trends and potentially help isolate the causes. In addition, data records facilitate the comparison of performance trends among different membrane units. (Note that a normal amount of instrument variation should be taken into account when conducting a statistical analysis comparing data among different membrane units.) A substantial amount of data collected over time may also enable operators to identify aberrations that do not necessarily indicate integrity problems. For example, if the turbidity is consistently higher than normal after a backwash, operators may be able to attribute this consistent aberration to air-entrainment error associated with the instrumentation. It is important that these types of aberrations are identified so that the system can be programmed not to trigger direct integrity testing during these events. Direct integrity testing data can also be a valuable tool for identifying trends; however, because continuous indirect integrity monitoring data are typically collected much more frequently and while the unit is online, they may often be the most practical means of tracking performance trends.

What Data Should Be Collected?

In addition to any regulatory requirements regarding data collection during operation, a utility should collect baseline data for each membrane unit (with respect to both direct integrity testing and continuous indirect integrity monitoring) before putting the plant into service. These data will serve as a reference baseline against which to evaluate membrane unit performance and also help refine a utility's strategy for collecting continuous indirect integrity monitoring data during regular operation. For example, for MF and UF systems the baseline data should demonstrate how long after a backwash event turbidity or particle count data might remain elevated. Based on this information, a utility can account for such data spikes that are known to not represent integrity problems, so that direct integrity testing and consequent loss of production are not unnecessarily triggered. Throughout operation, as well as during integrity testing conducted during routine maintenance or repair, it is recommended that data be collected in a spreadsheet or with software that establishes a database and allows data to be readily plotted in order to identify trends occurring over time.

What Are Some Methods for Reducing Continuous Indirect Integrity Monitoring Data?

As noted earlier, the LT2ESWTR defines continuous monitoring as a frequency of no less than once every 15 min. However, instruments used to collect integrity monitoring data—particle counters, particle monitors, turbidimeters, etc.—may allow data collection at much shorter frequency intervals. Thus in the absence of more specific requirements from the state, a utility may collect data as often as possible, if desired, provided that the minimum frequency is met or exceeded. Furthermore, in the absence of specific guidelines for how a large amount of data should be reduced for compliance and reporting purposes, a utility has the latitude to select a statistical method that it determines to be appropriate for its system. Some potential methods include

- maximum value,
- 95th percentile,
- average value, and
- singular timed measurement.

Using the example of continuous monitoring as defined under the LT2ESWTR (i.e., a minimum frequency of once every 15 min), each of these methods is described in context as follows.

Maximum Value

Using this method, the maximum value that occurs over every 15 min period represents the entire data set. Thus, direct integrity testing is triggered if even one measurement exceeds the UCL. This method is very conservative and could potentially result in excessive direct integrity testing and subsequent loss of filtered water production from any anomalous data spikes, which may be attributable to any number of factors aside from an integrity problem.

95th Percentile

With this method, the 95th percentile datum represents the entire 15 min data set, effectively screening out the largest 5% of data spikes. Direct integrity testing would be triggered if this datum exceeded the UCL. This method is less conservative than the maximum value approach and is more likely to screen anomalous data spikes that are not indicative of an integrity problem. The rationale behind this method is that if an integrity breach occurs, it may be likely that more than 5% of the data collected exceed the UCL. The premise for this method may be used with any percentile, and it may be advantageous for a utility using this technique to conduct a statistical analysis to determine an appropriate percentile to eliminate anomalous spikes without screening data that might indicate an integrity breach.

Average Value

Using this method, the average value represents the entire 15 min data set. This technique is roughly equivalent to a 50th percentile approach, and thus is less conservative than the 95th percentile method. However, the average value method lessens the need to artificially exclude anticipated integrity spikes that are known not to indicate an integrity problem (e.g., data collected immediate after a backwash with MF and UF systems), as the effect of these spikes may be sufficiently dampened that the average value is below the UCL.

Singular Timed Measurement

The singular timed measurement approach uses one reading collected exactly every 15 min for comparison to the UCL (i.e., for regulatory compliance), independent of how frequently data are collected between these compliance readings. Because many other noncompliance measurements collected during these 15 min intervals could exceed the UCL without triggering direct integrity testing, this method is one of the least conservative approaches. This technique represents the minimum requirement for compliance with the LT2ESWTR.

Note that because the LT2ESWTR does not specify a statistical reduction technique for data collected more frequently that at 15 min intervals, the methods described in this section are not specific compliance options under the rule. These methods are not meant to represent an exclusive or exhaustive list, and are simply examples of approaches that utilities could employ if not otherwise constrained by state requirements. A utility's IVP should clearly specify its data collection and analysis practices for both continuous indirect integrity monitoring and direct integrity testing.

Reporting

Utilities that use membrane filtration for compliance with the LT2ESWTR are required to submit a monthly operating report to the state summarizing the UCL exceedances for both direct integrity testing and continuous indirect integrity monitoring, as well as any corrective action that is taken in response (40 CFR 141.721(f)(10)(ii)). Because these monthly reports are directly linked to integrity testing results, it is important that reporting be incorporated into a comprehensive IVP. As with other sections in this appendix, the following discussion on reporting is organized into a series of critical questions that should be addressed in an IVP.

What Is the Purpose of Reporting?

As it relates to integrity verification, the primary purpose of reporting is to document the ability of a membrane filtration system to meet its required log removal or other performance-based objectives on an ongoing basis. Under the minimum requirements of the LT2ESWTR, this documentation generally consists of all direct integrity test and continuous indirect integrity monitoring results that exceed the respective UCLs and the subsequent corrective action that was taken in each case (40 CFR 141.721(f)(10)(ii)). Note that collecting, recording, and storing an abundance of integrity test data may be very beneficial for optimizing membrane filtration system performance, as discussed previously, even if the majority of the accumulated data are not necessary for complying with reporting requirements.

If membrane filtration is not applied specifically for LT2ESWTR compliance, reporting may not necessarily be directly related to integrity verification, depending on state requirements. For example, membrane filtration might be considered an alternative filtration technology (as provided for under the federal SWTR), in which case reporting requirements might simply include turbidity, similar to those for conventional media filters. In this case, including reporting requirements in an IVP may not be critical for a utility. Nevertheless, because verifying and preserving membrane integrity are critical to successful membrane filtration system operation, it is recommended that even turbidity, particle counts, and other required filtrate quality data always be considered continuous indirect integrity monitoring results, thus linking reporting requirements to membrane integrity and, subsequently, an IVP.

What Should an IVP Include with Respect to Reporting?

An IVP should specify any state requirements for reporting, including both the content and the reporting frequency. (Reporting requirements for utilities that employ membrane filtration for compliance with the LT2ESWTR are addressed in Chapters 4 and 5 for direct integrity testing and continuous indirect integrity monitoring, respectively.) Any utility-specific procedures for preparing the compliance report should also be included in an IVP, along with a sample report form. An IVP should also indicate the duration of time over which the utility is required to keep records related to reporting data. Under the LT2ESWTR, utilities must keep records of all treatment monitoring associated with membrane filtration and used for rule compliance (including both direct integrity testing and continuous indirect integrity monitoring results, as applicable) for a period of three years (40 CFR 141.422(c)).

Summary

Although not required under the LT2ESWTR, the development of a comprehensive IVP can be a valuable organizational tool to help a utility verify and maintain membrane system integrity. An IVP should essentially serve as a utility's system-specific how-to guide for all aspects of operation and maintenance that are related to system integrity, including (but not necessarily limited to) the following:

- regulatory requirements;
- voluntarily implemented system-specific practices;
- clear objectives for all IVP procedures;
- instructions for all IVP procedures;
- equipment listings, descriptions, and purposes;
- system troubleshooting tips;
- guidelines for interpreting test results;
- sample calculations (where applicable); and
- membrane manufacturer contact information.

A well-developed IVP containing these elements can make the process of integrity verification more effective and efficient, thus helping a utility maximize the benefit of its membrane filtration system for serving as a barrier to pathogens and other particulate matter.

APPENDIX K

Overview of Bubble Point Theory*

Introduction

The various methods of pressure-based direct integrity testing are predicated on capillary theory as described by the bubble point equation, which is derived from a balance of static forces on the meniscus in a capillary tube. The bubble point itself is defined as the threshold gas pressure required to displace liquid from the pores or capillary-like breaches of a fully wetted membrane. In the context of porous membranes, bubble point theory was originally used as the basis for developing a test to characterize pore sizes. Because the bubble point equation describes an inverse relationship between the applied pressure and the capillary (or pore) diameter, the pressure at which bubbles are first detected in a fully wetted membrane can be used to calculate the diameter of the largest pore (see Equation K.1). Accordingly, larger threshold pressures are indicative of membranes with smaller pores. A diagram illustrating a membrane pore as a capillary tube is shown in Fig. K-1.

Bubble point theory has also been applied to the detection of integrity breaches in the form of the various pressure-based direct integrity tests, such as the pressure or vacuum decay tests (Sections 4.7.1 and 4.7.2, respectively, in Appendix L of this book), the diffusive airflow test (Section 4.7.3), and the water displacement test (Section 4.7.4). Integrity breaches such as broken hollow fibers or holes in the surface of the membrane are analogous to pores, and larger test pressures enable the detection of smaller breaches. If the applied pressure is below the bubble point of the largest membrane pore and does not decay (pressure and vacuum decay tests), generate airflow (diffusive airflow test), or displace water (water displacement test) to a degree that exceeds normal tolerances over the duration of the direct integrity test, the membrane is determined to be integral at the level of the threshold pore or breach size corresponding to that applied pressure.

---

*Source: *Membrane Filtration Guidance Manual* (USEPA, 2005).

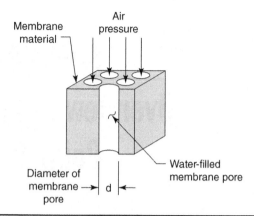

In the context of the Long Term 2 Enhanced Surface Water Treatment Rule (LT2ESWTR), this threshold pore or breach size is called the test resolution, as described in Section 4.2.

The purpose of this appendix is to provide a general overview of bubble point theory as it relates to direct integrity testing under the LT2ESWTR. In addition to the background provided in this introduction, subsequent sections of this appendix describe the bubble point equation and its parameters.

The Bubble Point Equation

The bubble point equation is derived from a balance of static forces on the meniscus in a capillary tube and is given as Equation K.1 (without specific units). A derivation of this equation is given in the literature by Meltzer (1987).

$$P_{bp} = \frac{4\sigma \cos\theta}{d_{cap}},$$ (Equation K.1)

where P_{bp} = bubble point pressure
σ = surface tension at the air-liquid interface
θ = liquid-membrane contact angle, and
d_{cap} = capillary diameter

Converting Equation K.1 to a form that utilizes convenient units for the various parameters yields Equation K.2:

$$P_{bp} = \frac{0.58\sigma \cos\theta}{d_{cap}},$$ (Equation K.2)

where P_{bp} = bubble point pressure (psi)

σ = surface tension at the air-liquid interface (dyn/cm)

θ = liquid-membrane contact angle (°), and

d_{cap} = capillary diameter (μm)

Because the pore structure of most membranes cannot be accurately represented by a perfectly cylindrical capillary, a shape correction factor κ can be included in this equation to account for nonideal conditions.

APPENDIX L

Direct Integrity Testing*

4.1   Introduction

In order for a membrane process to be an effective barrier against pathogens and other particulate matter, the filtration system must be integral, or free of any integrity breaches. Thus it is critical that operators be able to demonstrate the integrity of this barrier on an ongoing basis during system operation. Direct integrity testing represents the most accurate means of assessing the integrity of a membrane filtration system that is currently available.

Under the Long Term 2 Enhanced Surface Water Treatment Rule (LT2ESWTR), a direct integrity test is defined as a physical test applied to a membrane unit in order to identify and isolate integrity breaches. In order to receive *Cryptosporidium* removal credit for compliance with the rule, the removal efficiency of a membrane filtration process must be routinely verified during operation using direct integrity testing. The direct integrity test must be applied to the physical elements of the entire membrane unit, including membranes, seals, potting material, associated valves and piping, and all other components which could result in contamination of the filtrate under compromised conditions (40 CFR 141.719(b)(3)(i)).

There are two general classes of direct integrity tests that are commonly used in membrane filtration facilities: *pressure-based tests* and *marker-based tests*. The pressure-based tests are based on bubble point theory and involve applying a pressure or vacuum (i.e., negative pressure) to one side of a membrane barrier and monitoring for parameters such as pressure loss or the displacement of air or water in order to establish whether an integrity breach is present. The various pressure-based tests include the pressure and vacuum decay tests, the diffusive airflow test, and the water displacement test. Marker-based tests utilize either a spiked particulate or molecular

---

*Source: *Membrane Filtration Guidance Manual* (USEPA, 2005).

341

marker to verify membrane integrity by directly assessing removal of the marker, similar to a challenge test.

The LT2ESWTR does not require the use of a particular direct integrity test for rule compliance, but rather demands that any test used meet the specified performance criteria for *resolution, sensitivity,* and *frequency.* Thus a particular system may utilize an appropriate pressure- or marker-based test or any other method that both meets the performance criteria and is approved by the state. The performance criteria for direct integrity tests are summarized as follows:

- *Resolution.* The direct integrity test must be responsive to an integrity breach on the order of 3 μm or less (40 CFR 141.719(b) (3)(ii)).

- *Sensitivity.* The direct integrity test must be able to verify a log removal value (LRV) equal to or greater than the removal credit awarded to the membrane filtration process (40 CFR 141.719(b)(3)(iii)).

- *Frequency.* A direct integrity test must be conducted on *each membrane unit* at a frequency of no less than once each day that the unit is in operation. (The definition of a membrane unit under the LT2ESWTR is provided in Section 1.5.) Less frequent testing may be approved by the state if supported by demonstrated process reliability, the use of multiple barriers effective for *Cryptosporidium,* or reliable process safeguards (40 CFR 141.719(b)(3)(vi)).

In addition to the performance criteria, the rule also requires the establishment of a *control limit* within the associated sensitivity limits for the direct integrity test that is indicative of an integral membrane unit capable of achieving the *Cryptosporidium* removal credit awarded by the state (40 CFR 141.719(b)(3)(iv)). If the results of the direct integrity test exceed this limit, the rule requires that the affected membrane unit be taken off-line for diagnostic testing and repair (40 CFR 141.719(b)(3)(v)). The unit may be returned to service only after system integrity has been confirmed by a direct integrity test.

The objective of this appendix is to describe the various pressure- and marker-based direct integrity tests currently in use and the ways these tests can be applied to meet the performance criteria specified under the LT2ESWTR. Diagnostic tests, data collection, and reporting are also addressed.

This appendix is divided into the following sections:

- Section 4.2: Test Resolution. This section discusses the determination of pressure- and marker-based direct integrity test resolution for meeting the performance criteria required by the rule.

- Section 4.3: Test Sensitivity. This section discusses the determination of pressure- and marker-based direct integrity test sensitivity for meeting the performance criteria required by the rule, including general concepts and methods.

- Section 4.4: Test Frequency. This section reviews the direct integrity testing frequency requirements of the rule.

- Section 4.5: Establishing Control Limits. This section describes the mathematical and experimental determination of control limits for direct integrity testing.

- Section 4.6: Example. Establishing Direct Integrity Test Parameters. This section illustrates the calculation of some of the critical direct integrity test performance criteria, including test resolution, sensitivity, and control limits, for an example membrane filtration system.

- Section 4.7: Test Methods. This section provides an overview of the various types of pressure- and marker-based tests, including generic test protocols as well as some advantages and disadvantages of each.

- Section 4.8: Diagnostic Testing. This section describes some of the diagnostic tests that are used to identify and isolate integrity breaches following a failed direct integrity test.

- Section 4.9: Data Collection and Reporting. This section provides guidance on direct integrity test data collection and reviews the associated reporting requirements of the rule.

4.2 Test Resolution

Resolution is defined as the size of the smallest integrity breach that contributes to a response from a direct integrity test. Any direct integrity test applied to meet the requirements of the LT2ESWTR is required to have a resolution of 3 μm or less. This resolution criterion is based on the lower size range of *Cryptosporidium* oocysts and is intended to ensure that any integrity breach large enough to pass oocysts contributes to a response from the direct integrity test used. The manner in which the resolution criterion is met depends on whether the direct integrity test is pressure based or marker based, as described in the following subsections.

4.2.1 Pressure-Based Tests

In order to achieve a resolution of 3 μm with pressure-based direct integrity tests, the net pressure applied during the test must be great enough to overcome the capillary forces in a 3 μm hole, thus ensuring that any breach large enough to pass *Cryptosporidium* oocysts would also pass air during the test. The minimum applied test pressure necessary to achieve a resolution of 3 μm is calculated using Equation 4.1:

$$P_{test} = 0.193\kappa\sigma\cos\theta + BP_{max},\qquad\text{(Equation 4.1)}$$

where P_{test} = minimum test pressure (psi)
   0.193 = constant that includes the defect diameter (i.e., 3 μm resolution requirement) and unit conversion factors
   κ = pore shape correction factor (dimensionless)
   σ = surface tension at the air-liquid interface (dyn/cm)
   θ = liquid-membrane contact angle (°)
   BP_{max} = maximum back pressure on the system during the test (psi)

Equation 4.1 is based on bubble point theory and is derived from the balancing of capillary static forces. Note that the constant of 0.193 accounts for the LT2ESWTR resolution requirement of responding to a defect of at least 3 μm in diameter as well as the appropriate unit conversion factors, in order to simplify the equation for the purposes of rule compliance. The general form of Equation 4.1 includes the capillary diameter as a variable and represents an expression relating this diameter to the bubble point pressure. A discussion of bubble point theory and a derivation of Equation 4.1 is provided in Appendix B.

Values for several parameters in Equation 4.1 must be determined in order to calculate the minimum test pressure necessary to achieve a resolution of 3 μm when using a pressure-based integrity test. The parameters κ and θ are intrinsic properties of the membrane. In the absence of data supplied by the membrane manufacturer, conservative values of $\kappa = 1$ and $\theta = 0$ should be used. Appendix K provides an additional discussion of these parameters. The surface tension σ is inversely related to temperature; consequently, the surface tension at the coldest anticipated water temperature should be used to calculate a conservative value for the minimum required test pressure. As a point of reference, the surface tension of water at 5°C is 74.9 dyn/cm. Substituting these three values ($\kappa = 1$, $\theta = 0$, and $\sigma = 74.9$ dyn/cm) into Equation 4.1 yields the following simplified equation:

$$P_{test} = 14.5 + BP_{max},\qquad\text{(Equation 4.2)}$$

where P_{test} = minimum test pressure (psi)
   BP_{max} = maximum back pressure on the system during the test (psi)

Equation 4.2 indicates that the minimum test pressure necessary to achieve a 3 μm resolution is 14.5 psi plus the maximum back pressure on the system during application of a pressure-based direct integrity test (at a conservative temperature of 5°C). Ideally, there should be no hydrostatic back pressure on the system during the test. However, it is not always practical to perform the test without any hydrostatic back pressure, and in these cases the additional back pressure must be considered in establishing the minimum test pressure necessary to meet the resolution criterion. For example, there might be hydrostatic pressure on the undrained side of the membrane if a pressure-driven membrane module remains filled with water or if a vacuum-driven (i.e., immersed) membrane remains submerged underwater in a basin. Thus if the bottom of the membrane is under 7 ft of water, BP_{max} would be approximately 3 psi, yielding a P_{test} value of 17.5 psi to achieve a resolution of 3 μm.

Equations 4.1 and 4.2 assume that the applied pressure remains constant during the direct integrity test. However, in many cases there may be some baseline decay (e.g., attributable to diffusion) that is measurable over the duration of the test. In this case, it is important to account for this baseline decay in the resolution calculation. Thus, in order to ensure that the resolution requirement is satisfied throughout the duration of the test, the anticipated pressure at the end of the direct integrity test should be used to calculate the resolution. This value can be estimated using the initial applied pressure, the typical rate of baseline pressure decay for a fully integral system, and the duration of the test. If the baseline decay is small enough that the final test pressure is within approximately 5% of the initial applied pressure, the baseline decay can be assumed to be negligible, and the initial applied pressure may be used to calculate the test resolution.

The LT2ESWTR does not establish the minimum test pressure to be used during a pressure-based direct integrity test, but rather only requires that the test achieve a 3 μm resolution. If a membrane manufacturer has information to support the use of values other than $\kappa = 1$ and $\theta = 0$, and these less conservative values are approved by the state, then Equation 4.1 can be used to calculate the minimum required test pressure. It is essential that the use of values other than $\kappa = 1$ and $\theta = 0$ be scientifically defensible, since the use of inappropriate values could result in the use of a test pressure that does not meet the resolution criterion established by the rule. One approach for determining membrane-specific values for κ and θ is through direct experimental evaluation. Because these parameters can have a significant effect on the required direct integrity test pressure, it is strongly recommended that states require sufficient justification from a membrane manufacturer prior to approving the use of values other than $\kappa = 1$ and $\theta = 0$, such as independent third party testing results using a method accepted by the scientific community and demonstrating statistically significant data.

Although the rule does not include a frequency requirement for the recalculation of resolution, the resolution should be recalculated if the system back pressure during direct integrity testing is adjusted. Alternatively, if desired, direct integrity test instrumentation and the data recording system could be configured to calculate the resolution after each application of the direct integrity test using the applied test pressure, system back pressure, and surface tension corresponding to the temperature at which the test is conducted. Note that the liquid-membrane contact angle can also change over the life of a membrane module (e.g., as a result of the adsorption of organic matter by the membrane material), and these changes may not necessarily be uniform among the various modules in a unit (Childress et al., 1996; Jucker et al., 1994). Thus, if a value other than $\theta = 0$ (i.e., the most conservative value) is used, it may be appropriate to periodically recalculate the resolution based on a revised estimate of the actual value of θ—an exercise that may necessitate destructive testing of a representative sample of membrane modules in the system.

4.2.2   Marker-Based Tests

A marker-based direct integrity test can be viewed as a mini challenge study, in which a surrogate is periodically applied to the feed water in order to verify the integrity of a membrane filtration system. In order to meet the resolution criterion of the rule, the surrogate used in a marker-based test must have an effective size of 3 μm or smaller, as described in Section 3.9.2 of the *Membrane Filtration Guidance Manual*. A marker-based direct integrity test can use either particulate or molecular surrogates, but in either case it must be established that the surrogate meets the resolution criterion. Section 3.9 presents guidelines for the selection of a conservative surrogate for *Cryptosporidium* during challenge testing, and these same guidelines are applicable to selection of an appropriate surrogate for a marker-based direct integrity test. The effective size of the marker can be established through any accepted methodology, such as size distribution analysis of particulate markers or estimation techniques based on the molecular weight and geometry of molecular markers.

4.3   Test Sensitivity

Sensitivity is defined as the maximum log removal value that can be reliably verified by the direct integrity test (i.e., LRV_{DIT}). The sensitivity of the direct integrity test establishes the maximum log removal credit that a membrane filtration process is eligible to receive if it is less than or equal to that demonstrated during challenge testing (i.e., $LRV_{C\text{-}Test}$). For example, if the challenge test demonstrates an $LRV_{C\text{-}Test}$ of 5.5 log and the direct integrity test is capable of verifying an LRV_{DIT} of 4.5 log, the membrane filtration process would be eligible for removal credit up to 4.5 log. Although the sensitivity of the direct integrity test should not be expected to vary significantly over time, the determination of sensitivity as described in this section is designed to produce a conservative result that would remain applicable over the life of the membrane filtration system. However, if significant changes occur in terms of operational parameters, direct integrity test conditions, or any basic assumptions that might affect the value of the direct integrity test sensitivity, it is suggested that the sensitivity be reestablished to verify that it is at least equal to the removal credit awarded to the process.

The sensitivity of a direct integrity test is logarithmic in nature. For example, a test with an LRV_{DIT} of 5 log is 100 times as sensitive as a test with an LRV_{DIT} of 3 log. Thus when a higher sensitivity is required, the test must be capable of measuring very small changes in the direct integrity test response and distinguishing these results from background or baseline data. Data suggest that many direct integrity tests, as currently applied, have sensitivities in excess of 4 log; however, sensitivity must be determined on a case-by-case basis using the information provided by the membrane manufacturer and the guidance in this document. While determination of integrity test sensitivity can be complex, it provides a rational basis for awarding high removal credits to membrane processes that are commensurate with their abilities. As is the case with resolution, the manner in which sensitivity is determined depends on whether the type of direct integrity test used is pressure based or marker based.

Note that for systems that utilize multiple stages of membrane filtration, the sensitivity for each stage must be determined independently. The most common example of such an application would be a case in which a second stage is used to treat the backwash water from the first stage, after which the filtrate from the two stages is blended (i.e., concentrate staging). In this case, as well as in others involving multiple stages, if the filtrate from the various stages is blended, the stage using the membrane units with the lowest sensitivity would limit the maximum log removal credit that could be awarded to the overall process. However, if a second (or subsequent) stage is used strictly for residuals management, such that the filtrate is recycled to an upstream point in the overall treatment process, the LT2ESWTR would not be applicable to such ancillary stages.

4.3.1   Pressure-Based Tests

The discussion in this section regarding the calculation of sensitivity for pressure-based direct integrity tests is divided into three parts. First, the basic concepts that are applicable to all pressure-based tests are introduced. The subsequent subsection describes the calculation of sensitivity for pressure-based direct integrity tests based on this general conceptual framework. The third subsection discusses the determination of diffusive (or baseline) losses in a fully integral system during the application of a pressure-based direct integrity test.

4.3.1.1 Basic Concepts

The determination of sensitivity for pressure-based direct integrity tests is more complex than for marker-based tests. The equation used to determine the sensitivity of a pressure-based direct integrity test is specified in the LT2ESWTR (40 CFR 141.719(b)(3)(iii)(A)) and given here as Equation 4.3:

$$\text{LRV}_{\text{DIT}} = \log\left(\frac{Q_p}{\text{VCF} \times Q_{\text{breach}}}\right),$$ (Equation 4.3)

where LRV_{DIT} = direct integrity test sensitivity in terms of LRV (dimensionless)

Q_p = membrane unit design capacity filtrate flow (L/min)
VCF = volumetric concentration factor (dimensionless)
Q_{breach} = flow from the breach associated with the smallest integrity test response that can be reliably measured, referred to as the *critical breach size* (L/min)

Equation 4.3 represents a dilution model that assumes that water passing through the intact membrane barrier is free of the particulate contaminant of interest and that water flowing through an integrity breach has a particulate contaminant concentration equal to that on the high pressure side of the membrane. Under these assumptions, LRV_{DIT} is a function of the ratio of total filtrate flow to flow through the critical breach (i.e., Q_p/Q_{breach}), which quantifies the dilution of the contaminated stream passing through the breach as it mixes with treated filtrate. For a membrane unit of a given capacity (i.e., constant Q_p), LRV_{DIT} will increase as Q_{breach} decreases. This implies that a more sensitive direct integrity test capable of detecting a smaller breach can verify a higher log removal value and thus potentially increase the removal credit that a membrane filtration system is eligible to receive.

The volumetric concentration factor (VCF) is a dimensionless term that accounts for the increase in the suspended solids concentration that occurs on the feed side of the membrane for some hydraulic configurations. The VCF is important in the context of membrane integrity because the risk of filtrate contamination in the event of an integrity breach is increased for systems in which the influent suspended solids are concentrated on the feed side of the membrane. For example, for the same size integrity breach, systems with higher VCFs will allow increased passage of pathogens to the filtrate, reducing the verifiable removal efficiency. This effect is taken into account in the determination of sensitivity for pressure-based direct integrity tests in that systems with higher VCFs have proportionately lower test sensitivity. Thus in the denominator of Equation 4.3, Q_{breach} is multiplied by the VCF to account for the impact of this concentration effect.

The VCF is calculated as the ratio of the concentration of suspended solids maintained on the feed side of the membrane to that in the influent feed water, as shown in Equation 4.4:

$$\text{VCF} = \frac{C_m}{C_f},\qquad\qquad \text{(Equation 4.4)}$$

where VCF = volumetric concentration factor (dimensionless)
C_m = concentration of suspended solids maintained on the feed side of the membrane (number or mass/volume)
C_f = concentration of suspended solids in the influent feed water to the membrane system (number or mass/volume)

VCF generally ranges between 1 and 20; the value depends on the hydraulic configuration of the system. Membrane systems that operate in deposition mode do not increase the concentration of suspended solids on the feed side of the membrane, and thus have a VCF equal to 1. In contrast, membrane systems that operate in suspension mode, such as those that utilize a cross-flow hydraulic configuration, typically have a VCF in the range of 4 to 20, representing a 4- to 20-fold increase in the suspended particle concentration on the feed side of the membrane. The methods and equations used to calculate the VCF for various hydraulic configurations are provided in Section 2.5 of the *Membrane Filtration Guidance Manual*, and Table 2.4 presents equations for calculating both the average and maximum VCF for various hydraulic configurations. Alternatively, the VCF could be determined experimentally, as discussed in Section 2.5.4.

Note that the LT2ESWTR does not specify use of the maximum or average VCF value in calculating sensitivity, but the rule does require that the increase in suspended solids concentration on the high pressure side of the membrane, as occurs with some hydraulic configurations, be considered in the calculation. The maximum VCF typically ranges from 1 to 20 and provides the most conservative value for LRV_{DIT}, while the average VCF typically ranges from 1 to 7. In selecting between the maximum, average, or any other value for the VCF, consideration should be given to the concentration profile along the membrane surface in the direction of water flow and the implication of integrity breaches at various locations within the membrane unit. For example, although the maximum VCF does provide the most conservative value for LRV_{DIT} for systems in which the concentration varies with position this value represents only a very small portion of the concentration profile and thus is only representative of breaches that occur at the extreme end of the membrane unit. Similarly, some systems exhibit a concentration profile as a function of time within a filtration cycle, and the maximum VCF only occurs at the end of the filtration cycle immediately before a backwash event; prior to this time, the VCF is significantly lower.

4.3.1.2   Calculating Sensitivity

The sensitivity of a pressure-based direct integrity test can be calculated by converting the response from a pressure-based test that measures the flow of air (e.g., the diffusive airflow test) or rate of pressure loss (e.g., the pressure decay test) to an equivalent flow of water through an integrity breach during normal operation, as described in the following subsection. The second subsection outlines a general procedure for determining the threshold response of a pressure-based direct integrity test experimentally, if this information is not available from the membrane filtration system manufacturer.

Calculating Sensitivity Using the Air-Liquid Conversion Ratio   In order to calculate LRV_{DIT}, the flow through the critical breach for a direct integrity test (i.e., Q_{breach}) must be determined (as shown in Equation 4.3). Since most direct integrity tests do not directly measure Q_{breach}, it is necessary to establish a correlation between the direct integrity test response and the flow of water through the critical breach during system operation. In some of the most commonly used pressure-based direct integrity tests, including the pressure or vacuum decay test and the diffusive airflow test, air is applied to the drained side of a membrane and subsequently flows through any integrity breach that exceeds the test resolution. The response from such a test is typically measured as pressure decay or airflow. In order to relate the response from a pressure-based integrity test to Q_{breach}, it is necessary to establish a correlation between airflow and liquid flow through the critical integrity breach.

This correlation can be characterized through the air-liquid conversion ratio (ALCR), which is defined as the ratio of air that would flow through a breach during a direct integrity test to the amount of water that would flow through the breach during filtration, as defined in Equation 4.5:

$$ALCR = \frac{Q_{air}}{Q_{breach}}, \qquad \text{(Equation 4.5)}$$

where ALCR = air-liquid conversion ratio (dimensionless)
  Q_{air} = flow of air through the critical breach during a pressure-based direct integrity test (L/min)
  Q_{breach} = flow of water through the critical breach during filtration (L/min)

ALCR can be used to express the liquid flow through a breach in terms of corresponding flow of air, as shown in Equation 4.6:

$$Q_{breach} = \frac{Q_{air}}{ALCR}, \qquad \text{(Equation 4.6)}$$

where Q_{breach} = flow of water through the critical breach during filtration (L/min)

Q_{air} = flow of air through the critical breach during a pressure-based direct integrity test (L/min)

ALCR = air-liquid conversion ratio (dimensionless)

Substituting Equation 4.6 into the general expression for sensitivity (Equation 4.3) yields the following expression:

$$LRV_{DIT} = \log\left(\frac{Q_p \times ALCR}{Q_{air} \times VCF}\right), \qquad \text{(Equation 4.7)}$$

where LRV_{DIT} = direct integrity test sensitivity in terms of LRV (dimensionless)

Q_p = membrane unit design capacity filtrate flow (L/min)

ALCR = air-liquid conversion ratio (dimensionless)

Q_{air} = flow of air through the critical breach during a pressure-
based direct integrity test (L/min)

VCF = volumetric concentration factor (dimensionless)

Equation 4.7 can be used to directly calculate the sensitivity for any
pressure-based direct integrity test that is based on bubble point theory
and measures the flow of air (Q_{air}) through an integrity breach. The
four parameters that need to be determined to calculate sensitivity are
Q_p, VCF, ALCR, and Q_{air}. VCF can be established as previously
described. The value of Q_p is the design capacity filtrate flow approved
by the state from the membrane unit to which the direct integrity test is
applied. For a constant Q_{breach}, higher filtrate flows yield greater direct
integrity test sensitivity. Thus if different sizes of membrane units with
varying capacity are used in the same treatment system, the sensitivity
of the direct integrity test should be independently determined for
each unit size, with the lowest sensitivity establishing the maximum
log removal credit for the overall process. The flow of air Q_{air} is related
to the response from the direct integrity test. For tests that measure the
airflow through an integrity breach directly, such as the diffusive airflow test, Q_{air} is simply the airflow measured during application of the
test. On the other hand, methods such as the pressure and vacuum
decay tests yield results in terms of pressure loss per unit time, which
must be converted to an equivalent flow of air using Equation 4.8:

$$Q_{air} = \frac{\Delta P_{test} V_{sys}}{P_{atm}}, \qquad \text{(Equation 4.8)}$$

where Q_{air} = flow of air (L/min)

ΔP_{test} = rate of pressure decay during the integrity test (psi/min)

V_{sys} = volume of pressurized air in the system during the test (L)

P_{atm} = atmospheric pressure (psia)

Note that Equation 4.8 assumes that the temperature of the water and air are the same, since the air temperature should rapidly equilibrate with the water temperature. The value of V_{sys} encompasses the entire pressurized volume, including all fibers (typically the insides of the fibers are pressurized), piping, and other void space on the pressurized side.

Substituting Equation 4.8 into Equation 4.7 yields Equation 4.9, which can be used to calculate the sensitivity of a direct integrity test that measures the rate of pressure or vacuum decay.

$$LRV_{DIT} = log\left(\frac{Q_p P_{atm} \times ALCR}{\Delta P_{test} V_{sys} \times VCF}\right), \qquad \text{(Equation 4.9)}$$

where LRV_{DIT} = direct integrity test sensitivity in terms of LRV (dimensionless)

Q_p = membrane unit design capacity filtrate flow (L/min)

P_{atm} = atmospheric pressure (psia)

ALCR = air-liquid conversion ratio (dimensionless)

ΔP_{test} = smallest rate of pressure decay that can be reliably measured and associated with a known integrity breach during the integrity test (psi/min)

V_{sys} = volume of pressurized air in the system during the test (L)

VCF = volumetric concentration factor (dimensionless)

Regardless of whether the flow of air (Q_{air}) or pressure decay rate (ΔP_{test}) is measured during the direct integrity test, the smallest response from the test that can be reliably measured and associated with an integrity breach should be used in the sensitivity calculation. This should not be confused with the baseline integrity test response from an *integral* membrane unit, since there may be a small airflow or pressure decay due to diffusion of air through water in the wetted pores and/or the membrane material, even if there are no breaches in the system. In many cases, this smallest measurable response associated with an integrity breach may be provided by the membrane filtration system manufacturer. If this information is not available from the system manufacturer, it may be determined experimentally by progressively creating small integrity breaches of a known size in an otherwise integral membrane unit in order to determine the smallest measurable response from the direct integrity test that is distinguishable from the baseline response for an integral membrane unit. A general procedure for this experimental method is described in the next subsection.

The basic procedure for calculating the ALCR involves first making a reasonable assumption regarding whether the flow through the critical breach is laminar or turbulent for the particular membrane filtration system of interest. Then the ALCR can be calculated using the appropriate equation, as summarized in Table 4-1. Note that the

Module Type	Defect Flow Regime	Model	ALCR Equation	Appendix C Equation
Hollow fiber[a]	Turbulent[b]	Darcy pipe flow	$170Y\sqrt{\dfrac{(P_{test}-BP)(P_{test}+P_{atm})}{(460+T)TMP}}$	C.4
	Laminar	Hagen–Poiseuille[c]	$\dfrac{527\Delta P_{eff}(175-2.71T+0.0137T^2)}{TMP(460+T)}$	C.15
Flat sheet[d]	Turbulent	Orifice	$170Y\sqrt{\dfrac{(P_{test}-BP)(P_{test}+P_{atm})}{(460+T)TMP}}$	C.9
	Laminar[c]	Hagen–Poiseuille[c]	$\dfrac{527\Delta P_{eff}(175-2.71T+0.0137T^2)}{TMP(460+T)}$	C.15

[a]Or hollow fine fiber.

[b]Typically, characteristic of larger diameter fibers and higher differential pressures.

[c]The binomial in the Hagen–Poiseuille equation (C.15) approximates the ratio of water viscosity to air viscosity and is valid for temperatures ranging from approximately 32°F to 86°F. Additional details are provided in Appendix C.

[d]Includes spiral wound and cartridge configurations.

TABLE 4-1 Approaches for Calculating the ALC

equations in Table 4-1 assume that the flow regimes for the passage of air and water through an integrity breach are the same (i.e., both either laminar or turbulent). If this assumption is not considered appropriate for a specific system and/or application, such that inaccurate estimates for direct integrity test sensitivity may result, a hybrid approach may be considered, as described in Appendix C. Other assumptions utilized in the derivation of the ALCR equations in Table 4-1 may also not be valid for some membrane filtration systems. Section C.5 provides some examples of these types of systems, along with general guidance for modifying the ALCR formula derivations to accommodate these systems.

The various parameters given in the ALCR equations in Table N-1 include Y = net expansion factor for compressible flow through a pipe to a larger area (dimensionless), P_{test} = direct integrity test pressure (psi), BP = back pressure on the system during the direct integrity test (psi), P_{atm} = atmospheric pressure (psia), T = water temperature (°F), TMP = transmembrane pressure (psi), and ΔP_{eff} = effective integrity test pressure (psi).

Additional guidance for both calculating the ALCR and determining appropriate values for the component parameters of the respective ALCR equations is provided in section C of this Appendix, along with derivations of the ALCR equations. Further information regarding the net expansion factor Y may be found in various hydraulics references, including Crane (1988). Note that although the Darcy and orifice equations for the ALCR in Table N-1 are identical, the method for determining the net expansion factor Y is different for these two models, as described in section C of this Appendix.

The ALCR can also be determined via empirical means, which would be applicable to any flow regime or configuration of membrane material and is independent of a particular hydraulic model. Some manufacturers may have developed empirical models that could be used to determine the ALCR. If an empirical approach is preferred for determining the ALCR, and a valid empirical model is not available for the system, it may be necessary to develop one. One conceptual procedure for empirically deriving the ALCR for hollow fiber membrane filtration systems is the correlated airflow measurement (CAM) technique; the details of this procedure are presented in Appendix D of the *Membrane Filtration Guidance Manual*.

Measuring the Threshold Direct Integrity Test Response Experimentally   The smallest measurable response of a pressure-based direct integrity test that is associated with a known breach can be evaluated experimentally if this information is not available from the membrane filtration system manufacturer. For pressure-based tests, this response corresponds to the value of ΔP_{test} that should be used in the calculation of sensitivity. This experimental evaluation involves intentionally compromising system integrity in small, discrete, and quantifiable steps and monitoring the corresponding integrity test responses. In the case of microfiltration (MF) and ultrafiltration (UF) systems, several fiber-cutting studies conducted to evaluate the threshold response of various direct integrity tests have been documented in the literature (Adham et al., 1995; Landsness, 2001). In general, the procedure for measuring the threshold response experimentally involves the following steps:

1. The membrane system is determined to be integral through the application of a direct integrity test.

2. The investigator intentionally compromises a membrane to generate a known defect. Examples of such compromises include generating a hole in the membrane using a pin of a known diameter or cutting a hollow fiber at a predetermined location. In order to identify the threshold response, it is desirable to utilize a small integrity breach, such as a single cut fiber in a membrane unit.

3. After the membrane is compromised, the integrity of the membrane unit is measured using the designated direct integrity test.

4. The process is repeated with additional defects of progressively increasing size or quantity until a measurable response from the direct integrity test is detected. This minimum measurable response represents ΔP_{test} for the purposes of calculating the sensitivity of a direct integrity test.

4.3.1.3   Diffusive Losses and Baseline Decay

If it is determined to be appropriate for the membrane under consideration, the diffusive losses that constitute the baseline integrity test response may be subtracted from the smallest measurable response associated with an integrity breach for the purpose of determining sensitivity. For example, if a pressure decay rate of 0.05 psi/min is typical for an integral membrane unit and the limitations of the test are such that the smallest pressure decay rate that can be reliably associated with an integrity breach is 0.12 psi/min, the incremental response associated with an integrity breach is 0.07 psi/min, and this value may be used in the sensitivity calculation, as illustrated in Equation 4.10 (a variation of Equation 4.8):

$$Q_{air} = \frac{(\Delta P_{test} - D_{base})V_{sys}}{P_{atm}},\qquad \text{(Equation 4.10)}$$

where Q_{air} = flow of air (L/min)
ΔP_{test} = rate of pressure decay during the integrity test (psi/min)
D_{base} = baseline pressure decay (psi/min)
V_{sys} = volume of pressurized air in the system during the test (L)
P_{atm} = atmospheric pressure (psia)

In general, diffusive losses are most likely to be observed over the duration of the direct integrity test for thin-skinned, asymmetric membranes. The manufacturer should be able to provide information regarding whether or not diffusive losses are expected to be significant. If a high level of sensitivity is required for a membrane filtration system, baseline diffusive losses may be an important consideration. In case diffusive losses are determined to be significant and the membrane manufacturer cannot provide a value for the baseline decay, this section describes a means to estimate that decay.

For porous MF, UF, and membrane cartridge filtration (MCF) membranes, diffusive losses occur during direct integrity testing because a certain amount of compressed air used during the test dissolves into the water in the fully wetted pores and is transported across the membrane surface. In order to calculate the diffusive losses, it can be assumed that the water fills the pores of the membrane and forms a film of thickness z, and that diffusion directly through the membrane material itself is insignificant in comparison to the diffusion across the film of water. Using these assumptions, Equation 4.11 illustrates the relationship between diffusive losses and the various parameters that influence these losses:

$$Q_{diff} = 6\left(\frac{D_{aw}A_m(P_{test} - BP)H\varepsilon}{z}\right)\left(\frac{R_{gas}T}{P_{atm}}\right), \quad \text{(Equation 4.11)}$$

where Q_{diff} = diffusive flow of air through the water held in the membrane pores (L/min)

6 = unit conversion factor

D_{aw} = diffusion coefficient for air in water (cm^2/s)

A_m = total membrane surface area to which the direct integrity test is applied (m^2)

P_{test} = membrane test pressure (psi)

BP = back pressure on the system during the test (psi)

H = Henry's constant for air-water system (mol/psi-m^3)

ε = membrane porosity (dimensionless)

z = membrane thickness (mm)

R_{gas} = universal gas constant (L-psia/mol-K)

T = water temperature during direct integrity test (K)

P_{atm} = atmospheric pressure (psia)

Note that the dimensionless porosity ε is defined as the ratio of area of pores to the total membrane area in the unit. This term should not be confused with the pore size of porous MF, UF, and MCF membranes, which is given in terms of the limiting dimension of the openings in the membrane. The porosity of the membrane material can typically be provided by the manufacturer, if necessary. In addition, the diffusion flow path, which is affected by porosity, tortuosity, and the differential pressure across the membrane, is approximated by the membrane thickness z in Equation 4.11. Because membrane porosity and tortuosity may be difficult to measure, thus making it problematic to accurately quantify the actual length of the diffusion flow path, a more precise empirical method accounting for these two factors as a combined term has been developed by Farahbakhsh (2003). Both the diffusion coefficient D_{aw} and Henry's constant H vary with temperature, and Henry's constant also varies somewhat with the concentration of dissolved solids in the water. However, these effects may partially offset each other and may not be significant. Values for D_{aw} and H as a function of these variables (as applicable) may be found in standard tables in the literature.

The parameters given for Equation 4.11 are applicable to flat sheet porous membranes, such as those used in membrane cartridge configurations. For porous membranes in a hollow fiber configuration, such as most MF and UF systems, the following modifications are required:

$$A_m = \frac{A_2 - A_1}{\ln(A_2/A_1)},$$

where A_m = log mean total membrane area to which the direct integrity test is applied (m²)
A_1 = total membrane surface area to which the direct integrity test is applied based on the *inside* fiber diameter
A_2 = total membrane surface area to which the direct integrity test is applied based on the *outside* fiber diameter

and

$$z = r_2 - r_1,$$

where z = differential fiber radius (mm)
r_1 = inside radius of the hollow fiber
r_2 = outside radius of the hollow fiber

It is generally accepted that diffusion of air across the membrane can reduce the measured LRV_{DIT}. Under most circumstances, the amount of diffusion is small in comparison to the flow of air through a breach. However, if the membrane has a propensity to diffuse a significant amount of air (e.g., if the porosity is unusually high) or if a high level of sensitivity is required, it may be necessary to account for diffusive losses. Typically, there is only limited information available regarding the amount of diffusion that occurs for membrane processes used in water treatment under production conditions; however, MF and UF membrane manufacturers can typically provide a value for the baseline decay during a pressure-based direct integrity test for their specific proprietary systems, so it is generally not necessary to explicitly calculate diffusive losses using Equation 4.11.

In the absence of information provided by the manufacturer, it may be advantageous to directly measure the baseline pressure decay on an integral membrane unit, a process that should be conducted using clean membranes to avoid the possibility that fouling might artificially hinder diffusion. Because the diffusive loss is directly proportional to temperature (as shown in Equation 4.11), and the diffusion coefficient (which is also directly proportional to diffusive loss) also increases with temperature, it is important to characterize the baseline decay for the membrane filtration system at a typical water temperature in order to generate an appropriately representative value for diffusive loss. This temperature of evaluation should be recorded for future reference.

With semipermeable nanofiltration (NF) and reverse osmosis (RO) membranes, diffusive losses occur via the diffusion of air through the saturated membrane material itself. However, because NF and RO modules are manufactured separately from the accompanying filtration systems and inserted manually into pressure vessels, small seal leaks may occur that can be difficult to distinguish from baseline decay. Thus, it is recommended that baseline response for a pressure-based direct integrity test be evaluated for each unit in an NF or RO system on a site-specific basis.

If the membrane module manufacturer has information for typical diffusion airflow rates per unit of membrane area, the expected diffusive airflow for the entire membrane unit should be calculated and compared against the baseline observed during the unit-specific evaluation. If the observed baseline is significantly higher than the predicted diffusive losses, the result could be indicative of an integrity problem, and diagnostic testing (see Section 4.8) may be necessary to identify the source of additional airflow or pressure loss. If the membrane module manufacturer is only able to provide a diffusion coefficient for air through the membrane material, Equation 4.12 may be used to estimate the diffusive airflow for a membrane unit with a known membrane area, if necessary:

$$Q_{\text{diff}} = 6\left(\frac{D_{\text{am}}A_m(P_{\text{test}} - \text{BP})H}{z}\right)\left(\frac{R_{\text{gas}}T}{P_{\text{atm}}}\right), \quad \text{(Equation 4.12)}$$

where Q_{diff} = diffusive flow of air through a saturated semipermeable membrane (L/min)

$\quad\quad$ 6 = unit conversion factor

$\quad\quad D_{\text{am}}$ = diffusion coefficient for air through a saturated semipermeable membrane material (cm^2/s)

$\quad\quad A_m$ = total membrane surface area to which the direct integrity test is applied (m^2)

P_{test} = membrane test pressure (psi), BP = back pressure on the
system during the test (psi)
H = Henry's constant for air-water system (mol/psi-m^3)
z = membrane thickness (mm)
R_{gas} = universal gas constant (L-psia/mol-K)
T = water temperature during direct integrity test (K)
P_{atm} = atmospheric pressure (psia)

Note that the equations for the diffusion of air through porous and
semipermeable membranes—Equations 4.11 and 4.12, respectively—
are very similar. These equations differ only in that Equation 4.12 for
semipermeable membranes does not require the membrane porosity
and utilizes a diffusion coefficient for a composite membrane layer
consisting of both the membrane and the water of saturation bound in
the microscopic interstices of the membrane material.

Because spiral wound NF and RO membranes are typically composite structures consisting of two or more layers, it is important that
the membrane thickness z that is used correspond to the layer to which
the diffusion coefficient D_{am} provided by the membrane manufacturer
applies. For example, the diffusion coefficient may apply to the thin,
active, semipermeable layer, in which case the membrane thickness
would correspond to this layer only. Alternatively, if the diffusion coefficient is a composite representing the diffusion of air through all the
layers of the membrane taken as whole, then the thickness used should
be that of the entire membrane, including all layers.

Equations 4.11 and 4.12 show that diffusive airflow is directly
proportional to membrane area, the applied direct integrity test pressure, and the system back pressure. As a result, the decay should be
quantified for each membrane unit of different size in the system and
also recalculated if either the applied test pressure or system back
pressure is modified.

If the sensitivity is calculated using the ALCR approach
described in Section 4.3.1.2 and if diffusion is significant, Equation
4.6 can be modified to compensate for diffusive airflow as shown in
Equation 4.13:

$$Q_{breach} = \frac{Q_{air} - Q_{diff}}{ALCR},$$ (Equation 4.13)

where Q_{breach} = flow of water through the critical breach during filtration
(L/min)
Q_{air} = flow of air through the critical breach during a pressure-
based direct integrity test (L/min) Q_{diff} = diffusive flow
of air (L/min)
ALCR = air-liquid conversion ratio (dimensionless)

Note that the flow of air through the critical breach during a pressure-based direct integrity test (Q_{air}) includes diffusive losses and

thus should always be larger than the diffusive flow of air (Q_{diff}). Equation 4.13 should not be used if the baseline decay has already been accounted for by subtracting the baseline decay from the total rate of pressure decay (ΔP_{test}) observed during the integrity test using Equation 4.10, as described at the beginning of this section.

Combining Equation 4.13 with Equation 4.7 yields Equation 4.14, which enables the calculation of sensitivity using the ALCR approach, taking into account diffusive losses.

$$LRV_{DIT} = \log\left(\frac{Q_p \times ALCR}{VCF(Q_{air} - Q_{diff})}\right),\qquad \text{(Equation 4.14)}$$

where LRV_{DIT} = direct integrity test sensitivity in terms of LRV (dimensionless)
  Q_p = filtrate flow (L/min), ALCR = air-liquid conversion ratio (dimensionless)
  VCF = volumetric concentration factor (dimensionless)
  Q_{air} = flow of air (L/min)
  Q_{diff} = diffusive flow of air (L/min)

Note that Equations 4.13 and 4.14 are applicable to MCF, MF, UF, NF, and RO membrane filtration systems.

4.3.2   Marker-Based Tests

Sensitivity for marker-based direct integrity tests is determined via a straightforward calculation of the log removal value, similar to the determination of log removal values during a challenge study, as shown in Equation 4.15 (also Equation 3.7 of the *Membrane Filtration Guidance Manual*):

$$LRV_{DIT} = \log(C_f) - \log(C_p),\qquad \text{(Equation 4.15)}$$

where LRV_{DIT} = direct integrity test sensitivity in terms of LRV (dimensionless)
  C_f = feed concentration (number or mass/volume)
  C_p = filtrate concentration (number or mass/volume)

In using Equation 4.15 to calculate the sensitivity of a marker-based test, the LT2ESWTR specifies that the feed concentration C_f is the typical feed concentration of the marker used in the test, and that the filtrate concentration C_p is the baseline filtrate concentration of the marker from an integral membrane unit. If the marker is not detected in the filtrate, C_p must be set equal to the detection limit. Unlike the sensitivity calculations for pressure-based tests, Equation 4.15 does not incorporate a VCF. This term is not necessary to account for concentration effects on the feed side of the membrane in association with marker-based tests, because the concentration of the marker in the filtrate is measured. This measurement would directly account for any increase in the quantity of the marker passing through an integrity breach as a result of feed side concentration effects.

Due to variability in dosing the marker, the day-to-day LRV is likely to vary during system operation. Thus in order to establish the sensitivity of a marker-based direct integrity test using Equation 4.15, it is necessary to assume an appropriately conservative feed concentration, such that the LRVs determined on a day-to-day basis meet or exceed the LRV_{DIT} unless there is an integrity breach. An example of such a conservative approach would be the use of the lower bound of the anticipated concentration range of seeded marker as the feed concentration for the purposes of calculating the sensitivity of the test. Note that a marker-based direct integrity test must utilize a feed concentration sufficient to demonstrate the required *Cryptosporidium* LRV for rule compliance. Because a typical feed water will normally not contain a sufficient number of particles at the required 3 μm resolution, or else will not be sufficiently characterized to demonstrate this, seeding will usually be required for marker-based tests.

In order to optimize the sensitivity and reliability of a marker-based direct integrity test, it is important to use an accurate method for quantifying the feed and filtrate concentrations. Since the feed and filtrate concentrations will differ by orders of magnitude, an analytical method with a wide dynamic range is desired. If such a range is not available using a single device, two different instruments may need to be used to measure these respective concentrations. Regardless of the dynamic range of the instruments, it is likely that different analytical volumes will need to be used to deal with the different concentration ranges; however, the concentrations will have to be expressed in terms of equivalent volumes for the purpose of calculating an LRV. Some specific considerations regarding the use of particulate and molecular marker-based direct integrity tests are discussed in Section 4.7.5.

4.4   Test Frequency

Most currently available direct integrity tests require the membrane unit to be taken off-line for testing and thus are conducted in a periodic manner, requiring a balance between the need to verify system integrity and the desire to minimize system downtime and lost productivity. In addition, although some marker-based tests may be conducted while the membrane unit is online and in production, it is generally neither practical nor cost effective to implement these tests on a continual basis. Thus, the frequency at which direct integrity testing is conducted for membrane filtration systems represents a compromise between these competing objectives.

The LT2ESWTR requires that direct integrity testing be conducted on each membrane unit at least once each day that the membrane unit is in operation, unless the state approves less frequent testing (40 CFR

141.719(b)(3)(vi)). This minimum test frequency is intended to balance the need to verify system integrity as often as possible with cost and production considerations, and is based in part on a USEPA report indicating that daily direct integrity testing was a relatively common practice at membrane filtration facilities (USEPA 2001). It is important to note that the rule requires that each unit be subjected to direct integrity testing on a daily basis, even if the unit is operational for only a fraction of each day. It is also recommended that a given unit be tested at approximately the same time each day (site-specific facility operations permitting), in order to maintain a roughly consistent time interval between applications of direct integrity testing; however, this is not specifically required by the rule. The state may require either more or less frequent integrity testing for compliance with the rule at its discretion, although less frequent testing must be supported by demonstrated process reliability, the use of multiple barriers effective for *Cryptosporidium*, or reliable process safeguards (40 CFR 141.719(b)(3) (vi)). For example, in terms of process reliability, the state may opt to reduce the frequency of direct integrity testing if the membrane filtration system has a significant history of demonstrated operation either without the detection of an integrity breach or with only rare occurrences of integrity breaches, which have never been large enough to compromise the ability of the system to achieve the awarded removal credit. Alternatively, the state may reduce the required test frequency if the overall treatment scheme (including both membrane filtration and other processes) incorporates at least one additional process that is capable of achieving a substantial portion of the required *Cryptosporidium* treatment credit. In this case, even with a small integrity breach the state may be confident that the multiple barrier treatment process is fully capable of achieving compliance with the LT2ESWTR.

The state may also permit a reduced direct integrity test frequency if other safety factors are utilized to mitigate the risk associated with a potential integrity breach. One such strategy might be maintaining filtrate storage with a detention time equivalent to or longer than the time between direct integrity test events. If an integrity breach were detected, this storage would provide the utility with sufficient time to take the necessary mitigation measures to ensure that any contamination risks were addressed before the water entered the distribution system. Another possibility might be the use of continuous indirect integrity monitoring techniques with very high demonstrated sensitivities relative to other indirect monitoring methods. This strategy would represent a trade-off in that although the more sensitive (relative to the indirect monitoring methods) direct integrity test would be conducted less often, the ability to detect an integrity breach between applications of the direct test would be increased.

Note that any reduction in direct integrity test frequency is subject to the discretion of the state, which may utilize any of these suggested strategies as the basis for its decision or any other criteria it determines to be appropriate. Also, although unrelated to rule compliance, more frequent testing may be appropriate under certain specific circumstances, such as during initial facility start-up, as described in Chapter 8. More frequent direct integrity testing may also be advantageous for systems that rely on membrane filtration to achieve a high log removal of *Cryptosporidium*, since the risk associated with an integrity breach is greater. Conducting direct integrity testing on a membrane unit more frequently than once per day may be voluntarily implemented by a utility or required at the discretion of the state.

4.5   Establishing Control Limits

A control limit (CL) is defined as a response that, if exceeded, indicates a potential problem with the system and triggers a response. Multiple CLs can be set at different levels to indicate the severity of the potential problem. In the context of direct integrity testing, CLs are set at levels associated with various degrees of integrity loss. Under the provisions of the LT2ESWTR, a CL within the sensitivity limits of the direct integrity test must be established at the threshold test response that is indicative of an integral membrane unit capable of achieving the *Cryptosporidium* removal credit awarded by the state for rule compliance (40 CFR 141.719(b)(3)(iv)). Because utilities or states would have the option to implement a series of tiered CLs that may represent progressively greater levels of integrity loss leading up to the specific CL required under the rule, in this guidance manual the LT2ESWTR-mandated CL is referred to as an upper control limit (UCL). If the integrity test response is below the UCL, the membrane unit should be achieving an LRV equal to or greater than the removal credit awarded to the process. Alternatively, if the UCL is exceeded, the membrane unit must be taken off-line for diagnostic testing (as described in Section 4.8) and repair.

The same principles used to establish direct integrity test sensitivity are also used to establish the UCL. For *pressure-based tests*, the UCL may be calculated using the ALCR methodology. A modified version of Equation 4.7 yields an expression for the UCL, as shown in Equation 4.16:

$$UCL = \frac{Q_p \times ALCR}{10^{LRC} \times VCF},$$
(Equation 4.16)

where   UCL = upper control limit in terms of airflow (L/min)
Q_p = membrane unit design capacity filtrate flow (L/min)
ALCR = air-liquid conversion ratio (dimensionless)
LRC = log removal credit awarded (dimensionless)
VCF = volumetric concentration factor (dimensionless)

Similarly, Equation 4.9 can be rearranged to establish an expression for calculating the UCL in terms of a pressure decay rate, as shown in Equation 4.17:

$$\text{UCL} = \frac{Q_p P_{\text{atm}} \times \text{ALCR}}{10^{\text{LRC}} V_{\text{sys}} \times \text{VCF}},$$ (Equation 4.17)

where   UCL = upper control limit in terms of pressure decay rate (psi/min)

Q_p = membrane unit design capacity filtrate flow (L/min)

P_{atm} = atmospheric pressure (psia)

ALCR = air-liquid conversion ratio (dimensionless)

LRC = log removal credit (dimensionless)

V_{sys} = volume of pressurized air in the system during the test (L)

VCF = volumetric concentration factor (dimensionless)

Values for the parameters in Equations 4.16 and 4.17 should be the same as the analogous terms used to calculate sensitivity using Equations 4.7 and 4.9, respectively. Note that to the extent possible, these values should be selected to yield a conservative result for the UCL.

Equations 4.16 and 4.17 establish the maximum direct integrity test response that can be used as a UCL for the log removal credit (LRC) awarded by the state. The LRC, in turn, must be less than or equal to the lower value of either the log removal value determined during challenge testing ($\text{LRV}_{\text{C-Test}}$) or the sensitivity of the direct integrity test used (LRV_{DIT}).

In the context of Equations 4.16 and 4.17, CLs are expressed in terms of the actual response from the direct integrity test (i.e., flow of air or pressure decay rate, respectively). In this form the CLs may be most useful for operators, since these could be directly compared to integrity test results. However, it may also be useful to calculate the corresponding LRVs for both the CLs and the individual direct integrity test responses using the generic forms of Equations 4.7 (for tests yielding results in terms of the flow of air) or 4.9 (for tests yielding results in terms of pressure decay). In this case, these equations will simply yield an LRV corresponding to a particular direct integrity test result (i.e., a general LRV) rather than the sensitivity of the test (i.e., LRV_{DIT}). Many membrane systems have automated data acquisition equipment that could be programmed to calculate the LRV based upon the results of the most recent integrity test results and current operating conditions. These parameters may be displayed and trended to track system performance. For *marker-based tests*, which yield results in terms of log removal, the UCL is simply equal to the log removal credit awarded by the state.

Any CLs other than that mandated by the LT2ESWTR that are either voluntarily implemented by the utility or required by the state are referred to as lower control limits (LCLs). For example, an LCL may be established to provide operators with an indication that there may be an integrity breach before the breach becomes a compliance concern. In this scenario the LCL could be used in the context of preventative maintenance, and excursions above the LCL could prompt investigation and repair during scheduled downtime for the unit rather than require an immediate shutdown. The use of CLs and integrity testing in the context of a comprehensive integrity verification program is discussed further in Appendix A of the *Membrane Filtration Guidance Manual.*

Unlike the UCL that is established by the log removal credit awarded to the process, an LCL can be established based on the needs and objectives of the utility. However, any LCLs should be above the baseline integrity test response for an integral membrane and below the UCL in order to be useful. The baseline integrity test value for a membrane unit can be established during the commissioning of the facility, after the membrane system has been fully wetted and determined to be integral. The baseline level is described as the normal range of direct integrity test results that would occur for an integral membrane unit. A practical lower bound for any LCLs is the sensitivity of the direct integrity test (i.e., the lowest response that can be reliably measured that is indicative of an integrity breach).

As with the determination of membrane unit sensitivity, for systems that utilize multiple stages of membrane filtration, the UCL for each stage must be determined independently. Thus for pressure-based tests, the parameters used to calculated the UCL (e.g., log removal credit, ALCR, VCF, etc.) must be specific to the membrane units in each respective stage. This requirement is applicable to all stages to which the LT2ESWTR applies (i.e., those that produce filtrate for drinking water or postfiltration treatment rather than for recycling to an upstream point in the treatment process).

4.6 Example: Establishing Direct Integrity Test Parameters

Scenario

A submerged, vacuum-driven hollow fiber membrane system that operates in a deposition (i.e., dead-end) mode hydraulic configuration is required by the state to provide a total of at least 3-log removal for *Cryptosporidium* under the LT2ESWTR. A pressure decay test is applied to one of the membrane units in the system. Assume the plant is at sea level. Applicable parameters are as follows.

Operational Parameters

- The permitted design capacity of the membrane unit is 1,200 gpm.

- The maximum anticipated water temperature is 75°F (24°C).

- The minimum anticipated water temperature is 41°F (5°C).

- The maximum anticipated back pressure that might be exerted on the units during direct integrity testing is 75 in. of water.

- The minimum anticipated back pressure that might be exerted on the units during direct integrity testing is 60 in. of water.

- The back pressure measured prior to the most recent pressure decay test was 65 in. of water.

- The filtrate flow measured prior to the most recent pressure decay test was 1,000 gpm.

- The TMP measured prior to the most recent pressure decay test was 10 psi.

Direct Integrity Test Parameters

- The volume of pressurized piping during the test is 285 L.

- The initial applied test pressure is 16 psi.

- The duration of the pressure decay test is 10 min.

- Baseline (i.e., diffusive) decay is negligible over the duration of the test.

- The smallest verifiable rate of pressure decay under known compromised conditions is 0.10 psi/min.

- The most recent pressure decay test yielded a result of 0.13 psi/min.

- The temperature of both the water and the applied air was 68°F (20°C) during the most recent pressure decay test.

Unit and Membrane Characteristics

- The maximum rated TMP is 30 psi.

- The pore shape correction factor κ for the membrane material was not determined experimentally, and thus a conservative value of 1 is assumed.

- The membrane material is relatively hydrophilic and has a liquid-membrane contact angle (i.e., wetting angle) of 30°, as determined using a method acceptable to the state.

- The membrane surface variation (i.e., roughness) is 0.3 μm, as provided by the manufacturer.

- The hollow fiber lumen diameter is 500 μm.
- The depth of the membrane into the potting material is 50 mm.
- All flow through any integrity breach that may be present is assumed to be turbulent.

Requirements

Calculate

- the minimum direct integrity test pressure commensurate with the required resolution of 3 μm for the removal of *Cryptosporidium*,
- the sensitivity of the direct integrity test,
- the UCL for this system, and
- the LRV verified by the most recent direct integrity test.

Solution

1. Calculate the minimum direct integrity test pressure commensurate with the required resolution of 3 μm for the removal of *Cryptosporidium*.

$$P_{test} = 0.193\kappa\sigma\cos\theta + BP_{max} \qquad \text{(Equation 4.1)}$$

$\kappa = 1$ (from given information)
$\sigma = 74.9$ dyn/cm (the surface tension of water at 5°C)
$\theta = 30°$ (from given information)
$BP_{max} = 75$ in. (of water column, from given information)

$$P_{test} = 0.193 \times 1 \times 74.9 \times \cos(30°) + \frac{75}{27.7}$$

$$= 12.5 + 2.7$$

$$= 15.2 \text{ psi}$$

Because the problem scenario states that baseline diffusive losses are negligible, the pressure calculated represents the lowest permissible initial applied test pressure. If diffusive losses could not be neglected, P_{test} would represent the lower bound above which the pressure must be maintained to ensure that the resolution is maintained throughout the duration of the test. If this were the case, the applied pressure would have to be increased over P_{test} by the total anticipated pressure decay over the duration of the test. In this particular example, since the applied test pressure is given as 16 psi, the resolution requirement of the LT2ESWTR is satisfied.

2. Calculate the sensitivity of the direct integrity test.

$$LRV_{DIT} = \log\left(\frac{Q_p P_{atm} \times ALCR}{\Delta P_{test} V_{sys} \times VCF}\right) \qquad \text{(Equation 4.9)}$$

Q_p = 1,200 gpm (design capacity filtrate flow, from given information)

P_{atm} = 14.7 psia (atmospheric pressure at sea level, from given information)

The minimum back pressure that might be exerted during direct integrity testing is used to establish a conservative value for sensitivity.

ΔP_{test} = 0.10 psi/min (smallest verifiable decay rate, from given information)

V_{sys} = 285 L (from given information)

VCF = 1 (standard for deposition mode hydraulic configuration) ALCR must be determined

Consult Table 4.1 and Appendix C—use the Darcy pipe flow model for a hollow fiber membrane filtration system under conditions of turbulent flow (as specified):

$$ALCR = 170Y\sqrt{\frac{(P_{test} - BP)(P_{test} + P_{atm})}{(460 + T)TMP}}, \qquad \text{(Equation C.4)}$$

P_{test} = 16 psi (initial applied test pressure, from given information)

If diffusion through an integral membrane unit (i.e., baseline pressure decay) were significant, the cumulative decay over the duration of the test would be subtracted from the initial test pressure before applying this parameter to Equation C.4, to yield a conservative result for the ALCR.

BP = 60 in. (of water column, minimum back pressure, from given information)

P_{atm} = 14.7 psia (atmospheric pressure at sea level, from given information)

T = 75°F (maximum anticipated temperature, from given information)

TMP = 30 psi (maximum allowable TMP, from given information)

The values for system back pressure, temperature, and transmembrane pressure (TMP) were selected to establish a conservative (i.e., lower) value for the ALCR, which in turn yields a conservative value for sensitivity.

As indicated in Equation C.5 in Appendix C, the gas compressibility factor Y is a function of the applied test pressure P_{test}, the system back pressure during the test BP, atmospheric pressure P_{atm}, and a flow resistance coefficient K, as follows:

$$Y \propto \left[\frac{1}{\left(\dfrac{P_{test} - BP}{P_{test} + P_{atm}} \right)}, K \right].$$ (Equation C.5)

Smaller values for Y are generated with larger values of the first parameter in Equation C.5 and smaller values of K. The values for these variables should be selected to produce smaller values of Y, which in turn yield smaller values for the ALCR and a more conservative value for the test sensitivity.

P_{test} = 16 psi
BP = 75 in. (of water column)
P_{atm} = 14.7 psia

$$\frac{P_{test} - BP}{P_{test} + P_{atm}} = \frac{16 - \dfrac{75}{27.7}}{16 + 14.7} = 0.43$$

The flow resistance coefficient is defined by Equation C.6, as described in Appendix C:

$$K = f \frac{L}{d_{fiber}}.$$ (Equation C.6)

L = 50 mm (potting depth, from given information)
d_{fiber} = 0.5 mm (fiber diameter, from given information)
f (friction factor) must be determined

The friction factor can be obtained using the iterative method described in Appendix C.

f = 0.037 (friction factor, from iterative method)

$$K = f \frac{L}{d_{fiber}} = 0.037 \left(\frac{50}{0.5} \right) = 3.7$$

Using the appropriate chart on page A-22 from Crane (1988) with the values just calculated,

$$\frac{P_{test} - BP}{P_{test} + P_{atm}} = 0.43, \quad K = 3.7,$$

yields a value of 0.78 for Y.

With a value having been determined for Y, the ALCR can be calculated as follows:

$$\text{ALCR} = 170Y\sqrt{\frac{(P_{\text{test}} - \text{BP})(P_{\text{test}} + P_{\text{atm}})}{(460 + T)\text{TMP}}}$$

$$= 170(0.78)\left[\frac{\left(16 - \frac{75}{27.7}\right)(16 + 14.7)}{(460 + 75)30}\right]$$

$$= 21.1$$

Values can now be substituted into Equation 4.9 for sensitivity:

$$\text{LRV}_{\text{DIT}} = \log\left(\frac{(1,200 \times 3.785) \times 14.7 \times 21.1}{0.10 \times 285 \times 1}\right)$$

$$= 4.7$$

Therefore, the maximum removal value that this membrane filtration system is capable of verifying is 4.7 log.

3. Calculate the UCL for this system.

$$\text{UCL} = \frac{Q_p P_{\text{atm}} \times \text{ALCR}}{10^{\text{LRC}} V_{\text{sys}} \times \text{VCF}} \qquad \text{(Equation 4.17)}$$

$Q_p = 1,200$ gpm (design capacity filtrate flow, from given information)

$P_{\text{atm}} = 14.7$ psia (atmospheric pressure at sea level, from given information)

ALCR = 21.1 (as determined earlier)

LRC = 3 (from given information)

$V_{\text{sys}} = 285$ L (from given information)

VCF = 1 (standard for deposition mode hydraulic configuration)

$$\text{UCL} = \frac{(1,200 \times 3.785) \times 14.7 \times 21.1}{10^3 \times 285 \times 1}$$

$$= 4.9 \text{ psi/min.}$$

4. Calculate the LRV verified by the most recent integrity test.
   In addition to calculating the sensitivity, Equation 4.9 can be used to determine the LRV verified by the most recent direct integrity test via applying values for the variables specific to this test event.

$$\text{LRV} = \log\left(\frac{Q_p P_{\text{atm}} \times \text{ALCR}}{\Delta P_{\text{test}} V_{\text{sys}} \times \text{VCF}}\right) \qquad \text{(Equation 4.9)}$$

Q_p = 1,000 gpm (flow measured prior to testing, from given information)

P_{atm} = 14.7 psia (atmospheric pressure at sea level, from given information)

ALCR = 21.1 (as determined earlier)

ΔP_{test} = 0.13 psi/min. (measured test decay rate, from given information)

V_{sys} = 285 L (from given information)

VCF = 1 (standard for deposition mode hydraulic configuration)

Note that because the ALCR is designed to be a conservative value, it is not necessary to recalculate it for a specific pressure decay test in order to use Equation 4.9 for the purpose of determining the LRV verified by that test.

Values can be substituted into Equation 4.9 for sensitivity:

$$
\begin{aligned}
\text{LRV} &= \log\left(\frac{(1,000 \times 3.785) \times 14.7 \times 21.1}{0.13 \times 285 \times 1}\right) \\
&= 4.5
\end{aligned}
$$

This result demonstrates the effect that the membrane unit filtrate flow and integrity test response can have on the verifiable LRV at any point during operation. Although the sensitivity of this direct integrity test for this particular system under typical membrane unit operating conditions was determined to be an LRV of 4.7, operating the unit at about 83% of the design flow coupled with a measured pressure decay rate somewhat higher than the baseline detectable value under known compromised conditions reduced the verifiable LRV to 4.5. Because this reduction still results in a verifiable LRV that is significantly higher than the LRC of 3.0, operation under these conditions would still allow the system to maintain compliance with the LT2ESWTR.

4.7   Test Methods

The LT2ESWTR does not specify a particular type of direct integrity test for rule compliance, but instead allows the use of any test that meets the requirements of the rule (40 CFR 141.719(b)(3)). The two general classes of tests currently employed in municipal water treatment applications are pressure- and marker-based tests. Within these two categories, the particular types of tests most commonly used are described in the following subsections: pressure and vacuum decay tests, the diffusive airflow test, the water displacement test, and particulate and molecular marker tests. General procedures for conducting each of these tests are provided, along with a listing of some of the advantages and disadvantages of each method. The particular manner in which each of these tests is applied may vary according to manufacturer or site- or system-specific conditions.

The specific tests addressed in this guidance manual are not intended to represent a comprehensive list of all types of direct integrity tests that could be used to comply with the requirements of the LT2ESWTR. Any method that is both consistent with the definition of a direct integrity test under the rule and capable of meeting the specified resolution, sensitivity, and frequency requirements could be used for rule compliance at the discretion of the state.

4.7.1   Pressure Decay Test

The pressure-based pressure decay test is the most common direct integrity test currently in use and is generally associated with MF, UF, and MCF systems, which utilize porous membranes. In a pressure decay test, a pressure below the bubble point value of the membrane is applied, and the subsequent loss in pressure is monitored over several minutes. An integral membrane unit will maintain the initial test pressure or exhibit a very slow rate of decay. Note that the pressure decay test is applicable to currently available pressure-driven and vacuum-driven systems, since this test is conducted under positive pressure for both types of systems. A schematic illustrating a pressure decay test is shown in Fig. 4-1.

An outline of a generic protocol for a pressure decay test as is as follows:

1. Drain the water from one side of the membrane. For hollow fiber systems, typically the inside of the fiber lumen is drained, which may represent the feed or the filtrate, depending on whether the system is operated in an inside-out or outside-in mode, respectively.

2. Pressurize the drained side of the fully wetted membrane. The applied pressure must be lower than the bubble point pressure of the membrane (i.e., the pressure required to overcome the capillary forces that hold water in the membrane pores).

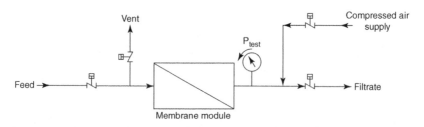

FIGURE 4-1   Schematic illustrating a pressure decay test.

Pressures ranging from 4 to 30 psi are typically applied during the pressure decay test, depending on the particular system. Membrane construction may limit the pressure at which a membrane can be tested. For compliance with LT2ESWTR requirements, the applied pressure must be sufficient to meet the resolution criterion of 3 μm based on Equation 4.1. For systems that utilize membranes submerged in an open basin, the test is typically applied to the filtrate side of the membranes without draining the basins. As a result, the hydrostatic pressure at the deepest part of the membrane unit at which feed water passes through the membrane must be considered in determining the resolution actually achieved by the test, as discussed in Section 4.2.

3. Isolate the pressure source and monitor the decay for a designated period of time. If there are no leaks in the membrane, process plumbing, or other pressurized system components, then air can only escape by diffusing through the water contained in the pores of the fully wetted membrane. Typically, this test is monitored over a period of 5 to 10 min. so that a stable rate of decay can be determined. The rate of pressure decay should be compared to the UCL (or any LCLs that may also be established) for the test to determine what, if any, subsequent action is triggered.

Advantages of the pressure decay test include

- the ability to meet the resolution criterion of 3 μm under most conditions;
- the ability to detect integrity breaches on the order of single fiber breaks and small holes in the lumen wall of a hollow fiber, depending on test parameters and system-specific conditions;
- its presence as a standard feature of most MF and UF systems;
- a high degree of automation;
- widespread use by utilities and acceptance by states;
- and simultaneous use as a diagnostic test to isolate a compromised module in a membrane unit in some cases (see Section 4.8.1).

Limitations of the pressure decay test include

- an inability to continuously monitor integrity;
- the necessity of measuring the volume of pressurized air in the system in order to calculate test method sensitivity (see Equation 4.8);

- the potential to yield false positive results if the membrane is not fully wetted (which may occur with newly installed and hydrophobic membranes that are difficult to wet, or when the test is applied immediately after a backwash process that includes air); and

- greater difficulty in application to membranes that are oriented horizontally, as a result of potential draining and air venting problems.

In addition to these disadvantages, another important concern associated with the pressure decay test is the potential for larger decay rates even within the UCL to affect the accuracy of the test. For example, if the total pressure decay over the duration of the test reduces the applied pressure on the membrane to a level below that sufficient to meet the resolution criterion, the test would not comply with the requirements of the rule. Consequently, parameters should be established so as to ensure that the pressure decay test meets the 3 µm resolution criterion throughout the duration of the test, as discussed earlier.

4.7.2   Vacuum Decay Test

The vacuum decay test is analogous to the pressure decay test with the exception that the test pressure is applied by drawing a vacuum on the membrane and monitoring the rate of vacuum (as opposed to pressure) decay over a period of time. An integral membrane unit will maintain the initial test vacuum or exhibit a very slow rate of decay. This test is generally associated with spiral wound NF and RO membranes. A schematic illustrating a vacuum decay test is shown in Fig. 4-2.

An outline of a generic protocol for a vacuum decay test as is as follows:

1. Drain the water from one side of the membrane. Typically, the filtrate side of a spiral wound NF or RO membrane is drained.

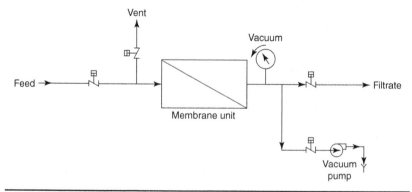

Figure 4-2   Schematic illustrating a vacuum decay test.

2. Apply a vacuum to the drained side of the fully wetted membrane. A vacuum of 20 to 26 in. Hg is typically applied during the test. Membrane construction may limit the vacuum with which a membrane can be tested. For compliance with LT2ESWTR requirements, the applied vacuum must be sufficient to meet the resolution criterion of 3 μm.

3. Isolate the vacuum source and monitor the decay for a designated period of time. If there are no leaks in the membrane, process plumbing, or other system components under vacuum, then the vacuum should not decay over the duration of the test. Typically, this test is monitored over a period of 5 to 10 min. to allow a stable rate of decay to be determined. The rate of pressure decay should be compared to the UCL (or any LCLs that may also be established) for the test to determine what, if any, subsequent action is triggered.

Advantages of the vacuum decay test include

- the ability to test spiral wound membranes or other systems that cannot be pressurized on the filtrate side of the membrane and
- the ability to meet the resolution criterion of 3 μm under most conditions.

Limitations of the vacuum decay test include

- an inability to continuously monitor integrity,
- a lack of current wide use for full-scale systems,
- difficulty in removing entrained air after the test has been completed, and
- the necessity of measuring the volume of air under vacuum in the system in order to calculate test method sensitivity (see Equation 4.8).

As with the pressure decay test, another important concern associated with the vacuum decay test is the potential for larger decay rates even within the UCL to affect the accuracy of the test. Thus if the total vacuum decay over the duration of the test reduces the applied vacuum on the membrane to a level below that sufficient to meet the resolution criterion, then the test will not comply with the requirements of the rule. Consequently, test parameters should be established so as to ensure that the vacuum decay test meets the 3 μm resolution criterion throughout the duration of the test, as discussed earlier.

4.7.3   Diffusive Airflow Test

The diffusive airflow test provides a direct measurement of the airflow through an integrity breach (i.e., Q_{air}). Like the pressure decay test, the diffusive airflow test is based on bubble point theory and generally associated with MF, UF, and MCF membranes. However, instead of the rate of pressure decay being measured, the test pressure is kept constant and the airflow through a breach is measured. If there are no breaches in the system, there will typically be a small flow of air resulting from the diffusion of air through water held in the fully wetted membrane pores. This diffusive airflow represents the baseline test response that is indicative of an integral membrane.

The diffusive airflow test has not been commonly used in municipal water treatment plants, at least partially as a result of the incorporation of pressure decay testing as a standard feature with most MF and UF systems. However, the diffusive airflow test has long been used for measuring the integrity of cartridge filters used in sterilization applications and has been described in the literature by a number of authors, including Meltzer (1987), Vickers (1993), and Johnson (1997). Note that this test is not applicable to submerged membrane filtration systems. A schematic illustrating a diffusive airflow test is shown in Fig. 4-3.

The outline of a generic protocol for a diffusive airflow test is similar to that for the pressure decay test, and is as follows:

1. Drain the water from one side of the membrane. For hollow fiber systems, typically the inside of the fiber lumen is drained, which may represent the feed or the filtrate, depending on whether the system is operated in an inside-out or outside-in mode, respectively.

2. Pressurize the drained side of the fully wetted membrane. The applied pressure must be lower than the bubble point pressure of the membrane (i.e., the pressure required to overcome the capillary forces that hold water in the membrane pores). As with the pressure decay test, pressures ranging from 4 to 30 psi are typically applied during the diffusive airflow test, depending on the particular system. Membrane construction may limit the pressure at which a membrane can be tested.

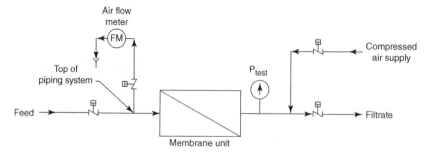

Figure 4-3   Schematic illustrating a diffusive airflow test.

For compliance with LT2ESWTR requirements, the applied pressure must be sufficient to meet the resolution criterion of 3 μm.

3. Maintain constant pressure and monitor the airflow for a designated period of time. If there are no leaks in the membranes, process plumbing, or other pressurized system components, then air can only escape by diffusing through the water contained in the pores of the fully wetted membrane. Typically, this test is monitored over a period of 5 to 10 min. to allow a stable airflow to be established. The flow of air should be compared to the UCL (or any LCLs that may also be established) for the test to determine what, if any, subsequent action is triggered.

Advantages of the diffusive airflow test include

- the ability to directly measure the flow of air through an integrity breach and
- the ability to meet the resolution criterion of 3 μm under most conditions.

Limitations of the diffusive airflow test include

- an inability to continuously monitor integrity;
- the requirement of equipment to measure flow of air that is generally not a standard inclusion with a membrane filtration system;
- a lack of current wide use for full-scale systems; and
- the potential to yield false positive results if the membrane is not fully wetted (which may occur with newly installed and hydrophobic membranes that are difficult to wet, or when the test is applied immediately after a backwash process that includes air).

4.7.4   Water Displacement Test

The water displacement test is similar to the diffusive airflow test, with the exception that the volume of water displaced by air flowing through an integrity breach is measured rather than the actual flow of air through the breach. Specifically, the air that passes across the membrane displaces a corresponding volume of water on the opposite side. This test can only be applied to membrane systems operated under positive pressure (as opposed to immersed systems that operate under vacuum). The volume of displaced water collected over a known period of time is measured and converted to a corresponding volume of air to determine the airflow (Q_{air}). If there are no breaches in the system, there will typically be a small flow of displaced water resulting from the diffusion of air. This flow represents the baseline test response that is indicative of an integral membrane. A larger volume (or flow) of displaced water would indicate an integrity breach. It is important to note that although this test measures the flow of water through an integrity breach, it does not constitute the flow of water through the critical breach (Q_{breach}) that is necessary for determining the direct integrity test sensitivity.

The water displacement test is described in an American Water Works Association Research Foundation (AWWARF) report prepared by Jacangelo et al. (1997). While this test is not commonly utilized in municipal drinking water treatment plants, it is relatively easy to use, requires minimal equipment, and can detect very small integrity breaches, on the order of a single broken fiber. Since the publication of that AWWARF report, a membrane filtration facility in Tauranga, New Zealand, has adopted the use of this integrity test. A schematic illustrating a water displacement test is shown in Fig. 4-4.

The outline of a generic protocol for a water displacement test is similar to that for the diffusive airflow test, and is as follows:

1. Drain the water from one side of the membrane. For hollow fiber systems, typically the inside of the fiber lumen is drained, which may represent the feed or the filtrate, depending on whether the system is operated in an inside-out or outside-in mode, respectively.

2. Pressurize the drained side of the fully wetted membrane. The applied pressure must be lower than the bubble point pressure of the membrane (i.e., the pressure required to overcome the capillary forces that hold water in the membrane pores). Pressures ranging from 4 to 30 psi are typically applied during the water displacement test, depending on

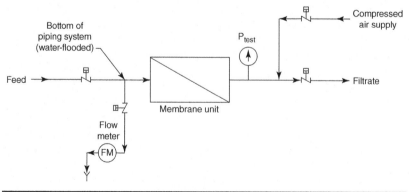

Figure 4-4 Schematic illustrating a water displacement test.

the particular system. Membrane construction may limit the pressure at which a membrane can be tested. For compliance with LT2ESWTR requirements, the applied pressure must be sufficient to meet the resolution criterion of 3 μm. Since one side of the membrane must remain flooded with water during the test, there is a potential for some hydrostatic back pressure. If this pressure is determined to be significant, it must be considered in the resolution calculations as discussed earlier.

3. Maintain constant pressure and monitor the flow for a designated period of time. If there are no leaks in the membrane, process plumbing, or other pressurized system components, then air can only escape by diffusing through the water contained in the pores of the fully wetted membrane. Typically, this test is monitored over a period of 5 to 10 min. to allow a stable flow to be established. The air displaces water that in turn is measured via either a flow meter or a graduated cylinder and timing device. Since the flow of water is equivalent to the flow of air, the resulting flow should be compared to the UCL (or any LCLs that may also be established) for the test to determine what, if any, subsequent action is triggered. Note that the flow meter and associated sample line used in this test must be configured to prevent the gravity flow of water.

Advantages of the water displacement test include

- the ability to directly measure the flow of air via a corresponding flow of water (provided that the back pressure on the system does not result in the compression of diffused air),
- the ability to meet the resolution criterion of 3 μm under most conditions,
- the ability to detect integrity breaches on the order of single fiber breaks and small holes in the fiber lumen wall of a hollow, and
- ease of measuring flow of water relative to that of air.

Limitations of the water displacement test include

- an inability to continuously monitor integrity,
- a lack of current wide use for full-scale systems, and
- the potential to yield false positive results if the membrane is not fully wetted (which may occur with newly installed and hydrophobic membranes that are difficult to wet, or when the test is applied immediately after a backwash process that includes air).

4.7.5   Marker-Based Integrity Tests

A marker-based test directly verifies the removal efficiency of a membrane filtration system by measuring the concentrations of a particulate or molecular marker in the feed and filtrate. The difference in the log of these concentrations represents the LRV demonstrated by the integrity test. It is important to note that a marker-based direct integrity test must utilize a feed concentration sufficient to demonstrate the required *Cryptosporidium* LRV for rule compliance. In addition, the full concentration of markers measured to demonstrate the required LRV must meet the 3 μm resolution criterion. Because a typical feed water will normally not contain a sufficient number of particles at the required 3 μm resolution or else will not be sufficiently characterized to demonstrate the fulfillment of this criterion, seeding will usually be required for marker-based tests.

Marker-based tests are advantageous in that they provide a direct assessment of the LRV achieved by the process. Both particulate and molecular marker-based direct integrity tests may be used to meet the requirements of the LT2ESWTR, provided the specified performance criteria for resolution, sensitivity, and frequency are satisfied (40 CFR 141.719(b)(3)). Since a marker-based direct integrity test is essentially a form of challenge testing, much of the guidance provided in Chapter 3 may be useful in designing a marker-based test. Because these tests are applied to water treatment equipment in active use, the particulate or molecular marker used must be inert and suitable for use in a water treatment facility—for example, approved by the Food and Drug Administration or certified by the National Sanitation Foundation as an NSF-60 approved material. In addition, because MF, UF, and MCF systems utilize porous membranes that would not remove a molecular marker, only particulate marker tests should be used with these systems. (Note that one possible exception is a system utilizing UF membranes with a low molecular weight cutoff, which could potentially remove larger molecular markers.) Conversely, because the semipermeable NF and RO membranes are not specifically designed to accommodate large particle loads and typically cannot be backwashed to removed particulate matter from the membrane surface, only molecular markers should be used with NF and RO systems. Schematics illustrating the particulate and molecular marker tests are shown in Figs. 4-5 and 4-6, respectively.

The generic protocols for both particulate and molecular marker tests are similar, and are as follows:

1. Select an appropriate marker and verify that it meets the resolution criterion. (Additional guidance is given in Section 3.9.)

2. Ensure that the expected feed and filtrate sampling concentrations fall within the limits for the measuring instrumentation used.

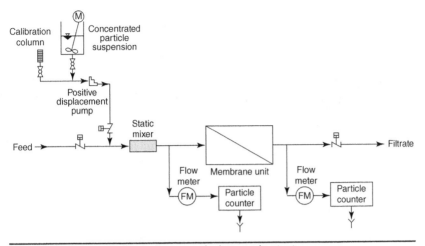

FIGURE 4-5 Schematic illustrating a particulate marker test.

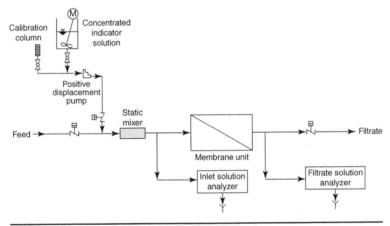

FIGURE 4-6 Schematic illustrating a molecular marker test.

3. Calculate the feed dosage rate and filtrate sample volumes. (Additional guidance is given in Section 3.10 of the *Membrane Filtration Guidance Manual*.)

4. Make sure that any operational processes such as chemical cleaning or backwashing are not initiated during the marker-based test.

5. Determine the amount of time the marker has to be applied to attain a steady state or equilibrium condition. (Additional guidance is given in Section 3.10 of the *Membrane Filtration Guidance Manual*.)

6. Feed the marker into the system.

7. Commence filtrate sampling when equilibrium is reached.

8. Continue simultaneous feed and filtrate sampling until results are attained.

9. Discontinue dosing.

Advantages of marker-based integrity tests include

- direct evaluation of the LRV of a membrane unit,

- the potential for application while the unit is online and producing water (i.e., during normal operation), and

- the potential to use instruments used for filtrate monitoring for continuous indirect integrity monitoring between applications of the direct integrity test at the discretion of the state.

Limitations of marker-based integrity tests include

- the potential for an improperly selected marker to cause fouling and interfere with normal system operation,

- the cost and calibration of instrumentation,

- the cost of the marker,

- the possible need for disposal of filtered water produced during the test,

- the possible need for special considerations for disposal of the filtered marker waste, and

- the typically discontinuous nature of measurement of marker concentrations.

Additional Considerations for Particulate Marker Tests

Typically, particulate marker tests require the use of particle counters to measure the feed and filtrate particle concentrations during the test, thus enabling the LRV to be directly calculated. Consequently, there are some important considerations regarding particle counters to consider in implementing a particulate marker test. First, it is important that the feed and filtrate particle counters continually monitor baseline concentrations between test events, so that background particle levels can be taken into account in the determination of the LRV as measured by the test during routine operation. The background levels recorded in both the feed and the filtrate should be subtracted from the respective particle count measurements during the test event in order to yield an accurate LRV.

In addition, particle counters are subject to coincidence error (i.e., two or more particles passing through the sensor simultaneously that are consequently measured as a single larger particle), particularly at high particle concentrations. This potential error may limit the maximum particle concentration that can be counted, which may in some cases be problematic for measuring the feed concentration during a particulate marker test. In order to compensate for possible coincidence error, some particle counters can allow for the sample to be diluted. However, because any particles in the dilution water would introduce error into the measurement, the quality of any dilution water used should also be taken into account in measuring the particle concentration of interest during a particulate marker test. It is also important that the particulate marker selected does not clump, which could exacerbate coincidence error or clog the instrument sensor.

All particle counters (or other instruments used to measure particulate concentrations during a particulate marker-based test) should be routinely calibrated at a frequency recommended by the manufacturer or required by the state. The calibration should be targeted toward counting particles at the specified size and concentration used in the particulate marker test. The particles used for the particulate marker test should be inert and compatible with the membrane in order to avoid irreversible fouling.

Additional Considerations for Molecular Marker Tests

Although the ambient levels of a molecular marker in the feed and filtrate may be negligible in many cases, these background concentrations should nevertheless still be measured and accounted for in the LRV calculation. The background concentrations of the molecular marker in the feed and filtrate should be subtracted from the respective measurements collected during the molecular marker-based direct integrity test in order to accurately quantify the LRV of the membrane unit. In addition, as with particle counters, all instruments that may be used to measure molecular marker concentrations should be calibrated regularly at a frequency recommended by the manufacturer or required by the state. Another important consideration is the potential for a molecular marker to adsorb onto the membrane surface or other system components, resulting in an inaccurate LRV determination; thus, any prospective molecular marker should be known to have negligible adsorption affinity for the membrane and other materials in use (Lozier et al., 2003).

4.8   Diagnostic Testing

Diagnostic tests are types of direct integrity tests that are specifically used to identify and isolate any integrity breaches that are detected, thus supplementing primary methods such as the pressure- or marker-based tests described in the previous section, which simply determine whether or a not a leak is present in a membrane unit. Note that there are no sensitivity or resolution requirements for diagnostic tests under the LT2ESWTR, since the objective of these tests is simply to isolate a particular compromised module and/or broken fiber. However, it is important that a diagnostic test have sufficient resolution to enable the detection of the integrity breach that caused the UCL to be exceeded.

Many diagnostic tests are most useful for identifying a particular compromised module within a unit; however, for hollow fiber MF and UF systems, some diagnostic tests may be used locate a specific broken fiber in a module. The LT2ESWTR requires that if the results of a direct integrity test exceed the UCL, the affected unit be immediately taken off-line for diagnostic testing and subsequent repair (40 CFR 141.719(b)(3)(v)). Although many different types of diagnostic tests may be available, five of the tests most commonly used are described in the following subsections: visual inspection, bubble testing, sonic testing, conductivity profiling, and single module testing. Note that not all of these tests are applicable to every type of membrane filtration system. Visual inspection, bubble testing, and sonic testing are generally applicable to MF, UF, and MCF systems, while conductivity profiling is used with NF and RO systems. Single module testing may be applicable to all types of membrane filtration systems.

4.8.1   Visual Inspection

Visually inspecting a membrane unit for leaks is the simplest form of diagnostic testing and is generally applicable to MF, UF, and MCF systems. Since many pressure-based direct integrity tests apply air to one side of the membrane while maintaining water on the other side, it may be possible to see air bubbles form in a compromised module. If a pressure-based direct integrity is used, visual inspection may be able to be conducted simultaneously with the direct integrity test to identify a compromised module. In order to perform a visual inspection, some component of the module housing must be transparent; accordingly, some pressure-driven membrane systems have inspection ports or clear tubing at the top of the membrane module to allow an operator to identify the particular compromised module. For vacuum-driven membrane systems submerged in basins, this test simply involves observing the water surface to identify the module that is the source of the bubbles rising to the surface.

Advantages of visual inspection include

- the ability to identify specific compromised fibers and leaking seals,
- ease of use and result interpretation,
- no need for additional equipment,
- equal applicability to systems using enclosed pressure vessels or immersed modules, and

- no requirement for membrane modules to be removed from the unit.

Limitations of visual inspection include

- manual application and
- the need for system design considerations for implementation (e.g., sight tubes, removable plates or carriages, etc.).

4.8.2   Bubble Testing

The bubble test is based on bubble point theory (as described in Appendix B) and is applicable to MF, UF, and MCF membranes. In conducting a bubble test, the module of interest is generally first removed from the membrane unit. The external shell of the module is then drained and pressurized to a level below the bubble point of the membrane but higher than the pressure required to achieve the required 3 μm resolution for compliance with the LT2ESWTR. The pressures applied in a bubble test are generally similar to those applied for the pressure-based direct integrity tests. The end cap is removed and a dilute surfactant solution applied to the open ends of the membrane fibers at the end of the module. The formation of bubbles in the surfactant solution can be used to pinpoint specific leaking fibers. The defective fiber may be repaired by plugging the lumen with a stainless steel pin or epoxy adhesive dispensed from a syringe. The membrane module manufacturer should be consulted for specific recommendations regarding module repair. Note that the bubble test is distinct from the bubble point test described in Section 3.6. The bubble test described in this section is applied to identify leaks, and thus the pressure must be kept below the bubble point of the porous membrane. By contrast, the bubble point test is applied for the specific purpose of determining the bubble point of the membrane, and thus the pressure must be gradually increased until the bubble point is achieved.

Advantages of the bubble test include

- the ability to identify specific compromised fibers and leaking seals,
- ease of use and result interpretation, and
- equal applicability to systems using enclosed pressure vessels or immersed modules.

Limitations of the bubble test include

- manual application and
- the possible need to remove a membrane module from the rack.

4.8.3 Sonic Testing

The principle underlying sonic testing is that water passing through broken fibers or other damaged system components will make a unique sound that can be detected using specialized equipment. The analysis is typically conducted by manually applying an accelerometer (an instrument used to detect vibrations) to one or more locations on each membrane module. Using headphones, the operator listens for vibrations generated by leaking air. Since the test is applied to modules on an individual basis, it is useful for identifying the module that has a potential integrity breach when the direct integrity test indicates that there may be a problem in a membrane unit. Sonic testing is generally applicable to MF, UF, and MCF systems.

A sonic test is most effectively administered by a skilled and experienced operator, particularly since the results are more subjective than the other forms of integrity testing, either direct or indirect. Nonetheless, it is a useful diagnostic tool that can help isolate a compromised membrane module. Adham and colleagues (1995) report that sonic testing was able to detect a breach as small as a 0.6 mm needle puncture in the wall of one hollow fiber out of a set of over 22,000.

An automated sonic testing system could have the potential to eliminate the subjectivity of this test and serve as an online, continuous, and direct means of integrity testing. The early stages of development and testing of such an automated system are described in a paper by Glucina and colleagues (1999). This acoustic monitoring system utilizes a sensor on each membrane module to detect changes in noise caused by pressure fluctuations in any compromised fibers. Test results indicated that the system was capable of detecting a single compromised fiber, although performance was affected by the level of background mechanical noise associated with the membrane filtration system. The system described in that paper has not been fully developed to date.

Advantages of sonic testing include

- the ability to identify a specific compromised module within a membrane unit and
- ease of use (assuming the test is conducted by a trained operator).

Limitations of sonic testing include

- manual application,
- the potential for subjective interpretations of results,
- the need to purchase additional equipment, and
- infeasibility for immersed membrane systems.

4.8.4   Conductivity Profiling

Conductivity profiling is a common practice associated with NF and RO systems to identify leaks in modules, O-rings, and seals. A long sample tube is inserted into the permeate port and used to withdraw a sample. The conductivity of the sample is monitored at various points among the modules in a pressure vessel and then indexed along the length of the permeate tube. Any integrity breaches are identified by significant changes in conductivity. The most common locations of leaks in a NF or RO system are at the module filtrate tube collector and end adaptor O-ring seals.

Advantages of conductivity profiling include

- the ability to identify a specific compromised module within a pressure vessel.

Limitations of conductivity profiling include

- manual application,
- the requirement of operator skill for the indexing probe,
- the potential for subjective interpretations of results, and
- the need to purchase additional equipment.

4.8.5   Single Module Testing

Single module testing may be applicable to all types of membrane filtration systems and involves removing potentially compromised modules from a membrane unit that is known to contain an integrity breach and testing each module on an individual basis. Although a number of different methods of integrity testing could potentially be utilized to screen individual modules with this diagnostic technique, single module testing generally refers to the use of a small apparatus to conduct a pressure or vacuum decay test on one module at a time. One advantage of this diagnostic test is that the pressure or vacuum decay associated with an integrity breach may be much more pronounced in a single compromised module at a given test pressure than in one compromised module in an entire membrane unit that is otherwise integral, enabling a breach to be more readily detected. Thus while the process of single module testing is labor intensive, it may be especially useful for isolating a compromised module in cases in which other methods of diagnostic testing have not been successful.

Advantages of the single module testing include

- an increased ability to detect compromised modules compared to in-situ testing applied to an entire membrane unit;
- the possibility of isolating compromised modules in cases in which other diagnostic tests have been unsuccessful; and
- general applicability to MF, UF, NF, RO, and MCF systems.

Limitations of single module testing include

- manual application,
- labor-intensive testing processes,
- the possible requirement that modules be removed from the membrane unit, and
- a possible need for separate testing apparatus.

4.9   Data Collection and Reporting

The LT2ESWTR requires that direct integrity testing be conducted on each membrane unit at a frequency of at least once each day of operation unless the state approves less frequent testing based on demonstrated process reliability, the use of multiple barriers effective for *Cryptosporidium*, or reliable process safeguards (40 CFR 141.719(b)(3)(vi)). At a minimum, the direct integrity test results that exceed the UCL and result in a membrane unit being taken off-line must be reported, along with the corrective action taken as a result of the UCL exceedance. This information must be reported to the state within 10 days following the end of each monthly monitoring cycle (40 CFR 141.721(f)(10)(ii)(A)). The state may exercise its discretion as to whether or not a report is required for a monthly monitoring period in which no UCL exceedances have occurred. Routine direct integrity test results that do not exceed the UCL are not required to be reported under the LT2ESWTR; however, the state may require that additional results be reported at its discretion. The state may also require that direct integrity test results be reported as measured (e.g., the rate of pressure decay rate for the pressure decay test or the flow of air for a diffusive airflow test) and/or as converted to equivalent LRVs, as calculated using the methodology described under Section 4.5. Note that it is often most advantageous for a utility to record the actual process value(s) measured (e.g., the rate of pressure decay), to track the results of the direct integrity test over time. Additional guidance regarding data collection and the use of these data for system optimization is provided in Appendix A in the context of developing a comprehensive integrity verification program.

All data collected in the process of conducting direct integrity testing as required under the rule must be kept by utilities for a minimum of 3 y (40 CFR 141.722(c)). These data include all direct integrity test results as well as UCL exceedances and any corrective action taken.

A sample summary report form for a hollow fiber system using the pressure decay test (the most common type of direct integrity test) is provided in Fig. 4-7. The form includes the following components:

- facility information, membrane unit number, and date (i.e., month and year);
- system parameters and test constants;
- test conditions and results; and
- UCL exceedances.

Month		Utility	
Year		Facility Name	
Membrane Unit No.		Test Duration	min
Volume of System (V_{sys})	L	VCF	
UCL		Total No. of UCL Violations	
Signed		Dated	

Day	Pressure (psi) Initial	Pressure (psi) Final	ΔP_{test} (psi/min)	Within UCL?	Corrective Action Taken (if required)	Filtrate Flow (gpm)	TMP (psi)	ALCR	LRV Verified
1									
2									
3									
4									
5									
6									
7									
8									
9									
10									
11									
12									
13									
14									
15									
16									
17									
18									
19									
20									
21									
22									
23									
24									
25									
26									
27									
28									
29									
30									
31									
Min									
Max									
Ave									

FIGURE 4-7   Sample summary report form for pressure decay testing.

The sample summary report form also contains a column for recording the LRV that is verified for each particular daily application of the direct integrity test, as well as for the flow, TMP, and ALCR parameters that are required to calculate the LRV. Note that these data are not required for reporting purposes under the LT2ESWTR, but are included to underscore their utility for tracking overall unit performance over time.

The LT2ESWTR also requires that utilities submit to the state the resolution, sensitivity, frequency, control limit, and associated baseline response (i.e., for an integral membrane filtration system) for the direct integrity test proposed to be used on the full-scale facility (40 CFR 141.721(f)(10)(i)(B)). Because the resolution and sensitivity are site and system specific, it will generally not be possible to accurately quantify these parameters until the full-scale system is fully installed and operational. Consequently, utilities could submit to the state estimates for these parameters based on information available from the membrane system manufacturer, and then subsequently submit more accurate, field-verified data after the full system is operational. However, the LT2ESWTR only specifies that these data must be submitted; the specifics of the submittal process, including the timing and procedures, are subject to the discretion of the state.

Calculating the Air-Liquid Conversion Ratio

C.1 Introduction

As described in Chapter 4, the regulatory framework for the Long-Term 2 Enhanced Surface Water Treatment Rule (LT2ESWTR) requires that the flow through the smallest integrity breach that generates a measurable response from the direct integrity test (i.e., the critical breach size Q_{breach}) be determined in order to establish the sensitivity of a pressure-based test method (40 CFR 141.719(b)(3)(iii)(A)). (Under the LT2ESWTR, sensitivity is defined as the maximum log removal value [LRV] that can be reliably verified by the direct integrity test—i.e., LRV_{DIT}.) However, because most pressure-based direct integrity tests yield results in terms of airflow or pressure decay, it may be necessary to convert these results to an equivalent value for the flow of water through the critical breach under typical filtration conditions. This conversion is necessary for calculating both the sensitivity and the upper control limit (UCL) for a pressure-based direct integrity test.

Although there are a number of methods for converting a direct integrity test response to a corresponding flow of water, each can be generally categorized as one of two types of approach: mathematical modeling or experimental determination. This appendix describes a mathematical approach based on a parameter called the air-liquid conversion ratio (ALCR), which is defined as the ratio of air that would flow through a breach during a direct integrity test to the amount of water that would flow through the breach during filtration, as shown in Equation C.1 (also Equation 4.5):

$$\text{ALCR} = \frac{Q_{air}}{Q_{breach}}, \qquad \text{(Equation C.1)}$$

where ALCR = air-liquid conversion ratio (dimensionless)

Q_{air} = flow of air through the critical breach during a pressure-based direct integrity test (volume/time)

Q_{breach} = flow of water through the critical breach during filtration (volume/time)

Because of the many variations in membrane configurations, breaches in the membrane may exhibit either turbulent or laminar flow characteristics, depending upon the location and size of the defect as well as the pressure differential between the feed and filtrate. In addition, there are fundamental differences between hollow fiber and flat sheet membrane breaches, since the most common breaches associated with hollow fiber modules exhibit pipe flow characteristics, while flat sheet breaches are best represented by an orifice model. Consequently, three different hydraulic models have been developed for determining the ALCR for a particular membrane system, depending on the configuration of the membrane material (hollow fiber vs. flat sheet) and the type of flow (laminar vs. turbulent) that is expected through the critical breach. These three models are the Darcy pipe flow model (for breaches in a hollow fiber or hollow fine fiber module under conditions of turbulent flow), the orifice model (for modules utilizing flat sheet membranes such as spiral wound and membrane cartridge configurations under conditions of turbulent flow), and the Hagen–Poiseuille model (for any configuration under conditions of laminar flow). Table C.1 summarizes the various approaches for calculating the ALCR based on these three models and the conditions under which the use of each model is appropriate.

Note that the various methods presented in this appendix for determining the ALCR implicitly assume that the flow regime for airflow through a breach during direct integrity testing is the same as that for liquid flow though a breach during filtration (i.e., either both laminar or both turbulent). If this assumption is determined to be inappropriate for a given membrane filtration system such that inaccurate and non-conservative estimates for sensitivity may result, then a hybrid approach may be considered. An example of such a hybrid approach is to assume laminar water flow and turbulent airflow, which could be modeled through the application of the Hagen–Poiseuille equation for water and the Darcy equation for air. The ALCR for such cases could then be

Module Type	Defect Flow Regime	Model
Hollow fiber[a]	Turbulent[b]	Darcy pipe flow
	Laminar	Hagen–Poiseuille
Flat sheet[c]	Turbulent	Orifice
	Laminar	Hagen–Poiseuille

[a]Or hollow fine fiber.
[b]Typically characteristic of larger diameter fibers and higher differential pressures.
[c]Includes spiral wound and cartridge configurations.

TABLE C.1 Approaches for Calculating ALCR

derived using a similar methodology to that described in Sections C.2, C.3, and C.4.

Procedures for calculating the ALCR and subsequently the sensitivity and UCL for applicable pressure-based direct integrity tests are given in Chapter 4, but the derivations of the various hydraulic models that form the basis for the respective ALCR equations are provided in the following sections of this appendix. Note that while the derivation of the ALCR equations relies on various hydraulic models that could be used to directly calculate the flow of air (Q_{air}) and water (Q_{breach}) through an integrity breach, direct application of these equations requires knowledge of the critical breach size, which is difficult and impractical to accurately quantify.

The advantage of the ALCR is that the terms relating to the size of the breach cancel out, yielding equations for the ALCR that are a function of either known and/or more easily determined parameters and are independent of the critical breach size or geometry. Thus, although the ALCR equations in this appendix are derived for conditions assuming the flow of air (Q_{air}) and water (Q_{breach}) through the critical breach, the ALCR is independent of integrity breach size (i.e., physical dimensions of the breach) and magnitude (i.e., number of distinct breaches), and thus is a scalable parameter. An additional section is also included at the end of this appendix discussing cases in which the ALCR equations derived in Sections C.2, C.3, and C.4 may not be applicable for some membrane filtration systems, as well as modifying the derivations to accommodate these systems.

For the derivation of the various models described in this appendix, airflow equations have been developed using a standard temperature and pressure of 68°F (528°Ra, 293 K, or 20°C) and 0 psi (1 atm or 14.7 psia). These standard conditions were selected to be consistent with the convention for airflow measurement devices. The equations can be modified to a different set of reference conditions through application of the ideal gas law expressed in terms of absolute temperature and pressure, if necessary. In addition, the temperature of the air used in a pressure-based direct integrity test is assumed to be the same as that of the water in the membrane filtration system, since these temperatures are expected to rapidly equilibrate.

Additional background on the hydraulic modeling developed in this appendix may be found in Crane's *Flow of Fluids through Valves, Fittings, and Pipes* (1988). All of the basic hydraulic equations used in the derivation of the ALCR equations are included in that text.

C.2 Darcy Pipe Flow Model

The Darcy pipe flow model is used to describe turbulent flow through an integrity breach with characteristics similar to a broken hollow fiber. Generally, turbulent flow may be expected through larger diameter broken fibers and at higher differential pressures. The Darcy equations for the flow of air and water through a pipe are given in Equations C.2 and C.3, respectively.

$$Q_{air} = 11.3 Y d_{fiber}{}^2 \sqrt{\frac{(P_{test} - BP)(P_{test} + P_{atm})}{(460 + T)K_{air}}}, \quad \text{(Equation C.2)}$$

where Q_{air} = flow of air at standard conditions (ft^3/s)
$\quad Y$ = net expansion factor for compressible flow through a pipe to a larger area (dimensionless)
$\quad d_{fiber}$ = fiber diameter (in.)
$\quad P_{test}$ = integrity test pressure (psi)
$\quad BP$ = back pressure on the system during the integrity test (psi)
$\quad P_{atm}$ = atmospheric pressure (psia)
$\quad T$ = water temperature (°F)
$\quad K_{air}$ = resistance coefficient of air (dimensionless)

$$Q_{breach} = 0.525 d_{fiber}{}^2 \sqrt{\frac{TMP}{K_{water} \rho_w}}, \quad \text{(Equation C.3)}$$

where Q_{breach} = flow of water through the critical breach during filtration (ft/s)
$\quad d_{fiber}$ = fiber diameter (in.)
$\quad TMP$ = transmembrane pressure (psi)
$\quad K_{water}$ = resistance coefficient of water (dimensionless)
$\quad \rho_w$ = density of water (lb/ft^3)

Assuming that the resistance coefficients for air and water are similar (i.e., $K_{air} \approx K_{water}$) and applying a value of 62.4 lb/ft^3 for the density of water, the ratio of Equation C.2 to Equation C.3 yields an expression for the ALCR, as given by Equation C.4:

$$\text{ALCR} = 170Y\sqrt{\frac{(P_{\text{test}} - \text{BP})(P_{\text{test}} + P_{\text{atm}})}{(460 + T)\text{TMP}}}, \qquad \text{(Equation C.4)}$$

where ALCR = air-liquid conversion ratio (dimensionless)
      Y = net expansion factor for compressible flow through a pipe to a larger area (dimensionless)
      P_{test} = direct integrity test pressure (psi)
      BP = back pressure on the system during the integrity test (psi)
      P_{atm} = atmospheric pressure (psia)
      T = water temperature (°F)
      TMP = transmembrane pressure during normal operation (psi)

The ALCR is used in the equations for determining both the sensitivity and the UCL for pressure-based direct integrity tests, as described in Chapter 4. Consequently, the values of the parameters in Equation C.4 should be selected to yield a lower, more conservative value for the ALCR. For example, the transmembrane pressure (TMP) in Equation C.4 comes from the expression for Q_{breach} (Equation C.3) during filtration, and thus the most conservative ALCR result would be generated from using the maximum anticipated TMP during normal operation.

The net expansion factor Y for compressible flow may be obtained from charts in various hydraulics references, such as Crane (1988, p. A-22). Using the appropriate chart for airflow, values for Y are given as a function of pressure and the flow resistance coefficient, as shown in Equation C.5 (a nonspecific expression illustrating the relationship between Y and its variables):

$$Y \propto \left[\frac{1}{\dfrac{P_{\text{test}} - \text{BP}}{P_{\text{test}} + P_{\text{atm}}}}, K\right], \qquad \text{(Equation C.5)}$$

where Y = net expansion factor for compressible flow through a pipe to a larger area (dimensionless) P_{test} = direct integrity test pressure (psi)
      BP = back pressure on the system during the integrity test (psi)
      P_{atm} = atmospheric pressure (psia)
      K = flow resistance coefficient (dimensionless)

The flow resistance coefficient K is a common fluid flow parameter described by most hydraulics texts, and is defined as shown in Equation C.6:

$$K = f\frac{L}{d_{\text{fiber}}}, \qquad \text{(Equation C.6)}$$

where K = flow resistance coefficient (dimensionless)
       f = friction factor (dimensionless)
       L = length of the defect (in.)
       d_{fiber} = fiber diameter (in.)

Using the conservative scenario of a fiber break at the point where the fiber enters the pot, the length of the defect L is represented by the length of the lumen encasement into the membrane pot. The friction factor f may be estimated from a Moody diagram or the corresponding tabulated values, both of which are readily in available in most hydraulics references. The relative roughness (e/d_{fiber}) that is required to estimate the value for the friction factor may be calculated either by obtaining a product-specific value for the specific roughness (e) from the manufacturer or by using the membrane pore size as an estimate of the specific roughness.

Note that the net expansion factor Y should remain constant over time for practical purposes if appropriately conservative values are used to calculate this parameter. Thus the determination of Y should represent a one-time, site-specific calculation. Also, because the ALCR is directly proportional to Y (as shown in Equation C.4), lower values for Y result in lower, more conservative values for the ALCR.

An iterative solution may be required to determine a value for the net expansion factor Y. A general outline for one such iterative process is given as follows. The use of a spreadsheet may help facilitate the various calculations required.

1. Select a reasonable value for the friction factor f.

2. Calculate the flow resistance coefficient K using Equation C.6.

3. Obtain a value for the Reynolds number Re from tabulated values for the friction factor f as a function of the Reynolds number Re and the relative roughness (e/d_{fiber}).

4. Calculate airflow Q_{air} from the equation for the Reynolds number Re as a function of equivalent diameter, air velocity, and dynamic viscosity (as referenced in fluid mechanics and fluid dynamics texts). For the purposes of determining the equivalent diameter and velocity (i.e., the flow Q_{air} divided by the cross-sectional area), assume that the applicable integrity breach may be represented by a pipe (e.g., a hollow fiber) flowing full with air. Use the maximum anticipated temperature and the minimum pressure applied over the duration of the direct integrity test (accounting for baseline decay) to generate a conservative (i.e., low) value for the dynamic viscosity and thus in turn a conservative (i.e., low) value for both airflow Q_{air} and ALCR.

5. Using tables available in hydraulics texts (e.g., page A-22 of Crane (1988)), apply the flow resistance coefficient K and the pressure ratio to determine a value for the net expansion factor Y, as shown in Equation C.5.

6. Calculate airflow Q_{air} using Equation C.2. Assume that $K \approx K_{air}$.

7. If the airflow Q_{air} calculated in steps 4 and 6 is approximately the same, then the net expansion factor Y determined in step 5 is correct. Otherwise, select a revised value for the friction factor f and repeat steps 1–7 in an iterative process until the two calculated values for airflow Q_{air} converge.

C.3 Orifice Model

The orifice flow model may be used to approximate turbulent flow through an integrity breach with characteristics similar to a hole in a flat sheet membrane that may be configured as a cartridge or a spiral wound module. The representative equations for airflow and water flow through an orifice are given as Equations C.7 and C.8, respectively.

$$Q_{air} = 11.3 Y d_{fiber}{}^2 C \sqrt{\frac{(P_{test} - BP)(P_{test} + P_{atm})}{460 + T}}, \quad \text{(Equation C.7)}$$

where Q_{air} = flow of air at standard conditions (ft/s)
Y = net expansion factor for compressible flow through a pipe to a larger area (dimensionless)
d_{fiber} = fiber diameter (in.)
C = coefficient of discharge (dimensionless)
P_{test} = direct integrity test pressure (psi)
BP = back pressure on the system during the integrity test (psi)
P_{atm} = atmospheric pressure (psia)
T = water temperature (°F)

$$Q_{breach} = 0.525 d_{fiber}{}^2 C \sqrt{\frac{TMP}{\rho_w}}, \quad \text{(Equation C.8)}$$

where Q_{breach} = flow of water through the critical breach during filtration (ft^3/s)
d_{fiber} = fiber diameter (in.)
C = coefficient of discharge (dimensionless)
TMP = transmembrane pressure (psi)
ρ_w = density of water (lb/ft^3)

The ratio of Equation C.7 to Equation C.8 yields an expression for the ALCR, as given in Equation C.9. Note that this equation incorporates a value of 62.4 lb/ft^3 for the density of water.

$$ALCR = 170 Y \sqrt{\frac{(P_{test} - BP)(P_{test} + P_{atm})}{(460 + T) TMP}}, \quad \text{(Equation C.9)}$$

where ALCR = air-liquid conversion ratio (dimensionless)
       Y = net expansion factor for compressible flow through a
           pipe to a larger area (dimensionless)
       P_{test} = direct integrity test pressure (psi)
       BP = back pressure on the system during the integrity test
           (psi)
       P_{atm} = atmospheric pressure (psia)
       T = water temperature (°F)
       TMP = transmembrane pressure during normal operation
           (psi)

Note that although the derivations are slightly different, the result-ing ALCR equations for the Darcy and orifice models (Equations C.4 and C.9, respectively) are identical. However, the two models utilize different methodologies for determining the net expansion factor for compressible flow Y. As described in Section C.2 for the Darcy model, the values of the parameters in Equation C.9 should be selected to yield a conservative value for the ALCR. For example, the TMP in Equation C.9 comes from the expression for Q_{breach} (Equation C.8) dur-ing filtration, and thus the most conservative ALCR result would be generated from using the maximum anticipated TMP during normal operation.

As with the Darcy model, the net expansion factor for compressible flow Y may be obtained from charts in various hydraulics references, such as Crane (1988, p. A-21). However, for the orifice model, Equation C.10 may also be used to calculate the net expansion factor, as follows:

$$Y = 1 - 0.293\left(1 - \frac{BP + P_{atm}}{P_{test} + P_{atm}}\right), \qquad \text{(Equation C.10)}$$

where Y = net expansion factor for compressible flow through a pipe
           to a larger area (dimensionless)
       BP = back pressure on the system during the integrity test (psi)
       P_{atm} = atmospheric pressure (psia)
       P_{test} = direct integrity test pressure (psi)

Because the ALCR is directly proportional to Y (as shown in Equation C.9), lower values for Y result in lower, more conservative values for the ALCR. Also, for practical purposes the net expansion factor Y should remain constant over time if appropriately conserva-tive values are used to calculate this parameter. Thus the determina-tion of Y should represent a one-time, site-specific calculation.

C.4 Hagen–Poiseuille Model

The Hagen–Poiseuille model is appropriate for small integrity breaches (such as a pinhole or a broken small diameter hollow fiber under low differential pressure) that would result in laminar flow.

Using this model, the equation for airflow through a small defect under laminar flow conditions is given by Equation C.11:

$$Q_{air} = \frac{49.5\pi d_{fect}{}^4 \Delta P_{eff} g}{L\mu_{air}(460+T)},$$ (Equation C.11)

where Q_{air} = flow of air at standard conditions (ft³/s)
  d_{fect} = defect diameter (in.)
  ΔP_{eff} = effective integrity test pressure (psi)
  g = gravitational constant (32.2 lbm-ft/lbf-s²)
  L = length of the defect (in.)
  μ_{air} = viscosity of air (lbs/ft-s)
  T = water temperature (°F)

Because air is a compressible fluid, the airflow is determined using the effective integrity test pressure ΔP_{eff}, calculated according to Equation C.12:

$$\Delta P_{eff} = (P_{test} - BP)\left[\frac{P_{test} + 2P_{atm} + BP}{2(BP + P_{atm})}\right]\left(\frac{BP + P_{atm}}{P_{atm}}\right),$$ (Equation C.12)

where ΔP_{eff} = effective integrity test air pressure (psi)
  P_{test} = direct integrity test pressure (psi)
  BP = back pressure on the system during the integrity test (psi)
  P_{atm} = atmospheric pressure (psia)

The elements of the effective integrity test pressure include three primary terms, as individually bracketed in Equation C.12. These three terms are, respectively:

- the differential pressure across the membrane during the integrity test,
- a term that accounts for the average velocity gradient of the compressed air as it passes across the membrane, and
- a multiplier that is necessary to convert the back pressure as it leaves the membrane to standard atmospheric conditions.

The Hagen–Poiseuille equation for liquid flow through a breach under conditions of laminar flow is shown in Equation C.13:

$$Q_{breach} = \frac{0.094\pi d_{fect}{}^4 g \times TMP}{L\mu_w},$$ (Equation C.13)

where Q_{breach} = flow of water through the critical breach during filtration (ft/s)
  d_{fect} = defect diameter (in.)
  g = gravitational constant (32.2 lbm-ft/lbf-s²)
  TMP = transmembrane pressure (psi)

L = length of the defect (in.)

μ_w = viscosity of water (lbs/ft-s)

The ratio of Equation C.11 to Equation C.13 yields an expression for the ALCR, given in Equation C.14:

$$\text{ALCR} = \frac{527 \Delta P_{\text{eff}} \mu_w}{\text{TMP} \times \mu_{\text{air}} (460 + T)}, \quad \text{(Equation C.14)}$$

where   ALCR = air-liquid conversion ratio (dimensionless)

ΔP_{eff} = effective integrity test pressure (psi)

μ_w = viscosity of water (lbs/ft-s)

TMP = transmembrane pressure during normal operation (psi)

μ_{air} = viscosity of air (lbs/ft-s), and T = water temperature (°F)

The ratio of the viscosity of water to the viscosity of air (μ_w / μ_{air}) may be combined and expressed as a single function of the water temperature that is derived by fitting a curve to discrete data points for the viscosity ratio. Equation C.15 simplifies the calculation of the ALCR to an expression that is a function of only measured pressures and the water temperature. Note that this form of the equation is only valid in the temperature range from 32°F to 86°F, in accordance with the limitations of the binomial fit for the viscosity ratio. If the temperature is outside of this range, the more general expression in Equation C.15 should be used:

$$\text{ALCR} = \frac{527 \Delta P_{\text{eff}} (175 - 2.71T + 0.0137T^2)}{\text{TMP}(460 + T)}, \quad \text{(Equation C.15)}$$

where ALCR = air-liquid conversion ratio (dimensionless)

ΔP_{eff} = effective integrity test pressure (psi)

T = water temperature (°F)

TMP = transmembrane pressure (psi)

As with the Darcy and orifice models, the values for the parameters ΔP_{eff} and TMP used in Equation C.14 or C.15 should be selected to yield a lower, more conservative value for the ALCR. For example, the TMP in Equation C.15 comes from the expression for Q_{breach} (Equation C.13) during filtration, and thus the most conservative ALCR result would be generated from using the maximum anticipated TMP during normal operation.

C.5 Applicability of ALCR Equations

The various expressions for the ALCR developed in this appendix are each derived under the assumption that water and air pass through the same integrity breach during normal operation and pressure-based direct integrity testing, respectively. This consistency allows the terms relating to breach characteristics to cancel out in the ALCR derivations, resulting in equations that are independent of any specific knowledge of the integrity breach. Although this assumption is consistent with the operation of most membrane filtration systems, there may be some systems that operate in a manner that renders this assumption invalid. In these cases, the ALCR equations described in this appendix are not directly applicable, and the derivation of the ALCR must be modified to match the specific operation of a particular membrane filtration system.

One example of such a scenario might be a hollow fiber membrane filtration system utilizing modules operating in an inside-out mode in which feed water flows into the fiber lumen from both ends. In this case, a single broken fiber represents two different pathways for water to flow through an integrity breach. If the Darcy model is applicable (i.e., turbulent flow conditions prevail), careful consideration must be given to the selection of the appropriate value for the net expansion factor for compressible flow Y. This parameter is a function of the defect length L, which will be different for each of the two flow pathways created by a single broken fiber (assuming the fiber is not severed precisely in the middle). In order to generate the most conservative (i.e., lowest) value for the ALCR (thus resulting in the most conservative direct integrity test sensitivity), the smallest potential value for L (the shortest length) should be used, which is generally represented by the length of the fiber encasement into the potting material. This value for L should always be utilized to yield a conservative ALCR, independent of the magnitude of the integrity breach. This approximation is reasonable since the flow through the short fiber length (i.e., the length of fiber embedded in the pot) will be substantially greater than the flow through the longer fiber length in most cases. If there is any question as to whether or not this approximation is appropriate for a particular system, the ALCR for different lengths could be calculated to evaluate the sensitivity of this important parameter to the length of the flow path.

A more complex scenario might involve a system similar to the one described in the previous example, but which applies pressurized air from only one end of the fiber during direct integrity testing. In this case, a single broken fiber would represent two separate pathways for water to flow through an integrity breach during normal operation, but only a single path for air during integrity testing. This difference must be accounted for in the derivation of an expression for the ALCR. As shown in Equation C.1, the ALCR is the ratio of the air flowing through an integrity breach during direct integrity testing (Q_{air}) to the flow of water through the same breach during normal operation (Q_{breach}). For the system described in this example, any number of broken fibers would generate double that number of pathways for feed water to bypass the membrane filtration process via integrity breaches. Thus, the term Q_{breach} must also be doubled in the ALCR expression for both the Darcy (i.e., turbulent flow) and the Hagen–Poiseuille (i.e., laminar flow) models.

The two examples addressed in this section are just two possible scenarios in which the ALCR equations developed in this appendix may not be directly applicable to some membrane filtration systems. Because this guidance manual cannot anticipate and address every such potential case, it is recommended that each membrane filtration system be evaluated on a site-specific basis to determine whether the given ALCR equations may be used or if the derivations must be modified to accommodate system-specific characteristics. Thus, any deviations from the assumptions used to derive the ALCR equations presented in this document may require a more complex treatment in which the flows of water and air through an integrity breach are determined separately using system-specific assumptions that are valid for each of these two respective flows.

Empirical Method for Determining the Air-Liquid Conversion Ratio for a Hollow Fiber Membrane Filtration System.

D.1 Introduction

As described in Chapter 4, the regulatory framework for the Long-Term 2 Enhanced Surface Water Treatment Rule (LT2ESWTR) requires that the flow through the smallest integrity breach that generates a measurable response from the direct integrity test (i.e., the critical breach size Q_{breach}) be determined in order to establish the sensitivity of a pressure-based test method (40 CFR 141.719(b)(3)(iii)(A)). (Under the LT2ESWTR, sensitivity is defined as the maximum log removal value [LRV] that can be reliably verified by the direct integrity test— i.e., LRV_{DIT}.) However, because most pressure-based direct integrity tests yield results in terms of airflow or pressure decay, for systems that utilize such tests it may be necessary to convert these results to an equivalent value for the flow of water through the critical breach under typical filtration conditions. Although there are a number of potential methods for converting a direct integrity test response to a corresponding flow of water, each can be generally categorized as one of two types of approach: mathematical modeling or empirical determination. This appendix describes an empirical approach based on a parameter called the air-liquid conversion ratio (ALCR), which is defined as the ratio of air that would flow through a breach during a direct integrity test to the amount of water that would flow through the breach during filtration, as shown in Equation D.1:

$$\text{ALCR} = \frac{Q_{air}}{Q_{water}}, \qquad \text{(Equation D.1)}$$

where ALCR = air-liquid conversion ratio (dimensionless)
  Q_{air} = flow of air through an integrity breach during a pressure-based direct integrity test (volume/time)

Q_{water} = flow of water through an integrity breach during filtration (volume/time)

While Appendix C describes the hydraulic models that could be used to calculate the ALCR for various types of membrane filtration systems and under different flow regimes, this appendix provides an example of an empirical method based on bubble point theory that can be used to determine the ALCR for a system using microporous hollow fiber membranes—the correlated airflow measurement (CAM) technique. The CAM technique measures the flow of air (Q_{air}) and water (Q_{water}) through a fiber break scenario to empirically determine the ALCR of a membrane filtration unit for a system utilizing a pressure-based direct integrity test. This method is specific to hollow fiber membrane processes in which the geometry of the fiber and associated module is known.

The CAM methodology involves first measuring the flow of water through a known integrity breach at various transmembrane pressures (TMPs) representative of normal operation, and subsequently measuring the flow of air through the same integrity breach during a pressure-based direct integrity test using a variety of potential test pressures. The data obtained from these measurements may be fitted with respective equations to establish empirical relationships between the applied pressures and the flow of water or air through an integrity breach. These functions can be used to determine the ALCR for any given operating TMP and direct integrity test pressure P_{test} (accounting for back pressure during the test, as per Equations 4.1 and 4.2). Note that the ALCR varies as a function of the TMP during operation, since a higher TMP results in a greater flow of water through the defect and thus a lower ALCR. In addition, these empirical relationships assume constant temperature, since changes in air or water temperature may change the functional relationships.

Although the CAM technique for determining the ALCR empirically is more labor intensive than calculating the ALCR using a hydraulic model, the procedure does have several advantages. First, because the measurement is empirical, it is more accurate than calculations based on general hydraulic models. In addition, the CAM procedure does not rely on assumptions that are necessary to estimate the ALCR from hydraulic models, but instead facilitates direct determination of the ALCR based on measured air and water flows through a known defect. Another advantage of the CAM procedure is that it allows the ALCR to be easily recalculated for any operating TMP and direct integrity test pressure (assuming constant temperature).

D.2 Methodology

The following general procedure is provided as a guide for conducting the CAM technique for experimentally determining the ALCR. Note that although the CAM technique is described below for use with a single bench- or full-scale module for convenience in conducting the procedure, the resulting ALCR is scalable and independent of the size of the module or the integrity breach, as discussed in Section C.1. Thus, the ALCR determined via this procedure would be applicable to an entire full-scale membrane unit.

1. Determine the baseline integrity test response for an integral bench- or full-scale membrane module. (See Section 4.3.1.3 for a discussion of diffusive flow through the wetted pores of an integral membrane.)

2. For reference, measure the flow of water through the integral membrane module at various TMPs representative of the potential range of operating conditions.

3. Cut a known number of fibers (e.g., between 1 and 100). (Note that for many hollow fiber membrane filtration systems, cutting a fiber at the point at which it enters the potting material represents the most conservative condition.)

4. Measure the water flow through the cut fibers (i.e., Q_{water}) over the range of potential operating TMPs. One potential approach for determining Q_{water} is to compare the flow through an integral membrane module (as determined in step 2) with that through the compromised module; the difference would represent the flow through the cut fibers for each discrete TMP assessed. (Note that other approaches for determining Q_{water} may also be used.)

5. Develop an equation for a fitted curve that represents the water flow through the cut fibers as a function of TMP.

6. Determine the minimum bubble point of the porous membrane material. (This information should generally be available from the manufacturer.)

7. Establish the direct integrity test pressure. As a general rule, the test pressure should be less than 80% of the bubble point pressure for the membrane and below the maximum TMP. However, the test pressure must be sufficient to meet the resolution requirement of the LT2ESWTR for the removal of *Cryptosporidium*, as expressed in Equations 4.1 and 4.2.

8. Measure the airflow from the cut fibers at a variety of potential applied direct integrity test pressures. This may be particularly useful if the hydrostatic back pressure may vary between different direct integrity test applications.

9. If the diffusive flow (i.e., the baseline response, as determined in step 1) is significant (i.e., more than 5% of the total airflow) at the target test pressure to be utilized during normal operation (P_{test}), then a lower test pressure should be considered, if possible. (The test pressure must enable the direct integrity

test to meet the resolution requirement.) Alternatively, the diffusive flow will have to be accounted for in determining the ALCR, as described in Section 4.3.1.3.

10. Determine the ALCR using Equation D.1:

$$\text{ALCR} = \frac{Q_{air}}{Q_{water}},$$

where ALCR = air-liquid conversion ratio (dimensionless)
  Q_{air} = flow of air through the broken fibers at the direct integrity test pressure (mL/min.)
  Q_{water} = flow of water through the broken fibers at the reference TMP (mL/min.)

Note that the reference TMP described in association with the variable Q_{water} refers to the TMP that is used in the determination of the ALCR for the purpose of establishing the direct integrity test sensitivity for regulatory compliance, as described in Section 4.3.1.2 and Appendix C. For example, the most conservative ALCR result would be generated from using the maximum anticipated TMP during normal operation.

11. Use Equation 4.7 to calculate the direct integrity test method sensitivity (LRV$_{DIT}$), incorporating the test result Q_{air} (either as directly measured with the diffusive airflow test or as converted from the pressure decay rate ΔP_{test} using Equation 4.8, if the pressure decay test is used) and the ALCR determined in the previous step, as described in Chapter 4. Note that the use of Equations 4.7 and 4.8 to calculate sensitivity require that the parameters Q_{air} and ΔP_{test}, respectively, represent the smallest integrity test response that can be reliably measured and associated with an integrity breach, as specified in Section 4.3.1.2.

Glossary

18-megohm water A high purity water that conducts electrical current poorly because of the lack of ionized impurities (electrolytes). It has an electrical resistivity of approximately 18 MΩ-cm (180,000 Ω-m) and a conductivity of 0.0556 mΩ/cm (0.00000556 S/m) at a specified temperature, typically 25°C (77°F). This type of water is also called *ultrapure water*.

Absolute filter rating A value associated with a filter that represents the size of the smallest particle completely retained. For membrane filters used in the water industry, an integrity test using a nondestructive method should be used to verify the absolute filter rating of the membrane to ensure compliance with regulatory standards.

Acidity A measure of the capacity of water to neutralize strong bases.

Alkalinity A measure of the capacity of a water to neutralize strong acids; in natural waters this capacity is usually attributable to bases such as bicarbonate (HCO_3^-) and carbonate (CO_3^{2-}).

Angstrom A unit of length equaling 10^{-10} m (10^{-4} pm, or 4×10^{-9} in.). Its symbol is Å.

Anion A negatively charged atom or molecule that migrates towards an anode.

Anionic Having a negative ionic charge.

Anisotropic membranes Microporous membranes that vary in pore size. The surface with the smaller pore size is used as the filtering surface.

Antiscalant A chemical agent added to water to inhibit the precipitation or crystallization of salt compounds; also referred to as **scale inhibitor**.

Antitelescoping device A rigid structure firmly attached to each end of a spiral wound NF or RO membrane module that prevents telescoping, unwinding, and other undesirable movement of the membrane module.

Array An assembly of cartridges in pressured membrane systems (also referred to as a **train**). NF and RO membrane systems can be described based upon the ratio of the number of pressure vessels in each stage that operate in parallel, e.g., a 24:12 (absolute) two-stage array or a 2:1 (relative) two-stage array.

Asymmetric   Having a varying consistency throughout (e.g., to describe a membrane that varies in density or porosity across its structure).

Backpulsing   The reversal of permeate flow through membranes to flush trapped particles from membrane pores and cavities.

Backwashing   A mode of operation in which the transport direction through the membrane is reversed by applying pressure in the opposite direction, so that the feed water enters the reversal of flow through a filtration medium. Often used as a cleaning operation that involves periodic reverse flow to remove foulants accumulated at the membrane surface element through what is normally the filtrate outlet. Also sometimes called *washwater* or *reverse filtration*.

Bacteria   Prokaryotic (no cell nucleus) cellular organisms, 0.5 to 2 μm in diameter; they have simple internal organization and reproduce by binary fission.

Bar   International unit of pressure; 1 bar = 1 Mdyn/cm^2 = 100,000 Pa = 0.987 atm.

Bed volume (BV)   The volume occupied by a bed of porous media or resin used for filtration, absorption, or ion exchange. The number of bed volumes is a unitless measure that indicates the capacity of a system to remove contaminants. The product of the number of bed volumes and the empty bed contact time is the operation time.

Binders   Chemicals used to hold or bind short fibers together in a cartridge filter.

Biofouling   Membrane fouling (and associated decreases in flux) that is attributable to the deposition and growth of microorganisms on the membrane surface and/or the adsorptive fouling of secretions from microorganisms.

Bleed   The continuous waste stream from an MF or UF system operated in a cross-flow hydraulic configuration.

Boundary layer   A thin layer of water at the surface of a semipermeable membrane containing the rejected contaminants from the filtrate (i.e., permeate) flow in higher concentrations than the bulk feed or brine stream (called **concentration polarization**), affecting the osmotic pressure and salt passage.

Brackish water   Saline water in which the dissolved solids content generally falls between those of drinking water and seawater (500 to 35,000 mg/L).

Brine   (1) A concentrated salt solution, generally containing sodium, chloride, and other ions typically having a concentration of 3% wt or more. (2) A concentrated salt solution remaining after desalting brackish or seawaters. For a brackish water membrane desalting system, the word **concentrate** or **reject** is commonly preferred over *brine*. (3) A concentrated potassium chloride (KCl) or sodium chloride (NaCl) solution used in the regeneration stage of either cation or anion exchange water treatment devices. Sodium chloride brine saturation in an ion exchange softening brine tank is about 26% NaCl by weight at 15.5°C (60°F).

Brine disposal   The process of ultimately discharging a reject stream, most often associated with desalination or membrane processes, to the environment. Brine disposal often involves ocean discharge, deep well injection, or discharge to another receiving medium whose mineral concentration is at least as high as that of the brine.

Brine seal   A rubber seal around the circumference of a spiral wound module between the module and the interior pressure vessel wall that separates the feed water from the concentrate stream, preventing the bypass of feed between the module and the inside of the pressure vessel wall.

Bubble point test   A nondestructive performance test commonly used to characterize the largest pore (or defect) in a membrane module; also the Applied air pressure required to evacuate the largest pores of a fully wetted porous membrane.

Calcium (Ca), a divalent cation   Calcium in water is a factor contributing to the formation of scale and insoluble soup curds that are a means of clearly identifying hard water.

Calcium carbonate ($CaCO_3$)   A colorless or white crystalline compound. $CaCO_3$ is a sparingly soluble salt, the solubility of which decreases with increasing temperature; it has the potential to cause scaling if concentrated to supersaturation.

Calcium carbonate ($CaCO_3$) equivalent   An expression of the concentration of specified constituents in water in terms of the equivalent value of calcium carbonate ($CaCO_3$). The hardness in water that is caused by calcium, magnesium, and other ions is usually described in terms of the calcium carbonate equivalent. For example, the concentration of calcium ions (i.e., $[Ca^{2+}]$) can be multiplied by 100/40 (the ratio of the molecular weight of $CaCO_3$ to the atomic weight of Ca^{2+}) to give $[Ca^{2+}]$ as equivalent $CaCO_3$.

Calcium hardness   The portion of total hardness caused by calcium compounds such as calcium carbonate ($CaCO_3$) and calcium sulfate ($CaSO_4$).

Calcium sulfate ($CaSO_4$)   A sparingly soluble salt, the solubility of which decreases with increasing temperature. It is called gypsum in its hydrated form. It is a potential source of scaling in desalting systems if it is concentrated to supersaturation.

Cartridge   A term commonly used to describe a disposable filter element.

Cassette A series of modules manifolded together into a single removable unit. Typically used in immersed membrane systems. Also referred to as **rack**.

Cation A positively charged ion or radical that migrates towards a cathode.

Cation exchange A process in which cation contaminants are removed from a liquid phase by contact with a synthetic, porous medium or resin that is coated with different cation species. The cations on the medium are exchanged for the cation contaminants.

Cation exchange resin A synthetic material possessing reversible ion exchange ability for cations.

Cation exchange water softener An equipment unit capable of reducing water hardness by the cation exchange process.

Cation membrane An electrodialysis membrane that allows the passage of anions (but not cations) and is practically impermeable to water under typical electrodialysis system working pressure.

Cationic Having a positive ionic charge.

Caustic (1) Caustic soda (NaOH) or any compound chemically similar to it. (2) An adjective usually applied to strong bases.

Caustic soda (NaOH) Sodium hydroxide, a strongly alkaline chemical used for pH adjustment, water softening, anion exchange demineralizer regeneration, and other purposes. It is sometimes called **caustic**.

Caustic soda softening A process to remove hardness from waters by using caustic soda (NaOH).

Challenge particulate The target organism or acceptable surrogate used to determine the log removal value during a challenge test.

Challenge test A study conducted to determine the removal efficiency—i.e., log removal value (LRV)—of a membrane material for a particular organism, particulate, or surrogate.

Chelating agent A chemical reagent, typically a water-soluble organic molecule such as citric acid or EDTA, that reacts with metal ions to keep them in an aqueous solution, therefore increasing the solubility of a metal in water.

Chemically enhanced backwash (CEB) A technique to clean a membrane in which sodium hypochlorite, caustic soda, or mineral acid is added to the backwash water.

Clean in place (CIP) The periodic application of a chemical solution (or series of solutions) to a membrane unit for the purpose of removing accumulated microbial or inorganic foulants and thus restoring permeability to baseline levels.

Clinoptilolite   A naturally occurring zeolite that exchanges ammonium ions (NH_4^+) in preference to sodium, magnesium, and calcium ions.

Colloid   A type of particulate matter ranging in size from approximately 2 to 1,000 nm in diameter that does not settle out rapidly.

Compaction   The compression or densification of a membrane as a result of exposure to applied pressure over a period of time, typically resulting in decreased productivity.

Composite   Being made from different materials (e.g., a membrane manufactured from two or more different materials in distinct layers).

Concentrate   The portion of a feed stream that does not permeate membrane media in a cross-flow filtration system, thus retaining impurities or contaminants such as organic compounds, ions, and colloidal materials that get rejected by the membrane; also referred to as **retentate** or **reject**.

Concentrate staging   A configuration of spiral wound NF and RO membrane systems in which the concentrate from each stage of a multistage system becomes the feed for the subsequent stage.

Concentration polarization   A phenomenon that occurs when dissolved and/or colloidal materials concentrate on or near the membrane surface.

Conductivity   A measure of the ability of an aqueous solution to conduct an electric charge; related to the amount of total dissolved solids (TDS). In the U.S. customary system, measured in $m\Omega/cm$; in SI, measured in S/m, s^3A^2/m, or $s^3A^2/m\text{-}kg$.

Control limit (CL)   A response from an integrity test which, if exceeded, indicates a potential problem with the membrane filtration system and triggers a response; synonymous with *upper control limit* (UCL) as used in the *Membrane Filtration Guidance Manual* to distinguish it from additional voluntary or state-mandated lower control limits (LCLs).

Cross-flow   (1) The application of water at high velocity tangential to the surface of a membrane to maintain contaminants in suspension. (2) A suspension mode hydraulic configuration that is typically associated with spiral wound NF and RO systems and a few hollow fiber MF and UF systems.

Dalton   A unit of mass equal to one-twelfth the mass of a carbon 12 atom (i.e., one atomic mass unit [amu]); typically used as a unit of measure for the molecular weight cutoff of a UF, NF, or RO membrane.

Day tank   A treatment chemical storage vessel that contains a diluted concentration in a feed volume suitable for a short period, typically from 1 to 3 days. For example, dry or viscous polymers are often diluted, or aged, in day tanks prior to application. A day tank is also called an *age tank*.

DBP precursors   Compounds such as natural organic matter (NOM) and humic substances (soluble and colloidal) that can be converted into disinfection by products during the disinfection process.

Dead-end or depth filtration   A mode of membrane filtration that has only one feed stream (influent) and one effluent stream (permeate, filtrate). Such systems do not have a cross-flow type; however, a cross-flow module can be operated in dead-end mode by shutting off the module outlet.

Demineralize   To reduce the concentrations of minerals and inorganic constituents from water by ion exchange deionization or desalting processes such as distillation, electrodialysis, and reverse osmosis. See **desalting**.

Demineralized stream   (1) A water stream from which minerals have been removed. (2) For an electrical-driven membrane treatment system (e.g., electrodialysis), the stream in a membrane stack in which the concentrations of ions have been reduced as a result of ion transfer through the membranes to the concentration stream.

Demineralizer   A device that demineralizes water.

Denitrification   The conversion of nitrite (NO_2^-) and nitrate (NO_3^-) to molecular nitrogen (N_2) and nitrogen dioxide (NO_2).

Deposition mode   A hydraulic configuration of membrane filtration systems in which contaminants removed from the feed water accumulate at the membrane surface and, in MF and UF systems, are subsequently removed via backwashing.

Desalination   The process of removing dissolved salts from water.

Desalting   A term often used instead of desalination that describes the removal of dissolved salts from water.

Diagnostic test   A precise direct integrity test that is specifically used to isolate any integrity breaches that may be initially detected via other means, such as coarser direct integrity tests that simply indicate the presence or absence of a breach within a membrane unit.

Differential pressure   The pressure drop across a membrane module or unit from the feed inlet to concentrate outlet; distinguished from transmembrane pressure (TMP), which represents the pressure drop across the membrane barrier.

Diffusion   A process by which a solute moves from high concentration to low concentration across a semipermeable membrane as a result of concentration differences (gradient). Diffusion stops when the gradient no longer exists.

Direct filtration   With respect to membrane filtration, a term commonly used describe the deposition mode hydraulic configuration of membrane filtration systems; synonymous with **dead-end filtration**.

Direct integrity test   A physical test applied to a membrane unit in order to identify and isolate integrity breaches.

Direct reuse   The beneficial use of reclaimed water for potable purposes with transfer from the reclamation plant to the reuse site.

Disinfection by-products (DBPs)   A group of chemical compounds—total trihalomethanes (TTHMs) and total haloacetic acids (THAAs)—formed during disinfection of waters containing bromine and natural organic matter.

Dissociation   The process by which a chemical combination breaks up into simpler constituents—such as atoms, groups of atoms, ions, or multiple different molecules—without any change of valence. Often this breakdown is reversible, as in the case of ionization. See also **ionization**.

Dissolved carbon dioxide   The carbon dioxide (CO_2) that is dissolved in a liquid medium, typically expressed in mg/L. The saturation concentration is dependent on several factors, including partial pressure, temperature, and pH. See also **partial pressure**.

Dissolved concentration   The amount per unit volume of a constituent of a water sample filtered through a 0.45-μm pore-diameter membrane filter before analysis.

Dissolved gases   The sum of gaseous components that are dissolved in a liquid medium. Typical dissolved gases found in water include oxygen (O_2), nitrogen (N_2), carbon dioxide (CO_2), methane (CH_4), and hydrogen sulfide (H_2S), among others. High concentrations of dissolved gases can result in filter air binding and pump cavitation.

Dissolved organic matter (DOM)   That portion of the organic matter in water that passes through a 0.45-μm pore-diameter filter. Dissolved organic carbon represents the carbon portion of this matter. For aquatic humic substances, the carbon level typically represents approximately 50% of the organic matter (the rest being hydrogen, oxygen, nitrogen, and sulfur).

Dissolved solids   In operational terms, the constituents in water that can pass through a 0.45-μm pore-diameter filter. See also **total dissolved solids**.

Divalent ion   An ion with two positive or negative electrical charges, such as ferrous (Fe^{2+}) or sulfate (SO_4^{2-}).

Effluent   Water that has been passed through a membrane; also referred to as **permeate**, **filtrate**, or **product water**.

Electrodialysis (ED)   A process in which ions are transferred through ion-selective membranes by means of an electromotive force from a less concentrated solution to a more concentrated solution.

Electrodialysis reversal (EDR)   A variation of the electrodialysis process in which the polarity of the electrodes is periodically reversed on a prescribed time cycle, thus changing the direction of ion movement, in order to reduce scaling.

Element   A term commonly used to describe an encased spiral wound membrane module.

Equivalent weight   A weight of a substance based on the formula weight and number of reactive protons associated with the substance. For an element, the equivalent weight is the atomic weight divided by the number of reactive protons. For example, calcium (Ca) has an atomic weight of 40 and has two reactive protons, so its equivalent weight is $40/2 = 20$. For a compound, the equivalent weight is the molecular weight divided by the number of reactive protons. For example, sodium chloride (NaCl) has a molecular weight of 58.5 and one reactive proton, so the equivalent weight is $58.5/1 = 58.5$. Calcium carbonate ($CaCO_3$) has a molecular weight of 100 and two reactive protons (in the Ca^{2+} ion), so the equivalent weight is $100/2 = 50$. Equivalent weight is also known as *combining weight*.

Feed channel spacer   A plastic mesh spacer that separates the various leaves in a spiral wound module, providing a uniform channel for feed water to reach the membrane surface and promoting turbulence in order to minimize the formation of a boundary layer at the membrane surface.

Feed stream   The input stream to a membrane array or train; also referred to as **feed water** or **influent**.

Feed water   The input stream to a membrane array or train; also referred to as **feed stream** or **influent**.

Feed-and-bleed mode   A term used to describe a variation of the suspension mode hydraulic configuration of membrane filtration systems in which a portion of the cross-flow stream is wasted (i.e., bled) rather than recirculated.

Filter cake   Accumulated particles on a membrane surface (or any filter surface).

Filtrate   Water that has been passed through a membrane; also referred to as **permeate**, **effluent**, or **product water**.

Flux   The volume of water that passes through a membrane per unit time and per unit surface area of the membrane. Flux is measured in either lmh or gfd. Flux is affected by the water temperature; it is often normalized to a standard temperature of 25°C (77°F).

Flux maintenance system   A term sometimes used to refer, collectively, to any equipment involved in relaxing, backwashing, or cleaning membranes to restore the system's flux rate.

Foulant   Any substance that causes fouling.

Fouling   The buildup of impurities (such as colloidal materials) on the membrane. Fouling reduces the flux through membranes and thus causes the increase of transmembrane pressure.

Granular media   A material used for filtering water, consisting of grains of sand or other granular material.

Hardness   A quality of water caused by divalent metallic cations and resulting in increased consumption of soap, deposition of scale in boilers, damage in some industrial processes, and sometimes objectionable taste. The principal hardness causing cations are calcium, magnesium, strontium, ferrous iron, and manganese. Hardness may be determined by a standard laboratory titration procedure or computed from the amounts of calcium and magnesium expressed as equivalent calcium carbonate ($CaCO_3$).

Hardness as calcium carbonate ($CaCO_3$)   The value obtained when the hardness forming salts are calculated in terms of equivalent quantities of calcium carbonate. This method of water analysis provides a common basis for comparison of different salts and compounds. See also **calcium carbonate ($CaCO_3$) equivalent**.

Head loss   A reduction of water pressure (head) in a hydraulic or plumbing system. Head loss is a measure of (1) the resistance of a medium bed (or other water treatment system), a plumbing system, or both to the flow of the water through it or (2) the amount of energy used by water in moving from one location to another. In water treatment technology, head loss is basically the same as pressure drop. **Heterogeneous** Composed of a combination of different materials (e.g., composite and some asymmetric membranes).

Hollow fiber   The physical form of some membranes, which are composed of very small diameter fibers that have an open core or lumen for the filtrate to pass into or out of.

Hollow fiber module   A configuration in which hollow fiber membranes are bundled longitudinally and either encased in a pressure vessel or submerged in a basin; typically associated with MF and UF membrane processes.

Hollow fine fiber (HFF) module   A relatively uncommon configuration in which very small diameter (approximately 50 μm inside diameter) semi-permeable hollow fiber membranes are bundled in a U shape and potted into a pressure vessel; typically associated with RO membrane processes.

Homogeneous   Composed of the same material throughout (e.g., symmetric and some asymmetric membranes).

Hydraulic configuration   The pattern of flow through a membrane process by which the feed contaminants are removed or concentrated (e.g., cross-flow, dead-end, etc.).

Hydrophilic   Attracting water.

Hydrophobic   Repelling water.

Immersed membrane   A membrane system in which the membranes are immersed in a tank of feed water and the filtrate is drawn through the membranes; also sometimes called *submerged membrane*.

Immersed membrane system   The process tank(s) with interconnecting piping for feed water and drains, membrane units with air scour and filtrate connections, and frames used to support the membrane units. For pressurized systems, process tanks and frames are replaced with pressure vessels used to contain the membranes and racks or frames on which the pressure vessels are mounted. For cross-flow UF systems, piping and connections for the concentrate will be provided.

Indirect integrity monitoring   Monitoring defined under the LT2ESWTR as some aspect of filtrate water quality that is indicative of the removal of particulate matter.

Indirect potable reuse (IPR)   The beneficial use of reclaimed water after releasing it for storage or dilution into natural surface waters or groundwater.

Influent   The input stream to a membrane array or train; also referred to as **feed stream** or **feed water**.

Inorganic compounds   Those compounds not derived from hydrocarbons. Usually of mineral origin.

Integral asymmetric membranes   Membranes cast in one process that consist of a very thin layer (less than 1 μm or 4×13^{-5} in.), referred to as the *skin*, and a thicker, porous layer (up to 100 μm or 4×10^{-3} in.) that adds support and is capable of high water flux; also referred to as **skinned membranes**.

Integrity breach   Any leak that will result in the contamination or deterioration of the product (effluent quality). Under the Safe Drinking Water Act, it is defined under the Long Term 2 Enhanced Surface Water Drinking Rule. Additionally, ASTM developed a draft *Standard Practice for Integrity Testing of Water Filtration Membrane Systems*, D6908.

Integrity monitoring system   A system provided to monitor for integrity breaches of a membrane. Typically, each manufacturer will have its own system to monitor for integrity breaches.

Ion   An atom that is electrically unstable because it has more or fewer electrons than protons; thus it is an electrically charged particle. A positive ion is called a cation, and a negative ion is called an anion. In aqueous solution, ions may not actually exist as isolated charged atoms but tend to form a variety of hydrated complexes.

Ionic bond   A type of chemical bond in which electrons are transferred.

Ionization   The splitting or dissociation (separation) of molecules into negatively and positively charged ions.

Irreversible fouling   Any membrane fouling that is permanent and cannot be removed by either backwashing (if applicable) or chemical cleaning.

Isotropic membranes   Microporous membranes that are uniform in pore size; also referred to as **symmetric membranes**.

Langelier Saturation Index (LSI)   An expression to predict the precipitation of calcium carbonate under certain conditions of temperature, pH, hardness, alkalinity, and total dissolved solids.

Leaf   A sandwich arrangement of flat sheet, semipermeable membranes placed back-to-back and separated by a fabric spacer (i.e., permeate carrier) in a spiral wound module.

Lime (CaO)   Calcium oxide, a calcined chemical material. Lime is used in lime softening and in lime–soda ash water treatment, but first it must be slaked to calcium hydroxide ($Ca(OH)_2$). Lime is also called *burnt lime, calyx, fluxing lime, quicklime,* or *unslaked lime.* A water treatment that makes use of lime softening followed by a reduction of noncarbonated hardness through the addition of soda ash (Na_2CO_3) to form an insoluble precipitate that is removed by filtration. This method of removing hardness by precipitation is sometimes used by municipalities, but it will leave 85 mg/L or more of residual hardness as calcium carbonate ($CaCO_3$).

Log reduction   A method of expressing the reduction of a number of particles or organisms. For example, if one particle per 100 mL is detected in the filtrate and 10^4 particles per 100 mL are detected in the feed in a given size range, the log reduction is $-\log_{10}(1/10^4) = \log$.

Log removal value (LRV)   The filtration removal efficiency for a target organism, particulate, or surrogate expressed as a common logarithm, i.e., $\log_{10}$(feed concentration), $-\log_{10}$(filtrate concentration).

Lower control limit (LCL)   A control limit (CL) that is not mandated by the Long Term 2 Enhanced Surface Water Treatment Rule but which is instead voluntarily implemented or may be required by the state at its discretion.

Lumen   The interior of a hollow fiber membrane.

Macrofouling   The buildup of impurities (such as colloidal materials) on the external membrane surface.

Mass-transfer coefficient (MTC)   Volume unit or mass transferred through a membrane based on driving force.

Maximum design flux   The average maximum flux for membrane filters when no train is in a CIP.

Maximum instantaneous flux   The maximum flux membranes will see under normal operation; typically occurs when one train is in CIP and another train is in backwash when the works output is at maximum flow.

Membrane   An intervening phase separating two phases and/or acting as an active or passive barrier to the transport of matter between phases adjacent to it.

Membrane bioreactor (MBR)   A combination of suspended-growth activated sludge biological treatment and membrane filtration equipment performing the critical solids/liquid separation function that is traditionally accomplished using secondary clarifiers.

Membrane cartridge filtration (MCF)   Filtration using any device that meets the definition of membrane filtration as specified under the Long Term 2 Enhanced Surface Water Treatment Rule.

Membrane cell   With respect to immersed membranes, a tank containing one train; also called **membrane filter tank**.

Membrane filter tank   With respect to immersed membranes, a tank containing one train; also called **membrane cell**.

Membrane filtration   A pressure- or vacuum-driven separation process in which particulate matter larger than 1 μm is rejected by an engineered barrier, primarily through a size exclusion mechanism, and which has a measurable removal efficiency of a target organism that can be verified through the application of a direct integrity test. The definition of a membrane filtration process and what it represents are further defined by the Membrane Filtration Guidance Manual, USEPA 2005 (as defined under the USEPA Long Term 2 Enhanced Surface Water Treatment Rule) a pressure- or vacuum-driven separation process in which particulate matter larger than 1 μm is rejected by an engineered barrier, primarily through a size exclusion mechanism, and which has a measurable removal efficiency of a target organism that can be verified through the application of a direct integrity test; includes common membrane classifications of MF, UF, NF, and RO, as well as any membrane cartridge filtration (MCF) device that satisfies this definition.

Membrane softening   A semipermeable membrane treatment process designed to selectively remove hardness (i.e., calcium, magnesium, and certain other multivalent cations) but allow significant passage of monovalent ions; typically used to describe the application of NF for hardness removal.

Membrane unit (cassette or rack)   An assembly of membranes intended to be removed from an immersed system as a unit.

Microbes   Parasites, bacteria, and viruses.

Microfiltration (MF)   A pressure-driven membrane filtration process that typically uses hollow fiber membranes with a pore size range of approximately 0.1 to 0.2 μm (3.9×10^{-6} to 7.9×10^{-6} in.).

Microfouling   The buildup of impurities (such as colloidal materials) inside a membrane's pores; also referred to as *pore fouling*.

Micron   A metric measurement equivalent to 10^{-6} m (3.9×10^{-5} in.). It is represented by the symbol μm and used to classify the pore size of a membrane.

Microporous membranes   Membranes cast from one material (homogenous); can either be either uniform in pore size (isotropic) or vary in pore size (anisotropic).

Mineral acid   An inorganic acid, especially hydrochloric acid (HCl), nitric acid (HNO_3), or sulfuric acid (H_2SO_4).

Mineral acidity   Acidity in water caused by the presence of a mineral as opposed to weak acidity caused by such acids such as carbonic acid (H_2CO_3) and acetic acid (CH_3COOH). Mineral acidity is usually expressed in water analysis as free mineral acidity.

Mineral salt   A chemical compound formed by the combination of a mineral acid and a base. Minerals from dissolved rock exist in water in the form of dissolved mineral salts. An excess of mineral salts can give water a disagreeable taste or even be harmful to human health.

Mineral-free water   Water produced by either distillation or deionization. The term is sometimes found on labels of bottled water as a substitute term for *distilled water* or *deionized water.*

Module   A collection of membranes and housing units intended to be mounted and replaced as a unit. Sometimes in a spiral wound membrane the membrane element itself is called a module.

Molecular weight (mol. wt., MW)   The sum of the atomic weights of the atoms in a molecule.

Molecular weight cutoff (MWCO)   The molecular weight of the smallest material rejected by a membrane, typically measured in daltons; also referred to as **nominal molecular weight cutoff.**

Molecule   The smallest particle of an element or compound that retains all of the characteristics of the element or compound. A molecule is made up of one or more atoms. A helium molecule, for example, has only one atom per molecule. Oxygen molecules (O_2) have two atoms; ozone molecules (O_3) have three atoms. Molecules found in chemical compounds often have many atoms of various kinds.

Nanofiltration (NF)   A pressure-driven membrane separation process that employs the principles of reverse osmosis to remove dissolved contaminants from water; typically applied for membrane softening or the removal of dissolved organic contaminants.

National Pollution Discharge Elimination System (NPDES)   A permit under the Clean Water Act of 1972 that qualifies the pollutants that a user can discharge to a given body of water.

Net driving pressure (NDP)   The pressure available to force water through a semipermeable NF or RO membrane, defined as the average feed side pressure (i.e., the average of the feed and concentrate pressures) minus the filtrate side back pressure and the osmotic pressure of the system.

Neutralization   A chemical reaction in which water is formed by mutual destruction of the ions that characterize acids and bases when both are present in an aqueous solution.

New blue water   A reuse source water processed by MF/UF–RO–advanced oxidation (AOP).

Nominal filter rating   An arbitrary value established by the filter manufacturer that is typically based upon removal efficiency of a challenge particulate in a specific size range. Nominal ratings vary from manufacturer to manufacturer and cannot be used to compare filters among suppliers.

Nominal molecular weight cutoff   The molecular weight of the smallest material rejected by a membrane, typically measured in daltons; also referred to as **molecular weight cutoff**.

Nondestructive performance test (NDPT)   A physical quality control test typically conducted by a manufacturer to characterize some aspect of process performance without damaging or altering the membrane or membrane module.

Nonionic Resin   An ion exchange resin without exchangeable charged ions (e.g., chloride (Cl^-) or hydrogen (H^+)). A nonionic resin can either be a macroreticular, nonionic, acrylic ester type or—without ester cross-linking—can have a less polar structure and take the form of a polystyrene ($[C_6H_5CH{:}CH_2]_n$) backbone cross-linked with divinylbenzene ($C_6H_4[CH{:}CH_2]_2$). A nonionic resin is used to isolate and concentrate natural organic matter. The hydrophobic fraction, defined as the organic compounds adsorbed on the acrylic ester polymer resin at acidic pH, represents the most commonly used operational definition of aquatic humic substances. Organic solutes that do not adsorb on the acrylic ester polymer resin at acidic pH are defined as the hydrophilic fraction. The acrylic ester polymer and polystyrene divinylbenzene have been used in tandem for the isolation of both humic substances and hydrophilic acids. See also, **hydrophilic, hydrophobic, natural organic matter, polymer,** and **resin.**

Nonpotable reuse   The beneficial use of reclaimed water other than for potable water supply augmentation.

Normalization   The process of evaluating membrane system performance at a given set of reference conditions (e.g., at standard temperature, per unit pressure, etc.), allowing the direct comparison and trending of day-to-day performance independent of changes to the actual system operating conditions.

Osmosis   The natural, spontaneous flow of water from a less concentrated solution to a more concentrated solution across a semipermeable membrane; the flow stops when equilibrium is attained.

Osmotic pressure   The potential energy difference between two solutions separated by a semipermeable membrane. Osmotic pressure is a factor used in designing RO membrane separation equipment.

Partial pressure   The pressure exerted by a gas in a mixture of gases, proportional to the percent, by volume, of the gas.

Particle counter   An instrument used to count the number of discrete particles in a solution and classify them according to size.

Pascal (Pa)   A unit of pressure; $1\ Pa = 1\ N/m^2$. Multiply Pa by 0.0001450 to obtain psi.

Permeability   The ability of a membrane barrier to allow the passage or diffusion of a substance; defined as the ratio of flux to transmembrane pressure.

Permeable   Allowing some material to pass through.

Permeate   Water that has been passed through a membrane; also referred to as **effluent, filtrate,** or **product water.**

Permeate (filtrate) carrier   The fabric spacer in between two sheets of membrane material in one leaf of a spiral wound module, serving to transfer the water that permeates through the membrane(s)—i.e., the filtrate—to a perforated central collector tube (i.e., the permeate tube).

Permeate staging   A configuration of spiral wound NF and RO membrane systems in which the permeate (or filtrate) from each stage of a multistage system becomes the feed for the subsequent stage.

Permeate tube   The perforated tube in the center of a spiral wound module that collects permeate (or filtrate) and transports it out of the membrane module.

pH   An indicator of the acidity or alkalinity of a solution; a value of 7 is neutral, with lower numbers representing acidic solutions and higher numbers, alkaline solutions. Strictly speaking, pH is the negative logarithm of the hydrogen ion concentration (in mol/L). For example, if the concentration of hydrogen ions is 10^{-7} mol/L, the pH is 7.0. As a measure of the intensity of a solution's acidic or basic nature, pH is operationally defined relative to standard conditions that were developed so that most can agree on the meaning of a particular measurement. The pH of an aqueous solution is an important characteristic that affects many features of water treatment and analysis.

Plate and frame module   A relatively uncommon configuration consisting of a series of flat sheet membranes separated by alternating filtrate spacers and feed/concentrate spacers; used with electrodialysis reversal (EDR) membrane systems.

Plugging   The physical blockage of the feed side flow passages of a membrane or membrane module (e.g., a blockage in the lumen of a hollow fiber module operated in inside-out mode or in the spacer of a spiral wound module).

Pore   An opening in a membrane. The pore size in a membrane provides an idea of the average or smallest particle retained by a membrane.

Pore size   The size of the openings in a porous membrane, expressed either as nominal (average) or absolute (maximum), typically in microns.

Porosity   For a membrane material, the ratio of the volume of voids to the total volume.

Posttreatment   Any treatment applied to the filtrate of a membrane process in order to meet given water quality objectives.

Potable reuse The beneficial use of highly treated reclaimed water towards the augmentation of potable water supply.

Potable water Water that is safe and satisfactory for drinking and cooking.

Preliminary treatment Treatment steps including comminution, screening, grit removal, preaeration, and/or flow equalization which prepare wastewater influent for further treatment.

Pressure membrane A membrane system in which the membranes are contained in pressure vessels and filtrate is pushed through membranes; also called *encased membrane*.

Pressure system frame The structure that supports a group of pressure vessels that are interconnected by common feed water, filtrate, and drain blocks or piping. Feed water is pumped into the system and the residual pressure is used to convey the filtrate to storage or to a filtrate pumping station if higher pressure is required. For systems with multiple frames, the process connections from each frame are connected into larger piping manifolds.

Pressure vessel The cylindrical outer housing for pressure membrane modules.

Pretreatment Any treatment applied to the feed water of a membrane process to achieve desired water quality objectives and/or protect the membranes from damage or fouling.

Primary treatment Treatment steps including sedimentation and/or fine screening to produce an effluent suitable for biological treatment.

Product water Water that has been passed through a membrane; also referred to as **permeate, effluent,** or **filtrate**.

Productivity The amount of filtered water that can be produced from a membrane module, filtration unit, or system over a period of time, accounting for the use of filtrate in backwash and chemical cleaning operations as well as otherwise productive time that a unit or system is off-line for routine maintenance processes such as backwashing, chemical cleaning, integrity testing, or repair.

Protozoa Unicellular, eukaryotic organisms between 2 μm and 1 mm in size; they are nonphotosynthetic, with flexible cell membranes instead of cell walls. Protozoa come in a wide range of sizes and shapes and produce hardy cysts. Groups include flagellates (*Giardia*), amoebae, ciliates, sporozoa (*Cryptosporidium*), and microsporidia.

Quality control release value (QCRV) A minimum quality standard of a nondestructive performance test established by the manufacturer for membrane module production that ensures that the module will attain the targeted log removal value (LRV) demonstrated during challenge testing in compliance with the Long Term 2 Enhanced Surface Water Treatment Rule.

Quaternary treatment   The use of a double membrane process in the repurification of water; may involve MF or UF for pretreatment before RO or NF.

Rack   An assembly of membranes intended to be removed from an immersed system as a unit; also referred to as a **membrane unit** or **cassette**.

Rated capacity   (1) In filtration or adsorption, the manufacturer's statement regarding the expected number of days the equipment will be in service or the expected volume of product water delivered before backwashing and rinse down are expected to occur. (2) In softening or ion exchange, a value stated by the manufacturer, given in relation to the period between regenerations, of the expected number of days the equipment will be in service, the expected volume of product water delivered, or the weight of total hardness removed. The capacity of an ion exchange system varies, within limits, with the amount of regenerant used.

Reagent grade water   Very high purity water produced to meet the standards outlined by ASTM in its standard D1193-77, *Standard Specification for Reagent Water*. Four grades, types I through IV, are specified, and three levels of maximum total bacterial count, types A through C, are listed depending on intended use. Reagent grade water is used for chemical analysis and physical laboratory testing.

Reclaimed water   Wastewater that has been treated to a level that allows its reuse for a beneficial purpose.

Recovery   The ratio of the permeate flow to the feed flow; it is expressed, in general, as a percentage. Membrane bioreactor systems do not refer to recovery.

Regeneration   Restoration of the properties of a granular medium used for removing contaminants by surface interactions. For example, ion exchange resins are regenerated by passing through the medium a solution that exchanges ions of similar charge for those that have been removed. After regeneration, the medium is put back into service to remove the target contaminants.

Reject   The portion of a feed stream that does not permeate membrane media in a cross-flow filtration system, thus retaining impurities or contaminants such as organic compounds, ions, and colloidal materials that get rejected by the membrane; also referred to as **concentrate** or **retentate**.

Rejection   The term used in a cross-flow membrane system to express the retention of contaminants at the membrane that are larger than the pore size of the membrane.

Repurification   The use of advanced technologies to treat secondary, tertiary, and quaternary effluents to a very high quality, often exceeding potable water standards.

Resistance   The degree to which the flow of water is impeded by a membrane material or fouling.

Resolution   Under the Long Term 2 Enhanced Surface Water Treatment Rule, the size of the smallest integrity breach that contributes to a response from a direct integrity test; also applicable to some indirect integrity monitoring methods.

Retentate   The portion of a feed stream that does not permeate membrane media in a cross-flow filtration system, thus retaining impurities or contaminants such as organic compounds, ions, and colloidal materials that get rejected by the membrane; also referred to as **concentrate** or **reject**.

Reverse osmosis (RO)   A process whereby water is cleaned by being forced through an ultrafine, semipermeable membrane that allows only water to pass through and retains contaminants; RI filters are sometimes used in tertiary treatment and in pretreating water in chemical laboratories.

Salinity   The amount of salt in a solution; usually used in association with salt solutions in excess of 1,000 mg/L and synonymously with the term **total dissolved solids (TDS)**.

Salt passage   The transport of a salt through a semipermeable membrane; typically expressed either as a percentage or as mass of salt per unit of membrane area per unit time.

Salt rejection   The amount of salt in the feed water that is rejected by a semipermeable membrane, expressed as a percentage; also referred to as *solids rejection*.

Salt water   Water containing a relatively high concentration of salts or dissolved solids, usually over 1,000 mg/L.

Salt-splitting capacity test   A test that measures the performance of a used ion exchange resin to determine the capacity of the used resin versus the standard rated capacity of the resin when fresh. This allows a calculation of the degree of exhaustion of the resin.

Scale inhibitor   A chemical agent added to water to inhibit the precipitation or crystallization of salt compounds; also referred to as **antiscalant**.

Scaling   The buildup of precipitated solids (salts) on a surface, such as a membrane.

Secondary effluent   The treated water (effluent) from a secondary clarifier in a wastewater treatment plant—usually activated sludge.

Secondary treatment   The treatment of wastewater through biological oxidation after primary treatment.

Selective ion exchanger   An ion exchange medium that shows selectivity, e.g., a chelating ion exchange resin that will remove only gold ions from solution.

Semipermeable membrane   A membrane that allows only water to pass through while rejecting certain impurities, such as colloidal or dissolved materials.

Sensitivity   The maximum log removal value (LRV) that can be reliably verified by a direct integrity test; also applicable to some continuous indirect integrity monitoring methods.

Sequestration   To keep a substance (e.g., iron or manganese) in solution through the addition of a chemical agent (e.g., sodium hexametaphosphate) that forms chemical complexes with the substance. In the sequestered form, the substance cannot be oxidized into a particular form that will deposit on or stain fixtures. Sequestering chemicals are aggressive compounds with respect to metals, and they may dissolve precipitated metals or corrode metallic pipe materials.

Silica (SiO_2)   A common name for silicon dioxide; silica occurs widely in nature as sand, quartz, flint, and diatomite. It has various industrial uses, including in the manufacture of glass and ceramics and in water filtration.

Silt Density Index (SDI)   A test used to assess the treatability of a specific feed water with NF and RO membranes. It consists of measuring the time required to filter 500 mL (30.5 in.^{3}) of water through a 0.45-μm (1.8×10^{-5} in.) filter. If the SDI of a feed water is too high, pretreatment before NF or RO will be required.

Size exclusion   The removal of particulate matter by a sieving mechanism.

Skid   A group of pressure vessels that share a common valving and can be isolated as a group for testing, cleaning, or repair.

Skinned membranes   Membranes cast in one process that consist of a very thin layer (less than 1 μm or 4.0×10^{-5} in.), referred to as the *skin*, and a thicker, porous layer (up to 100 μm or 4.0×10^{-3} in.) that adds support and is capable of high water flux; also referred to as **integral asymmetric membranes**.

Sodium (Na)   A metallic element found abundantly in compounds in nature but never existing alone. Sodium compounds are highly soluble and do not form curds when used with soaps or detergents. Many sodium compounds are used in the water treatment industry, notably sodium chloride (NaCl), used as a regenerant in the cation exchange water softening process, and sodium hypochlorite (NaOCl), used as a disinfectant.

Sodium bicarbonate ($NaHCO_3$)   A substance that can be added to drinking water to increase alkalinity and provide additional buffering intensity. It is generally supplied as a powder and is used as part of a corrosion control strategy or in the coagulation treatment process; also called *baking soda*.

Sodium carbonate (Na_2CO_3)   A salt in water treatment used to increase the alkalinity or pH value of water or to neutralize acidity; also called *soda ash*.

Sodium chloride (NaCl)   The chemical name for common table salt. Sodium chloride is widely used for regeneration of ion exchange water softeners and in some dealkalizer systems.

Sodium hexametaphosphate   A substance that has a molecular ratio of 1.1 parts of sodium monoxide (Na_2O) to 1 part of phosphorus pentoxide (P_2O_5), with a guaranteed minimum of 76% P_2O_5. Several specialized compositions are available. In general, it is used as a sequestering, dispersing, and deflocculating agent. It is soluble in water and insoluble in inorganic solvents; also called *sodium polyphosphate* or *glassy sodium phosphate*.

Sodium silicate (Na_2SiO_3)   A caustic solution of silica used in water treatment to prevent corrosion or, after a treatment known as activation, to assist coagulation.

Softened water   Any water that has been processed in some manner to reduce the total hardness to 17 mg/L (1 gr./ga) or less, expressed as equivalent calcium carbonate ($CaCO_3$).

Softening   The removal of hardness (i.e., divalent metal ions, primarily calcium and magnesium) from water.

Solutes   The materials (such as chemicals) contained in a solution.

Solvent   The water (or any liquid) in a solution that contains dissolved matters or total dissolved solids. A solution is made up of the solvent and the solutes.

Spacer   A meshlike material used in flat sheet modules, such as in spirals, plates, and pleated sheets, to separate successive layers of membranes. Spacers are very important in membrane systems, as they control the feed channel dimensions in membrane modules.

Specific flux   The membrane flux normalized for pressure and temperature.

Spiral wound membrane modules (or elements)   A membrane configuration in which semipermeable membrane sheets are put together with a support matrix and a spacer and then wrapped around a central tube to collect the filtrate. They are primarily used in NF and RO membrane processes.

Stage   One portion of a train or array that includes membranes operating in a series.

Sulfuric acid (H_2SO_4)   A very strong, corrosive, and hazardous acid used as a regenerant for the cation stage of an ion exchange deionization system. Sulfuric acid is also used occasionally to lower the pH of highly alkaline water. When high concentrations of sulfuric acid are combined with high concentrations of calcium, calcium sulfate ($CaSO_4$) crystals precipitate, creating tenacious fouling of media particles. Sulfuric acid was once commonly called *oil of vitriol*.

Surrogate   A challenge particulate that is a substitute for the target microorganism and which is removed to an equivalent or lesser extent by a membrane filtration device.

Suspension mode   A hydraulic configuration of membrane filtration systems in which contaminants are maintained in suspension through the application of an external force, typically either air or water tangential to the membrane surface.

Symmetric   Having the same consistency throughout.

Symmetric membranes   Microporous membranes that are uniform in pore size; also referred to as **isotropic membranes**.

Synthetic ion exchange resin   A manufactured ion exchange resin, commonly made with cross-linked polymers having exchangeable functional groups.

Synthetic organic chemicals (SOCs)   Organics chemicals made commercially and include pesticides, solvents, monomers, pharmaceuticals, endocrine disruptors, and hormones.

System arrays   The trains or arrays needed to produce the design flow of a plant.

Telescoping   The physical deformation of a spiral wound membrane module due to high differential pressure, in which the membrane, support, and spacer layers are displaced axially (i.e., in the direction of the feed flow) from the center, causing membrane fracture and element failure.

Tertiary treatment   The use of physical, chemical, or biological means to improve secondary wastewater effluent quality. Also, the treated water or effluent from a sand filter or mixed media filter used to treat secondary effluent.

Thin-film composite (TFC) membranes   Membranes made by bonding a thin cellulose acetate, polyamide, or other acetate layer—typically 0.15 to 0.25 μm (6.0×10^{-6} to 9.8×10^{-6} in.) thick—to a thicker porous substrate, which provides structural stability.

Total dissolved solids (TDS)   The concentration of solids in the water filtered through a glass fiber filter to remove suspended solids and colloids. The filter medium is usually 1 or 0.45 μm pore size. Includes sulfates and chlorides.

Train   Multiple interconnected stages in series, or an assembly of cartridges in pressured membrane systems; also called **array**.

Transmembrane pressure (TMP)   The difference between the average feed concentrate pressure and the permeate pressure; the driving force, or hydraulic head loss, associated with any given flux. The TMP of a membrane system is an overall indication of the feed pressure requirement; it is used, with the flux, to assess membrane fouling and the need for chemical cleaning. In a cross-flow membrane system, TMP is measured as the average of the inlet and outlet pressures, minus the permeate back pressure.

Tubular module   A relatively uncommon configuration (in drinking water applications) similar to a hollow fiber module but utilizing membranes of much larger diameter; may be associated with either MF and UF or NF and RO membrane processes.

Tubular systems   Membrane systems in which the membranes are cast on the inside of a support tube and then placed into a pressure vessel. The feed water is pumped through the feed tube and the product water is collected on the outside of the tubes, while the concentrate continues to flow through the feed tube.

Turbidimeter   An instrument used to measure turbidity via the scattering of a light beam through a solution that contains suspended particulate matter.

Turbidity   (1) A condition in water caused by the presence of suspended matter, resulting in the scattering and absorption of light. (2) Any suspended solids that impart a visible haze or cloudiness to water and can be removed by treatment. (3) An analytical quantity, usually reported in nephelometric turbidity units (ntu), determined by measurements of light scattering. The turbidity in finished water is regulated by the US Environmental Protection Agency.

Ultrafiltration (UF)   A pressure-driven membrane filtration process that typically employs hollow fiber membranes with a pore size range of 0.01 to 0.05 μm (4.0×10^{-7} to 2.0×10^{-6} in.).

Upper control limit (UCL)   A control limit (CL) for a membrane filtration system that is required by the Long Term 2 Enhanced Surface Water Treatment Rule; used in the *Membrane Filtration Guidance Manual* to distinguish CLs mandated by the rule from additional lower control limits (LCLs) that either are voluntarily implemented or may be required by the state at its discretion.

Vessel   A pressurized tube that contains several membrane elements in series.

Viruses   The smallest, simplest organisms, 0.02 to 0.03 μm in diameter; made up of nucleic acid plus a protein coat and a lipoprotein envelope.

Volumetric concentration factor (VCF)   The concentration of suspended solids on the high pressure side of the membrane relative to that of the ambient feed water.

Water   (1) A transparent, odorless, tasteless compound of hydrogen and oxygen (H_2O). At a pressure of 1 atm (101.3 kPa), water freezes at 32°F (0°C) and boils at 212°F (100°C). Water, in a more or less impure state, constitutes rain, oceans, lakes, rivers, and other such surface water bodies, as well as groundwater. It contains 11.188% hydrogen and 88.812% oxygen by weight. It may exist as a solid, liquid, or gas. As normally found in the lithosphere, hydrosphere, and atmosphere, it may have other solid, gaseous, or liquid materials in solution or suspension. (2) To wet, supply, or irrigate with water.

Water reclamation   The restoration of wastewater to a state that will allow its beneficial reuse.

Water recycling   The reclamation of effluent generated by a given user for reuse by the same user. In different parts of the world, *water reuse, water reclamation*, and *water recycling* are used synonymously. In Singapore, for example, the usual term is *water reclamation*, while in Australia and in some parts of the United States the common term is *water recycling*.

Water scarcity   Limited water supplies for meeting prevalent or projected water scarcity.

Water softener   (1) A chemical compound that, when introduced into water used for cleaning or washing, counteracts the effects of the hard water minerals (calcium and magnesium) and produces the effect of softened water. For example, detergent additives and polyphosphates are water softeners. See also **sodium hexametaphosphate.** (2) A pressurized water treatment device in which hard water is passed through a bed of cation exchange media (either inorganic or synthetic organic) for the purpose of exchanging calcium and magnesium ions for sodium or potassium ions, thus producing a softened water that is more desirable for laundering, bathing, and dishwashing. This cation exchange process was originally called *zeolite water softening*. Most modern water softeners use a sulfonated bead form of styrene $(C_6H_5CH{:}CH_2)$/divinyl benzene $(C_6H_4[CH{:}CH_2]_2)$ cation resin.

References

40 CFR 141.719(b)(3)(ii-v)

49 CFR 144.82

Adham, S., *et al.* (2005). *Development of a Microfiltration and Ultrafiltration Knowledge Base.* AWWARF Report 91059, American Water Works Association, Denver, CO.

AlanPlummer Associates, Inc. (2010). Final Report: State of Technology of Water Reuse, Prepared for the Texas Water Development Board, Austin, Texas.

Alfrey, T., Jr. (1985). "Structure-Property Relationships in Polymers," in *Applied Polymer Sciences*, 2nd ed., ACS Symposium Series 285, ed. R. W. Tess and G. W. Poehlein (Washington, DC: American Chemical Society), 241–52.

American Public Health Association, American Water Works Association, and Water Environment Federation (1995). *Standard Methods for the Examination of Water and Wastewater*, 20th ed. Washington, DC: American Public Health Association.

American Public Health Association, American Water Works Association, and Water Environment Federation (2005). *Standard Methods for the Examination of Water and Wastewater*, 20th ed. Washington, DC: American Public Health Association.

American Society for Testing and Materials (2003). *Practice A, Pressure Decay (PDT) and Vacuum Decay (VDT) Tests.* ASTM D-6903. West Conshohocken, PA: ASTM International.

American Society for Testing and Materials (1995). ASTM Standard 4189-95. ASTM D-6903. West Conshohocken, PA: ASTM International.

American Society for Testing and Materials (1995). ASTM Standard 6908-03-06, Standard Practice for integrity Testing of Water filtration Systems. West Conshohocken, PA: ASTM International.

American Society for Testing and Materials (2008). ASTM Standard 4189-95. ASTM D-6903. West Conshohocken, PA: ASTM International.

American Water Works Association (2005). *Microfiltration and Ultrafiltration Membranes for Drinking Water*, 1st ed., AWWA Manual M53. Denver, CO: American Water Works Association.

American Water Works Association (1990). *Water Quality and Treatment*, 4th ed., Denver, CO: American Water Works Association.

American Water Works Association (2003). "Residuals Management for Low-Pressure Membranes." Committee report. *Journal American Water Works Association* 95 (6).

American Water Works Association Research Foundation, Lyonnaise des Eaux, Water Research Commission of South Africa, eds. (1996). *Water Treatment Membrane Process.* New York, NY: McGraw-Hill.

Arizona Department of Environmental Quality (2011). Water Quality Division. Permits Reclaimed Water.

Arkhangelsky, E., *et al* (2007). "Hypochlorite Cleaning Causes Degradation of Polymer Membranes." *Tribology Letters*, no. 28: pp 109–116.

Asahi-Kasei Chemicals Corporation (2005). Microza Bulletion MUNC–620A MF MBR Module, Microza Division, Tokyo, Japan.

Asano, T., *et al*, and Metcalf & Eddy (2007). *Water Reuse, Issues, Technologies, and Applications.* New York, NY: McGraw-Hill.

Asian Water (2011) "Recent Cast Trends in Seawater Desalination," Vol 26, No 09, 2011, p 14–17.

Atasi, Khalil, *et al*, and WEF Press (2006). *Membrane Systems for Wastewater Treatment.* New York, NY: Water Environment Federation and McGraw-Hill.

AWWA Subcommittee on Periodical Publications of the Membrane Process Committee, "Microfiltration and Ultrafiltration Membranes for Drinking Water, Journal AWWA 100:12. AWWA, Denver, CO.

AWWA Subcommittee on Periodical Publications of the Membrane Process Committee, "Microfiltration and Ultrafiltration Membranes for Drinking Water, Journal AWWA 100:12: pp 94–96. AWWA, Denver, CO.

B. Pellegrin, E. Gaudichet-Maurin, C. Causserand (2011). "Aging and Characterization of PES Ultrafiltration Membranes Exposed to Hyperchlorite." Poster. IWA MTC Aachen, Toulouse, France.

Brantley, John D., and Jerry M. Martin (1997). "Integrity Testing of membrane Filters to Assure Production of Sterile Effluent. "Genetic Engineering News, Vol 17, No. 10, Mary Ann Liebert, Inc., New York, May 15.

Business Wire (1999). "Koch Membrane Systems to Provide Clean Water Solutions in Midwest." March 23. http://www.thefreelibrary.com/Koch+Membrane+Systems+to+Provide+Clean+Water+Solutions+in+Midwest.-a054190855.

California Department of Public Health (2008). Groundwater Recharge Regulation (Draft) Sacramento, CA.

California Department of Public Health (2009). Regulations Related to Recycled Water. California Code of Regulations, Title 22, Division 4, Chapter 3, Water Recycling Criteria, Sacramento, CA.

Causserand, C., *et al* (2006). *Degradation of Polysulfone Membranes Due to Contact with Bleaching Solution.* Laboratoire de Genie Chimique, CNRS UMR 5503, University Paul Sabatier, Toulouse Cedex, France.

Chellam, S., C. A. Serra, and M. R. Wiesner (1998). "Estimating the Cost of Integrated Membrane Systems." *Journal American Water Works Association* 90 (11): 96–104.

Cheryan, M. (1998). *Microfiltration and Ultrafiltration Handbook.* Lancaster, PA: Technomic Publishing Co., Inc.

Childress, A., *et al* (2005). *Mechanical Analysis of Hollow Fiber Integrity in Water Reuse Applications.* Reno, NV: University of Nevada.

Clements, J., *et al* (2006). "Ceramic Membranes and Coagulation for TOC Removal from Surface Water," in *Proceedings of Water Quality Technology Conference,* American Water Works Association. Denver, CO: AWWA.

Colorado Department of Health and Environment (2007). Reclaimed Water Control Regulation. 5 CCR1002-84, Colorado Department of health and Environment, Denver, CO.

David R. Lide, Editor-in-Chief (1996–1997). CRC Handbook of Chemistry and Physics, 77th ed. CRC Press, Cleveland, OH.

Dempsey, B. A., R. M. Ganho, and C. R. O'Melia (1995). "The Coagulation of Humic Substances by Means of Aluminum Salts." *Journal American Water Works Association* 76(4): 141–50. American Water Works Association, Denver, CO.

Dennett, K. E., *et al* (1996). "Coagulation: Its Effect on Organic Matter." *Journal American Water Works Association* 88(4): 129–42. American Water Works Association, Denver, CO.

Design of Municipal Wastewater Treatment Plants (1992). WEF Manual of Practice No. 8, ASCE Manual and Report on Engineering Practice No. 76, Water Environment Federation, Alexandria, Virginia, American Society of Civil Engineers, New York.

Dwyer (2003). University of New Hampshire, Durham, NH.

Ebbing, D. D. (1996). *General Chemistry.* Houghton Mifflin Company: Boston, MA.

Escobar-Ferrand, Lui, *et al.*, "Detailed Analysis of the Silt Density Index (SDI) Results on Desalination and Wastewater Reuse Applications for Reverse Osmosis Technology Evaluation," Pall Corporation, Port Washington, New York.

Garcia-Aleman, J., and J. Lozier (2005). "Managing Fiber Breakage: LRV Operations and Design Issues Under the LT2 Framework," in *Proceedings of Membrane Technology Conference, Phoenix, AZ*. American Water Works Association, Denver, CO.

Global Water Intelligence (2010). *Global Water Report 2011*. Oxford, UK: Media Analytics, Ltd., The Jam Factory.

Global Water Summit (2010). *Transforming the World of Water/Promoting Water Reuse*. Paris, France: Global Water Intelligence and International Desalination Association.

Gregory, J. (1993). "The Role of Colloid Interactions in Solid-Liquid Separation." *Water Science and Technology* 27 (10): 1–17.

Grozes, G., P. White, and M. Mashall (1995). "Enhanced Coagulation: Its Effect on NOM Removal and Chemical Costs." *Journal American Water Works Association* 87 (1): 78–89, American Water Works Association, Denver, CO.

Griss, P., and J. Dihrich (2000). "Microfiltration of Municipal Wastewater Disinfection and Advanced Phosphorus Removal: Results from Trials with Different Small Scale Pilot Plants." *Water Environment Research* 72 (5): 602–09.

Hagstrom, J. P. (2007). "MBR, Membrane Basics and Fundamentals: What You Need to Know Before Embarking on Your MBR Project." Presented at Technical Workshop 202, October 14, San Diego, CA.

Hertzberg, R. W., and J. A. Manson (1980). *Fatigue of Engineering Plastics*. New York, NY: Academic Press.

Hofmann, F. (1984). "Integrity Testing of Microfiltration Membranes." *Journal of Parenteral Science and Technology* 38 (4).

Howe, K. J., *et al* (2002). *Coagulation Pretreatment for Membrane Filtration*. Denver, CO: AWWA Research Foundation.

Howe K. J., and M. M. Clark (2002). "Fouling of microfiltration and ultrafiltration membranes by natural waters," *Environmental Science and Technology* 36: 3571–3576, American Chemical Society.

International Union of Pure and Applied Chemistry (1985). "Reporting Physisorption Data," Pure and Applied Chemistry 57:603, Research Triangle Park, NC.

Jacangelo, J., S., Adham, and J.-M. Laine (1997). Membrane Filtration for Microbial Removal. American Water Works Research Foundation, Denver, CO.

Kumar, A., and R. K. Gupta (1996). *Fundamentals of Polymers*. New York, NY: McGraw-Hill.

Lang, Heather, and Jablanka, Azelac (2011). *Global Water Market 2011*. Oxford, UK: Media Analytics, Ltd., The Jam Factory.

Lee, S., and J. Cho (2004). "Comparison of Ceramic and Polymeric Membranes for Natural Organic Removal." *Desalination* 160 (3): 223–232.

Lehmann, *et al* (2009). *Application of New Generation Ceramic Membranes for Challenging Waters*. Arcadia, CA: Applied Research Department, MWH Americas.

Letterman, R. D., A. Amirtharajah, and C. R. O'Melia (1999). "Coagulation and Flocculation," in *Water Quality and Treatment: A Handbook for Community Water Supplies*, 5th ed., ed. R. D. Letterman. New York, NY: McGraw-Hill.

Liu, C. (1998). *Developing Integrity Testing Procedures for Pall Microza MF and UF Modules, I: Diffusion Flow and Pressure Hold Methods*. SLS Report 7426. Port Washington, NY: Pall Corporation.

Liu, C. A., M. Wachinski, and D. Vial. (2006). "Factors Affecting NOM Removal by Coagulation-Membrane Filtration," Pall Corporation, Port Washington, New York.

Liu, C. (2007). "Mechanical and Chemical stabilities of Polymeric Membranes," in *Proceedings of Membrane Technology Conference*, Charlotte, NC.

Liu, C., and A. M. Wachinski (2010). "Attributes of High Performance Membranes," Pall Corporation, Port Washington, New York.

Liu, Charles (2012). *Integrity Testing for Low-Pressure Membranes*. Denver, CO: American Water Works Association.

Logsdon, G., A. Hess, and M. Horsley (1999). "Guide to Selection of Water Treatment Processes," in *Water Quality and Treatment: A Handbook for Community Water Supplies*, 5th ed., ed. R. D. Letterman. New York, NY: McGraw-Hill.

Macpherson, Linda (2011). "Water: Changing Mental Models from Used to Reusable," in *Australian Water Association Conference*, Melbourne, Australia.

Maletzko, Christian (2003). "PESU Membranes in Municipal Water Treatment Applications." *Desalination and Water Reuse* 19 (3): 22–26.

Markhoff, Heiner (2010). "Challenges and Opportunities for Reuse." Presented at Global Water Summit 2010, Transforming the World of Water IDA. Paris, France.

Metcalf and Eddy (2003). *Wastewater Engineering Treatment & Reuse*, 3rd ed., Revised, New York, NY: McGraw-Hill.

Metcalf and Eddy (2004). *Wastewater Engineering Treatment & Reuse*, 4th ed. New York, NY: McGraw-Hill.

Moore, Tara (2011). "Beating the Coming Water shortage," *Fortune*, October 17.

Mulder, M. (1991). *Basic Principles of Membrane Technology*, Kluwer Academic Publishers, Dordrecht, Germany.

Omori, A., *et al* (2010). High Efficiency Absorbent System for Phosphorous Removal." Asahi Kasei Chemicals Corporation, Tokyo, Japan.

Osmosis Filtration and Separation Spectrum, www.osmonics.com/library/filspc .htm.

Owen, D. M., *et al* (1995). "NOM Characterization and Treatability." *Journal American Water Works Association* 87 (1): 46–63.

Pearce, Graeme (2007). "Introduction to Membranes: Manufacturer's Comparison, Parts 1–2." *Filtration and Separation* 44 (10) (October): 36–38.

Pall Corporation (2004). Port Washington, New York.

Pearce Graeme K (2011). *UF/MF Membrane Water Treatment, Principles and Design*, Water Treatment Academy, TechnoBiz Communications Co., Ltd., Bangkok, Thailand.

Randtke, S. J. (1988). "Organic Contaminants Removal by Coagulation and Related Processes." *Journal American Water Works Association* 80 (5): 40–56.

Reich, L., and S. S. Stivala (1971). *Elements of Polymer Degradation*. New York, NY: McGraw-Hill.

Sakaji, R. H., ed. (2001). *California Surface Water Treatment Alternative Filtration Technology Demonstration Report* [Draft]. Berkley: California Department of Health.

Stumm, W., and J. J. Morgan (1996). *Aquatic Chemistry: Chemical Equilibria and Rates in Natural Water*, 3rd ed. New York, NY: John Wiley and Sons.

United States Environmental Protection Agency (1979). *Estimating Water Treatment Costs: Volume II: Cost Curves Applicable to 1 to 200 MGD Treatment Plant*. Cincinnati, OH: Office of Research and Development, US Environmental Protection Agency.

United States Environmental Protection Agency (1979). "National Interim Primary Drinking Water Regulations: Control of Trihalomethanes in Drinking Water: Final Rule. *Fed Register* 44(231): 68624-68707.

United States Environmental Protection Agency (1989). "National Primary and Secondary Drinking Water Regulations." *Federal Register* 54: 22062–22160.

United States Environmental Protection Agency (1993). *Manual: Nitrogen Control*. EPA-625/R-93-010. Washington, DC: US Environmental Protection Agency.

United States Environmental Protection Agency (1998). "National Drinking Water Regulations: Disinfectants and Disinfection Byproducts: Final Rule." *Federal Register (Part IV)* 63 (241): 69390–69476.

United States Environmental Protection Agency (1999). *Enhanced Coagulation and Enhanced Precipitation Guidance Manual*. Washington, DC.

United States Environmental Protection Agency (2001). *Low Pressure Membrane Filtration for Pathogen Removal: Application Implementation and Regulatory Issues*. EPA 815-C-01-001. Washington, DC: US Environmental Protection Agency.

United States Environmental Protection Agency (2003). "National Drinking Water Regulations: Long Term 2 Enhanced Surface Water Treatment Rule, Proposed Rule." *Federal Register* 68 (154): 47640–47795.

United States Environmental Protection Agency (2003c). "National Drinking Water Regulations: Stage 2 Disinfectants and Disinfectant Byproducts: Final Rule." *Federal Register* 68 (159): 49547.

United States Environmental Protection Agency (2004). *Guidelines for Water Reuse*, EPA 625-R-04-108. Washington, DC: US Environmental Protection Agency.

United States Environmental Protection Agency (2005). *Membrane Filtration Guidance Manual*. EPA 815-R-06-009. Washington, DC: US Environmental Protection Agency.

United States Environmental Protection Agency (2009a). "National Drinking Water Regulations: Stage 2 Disinfectants and Disinfectant Byproducts: Final Rule. *Federal Register* 71: 388.

United States Environmental Protection Agency (2009b). "National Drinking Water Regulations." EPA/816F-09/04.

United States Geological Survey (2000). *Hydrological Units of the United States*. USGS: Washington, DC.

Van der Bruggen, B. (2003). "Pressure Driven Membrane Processes in Process and Wastewater Treatment and in Drinking Water Production." *Environmental Progress* 22 (1): 46–56.

Vieth, W. R. (1991). *Diffusion in and Through Polymers: Principles and Applications*. New York, NY: University Press.

Wachinski, A. M. (1985). Class Notes, Chemistry Review and Sampling, in Department of Civil Engineering Course, Wastewater Treatment Plant Design, USAF Academy Colorado.

Wachinski, A. M. (2002). *Handling and Disposal of Process Wastewaters for Microfiltration Systems*. Port Washington, NY: Pall Corporation.

Wachinski, A. M. (2003). *Water Quality*, 3rd ed. Denver, CO: American Water Works Association. Chapter 1, pp 6–23.

Wachinski, A. M. (2006). *Ion Exchange Treatment for Water*. Chapter 1: Fundamental Concepts of Water Chemistry for Ion Exchange. Denver, CO: American Water Works Association.

Wachinski, A. M. (2007). "Industrial Water Reuse Makes Cents," Environmental Protection. Environmental Protection Magazine. Dallas Texas. EP online.

Wachinski, A. M. *et al* (2009). Pall Corporation video.

Wachinski, A. M., and J. E. Etzel (1997). *Environmental Ion Exchange Principles and Design*. Boca Raton, FL: Lewis Publishers.

WEF Press (2006). *Biological Nutrient Removal (BNR) Operation in Wastewater Treatment Plants*, WEF Manual of Practice No. 29, ASCE/EWRI. Manuals and Reports on Engineering Practice No. 109, McGraw-Hill, New York.

White, M. C.,Thompson, J. D., Harrington, G. W., *et al* (1997). "Evaluating Criteria for Enhanced Coagulation Compliance." *Journal American Water Works Association* 89(5). pp 64–67. American Water Works Association, Denver, CO.

Wideman, S. (2009). "Aldermen to review 'brutal facts' on fix for water plant." *Appleton Post-Crescent*, August 4. http://www.postcrescent.com.

Williams, M. E. (2003). "A Brief Review of Reverse Osmosis Technology," EET Corporation and Williams Engineering Services Company, Inc.

Wingfield, Tom, and James, Schaeffer (2001). "Making Water Work Harder, Environmental Protection." Dallas, TX, November.

Wingfield, Tom, and James, Schaeffer (2002). "Cleaner Purer Water." *Pollution Engineering*. September.

www.pall.com.

www.ROTOOLS.com.

Zondervan (2007). "Statistical Analysis of Data from Accelerated Ageing Tests of PES UF Membranes," *Journal of membrane science*, Vol. 300, No. 1, London, Amsterdam.

Index

18 megohm water, 239